深智數位
股份有限公司

深智數位
股份有限公司

序

「想要打造看看派對遊戲。」

這個點子從幾年前就一直在腦中打轉，一開始嘗試用 MCU 打造如 joy-con 的搖桿，雖然不是不行，但是設計電路、選電子零件、建構外殼到組裝結構等等步驟實在太麻煩，所以很快就棄坑了。(>´ω`<)

（絕對不是因為沒有朋友一起玩才棄坑，絕對不是。 ㄒ▽ㄒ）

直到後來學了網頁開發，看到網頁技術的高速發展，同時看到手中有著多種感測器的手機，腦中閃過了新點子：「手機不就是一個現成的joy-con了嗎？」，於是打造派對遊戲這個點子有了新方向。

在各種偷懶與拖延的干擾下，派對遊戲還是成功誕生了！✧*.९(ˊᗜˋ*)ɞ✧*。

感謝深智數位的邀請，讓這個原先零散的專案擴增並集結成本書，還有總是鼓勵我搞些奇怪東東的父母與一直很支持我的老婆大人，即使我把家裡搞到跳電也沒把我轟出去，最後希望女兒趕快長大，趕快陪老爸玩 Switch！(/≧▽≦)/

感謝翻開本書的各位讀者，希望你們會喜歡。(ˊᗜˋ)/

可以掃描下方 QR Code 觀看介紹影片，
更了解本書專案成果

行前準備

本書主要使用以下環境開發：

- Windows 的 Chrome 瀏覽器

- Android 的 Chrome 瀏覽器

其他環境可以試試看，但是不保證所有功能可以支援。(´・ω・`)

除了圖解說明外，本書所有的範例與開發環境等等，都會在 GitLab 提供下載，讓讀者省去逐字輸入程式的困擾，將腦力集中在理解邏輯，更無負擔的學習。

專案連結：

- https://gitlab.com/deepmind1/animal-party

鱈魚：「另外我還找了一個助教一起協助解說。✧(ˋ‿ˊ)」

助教：「肥魚講太多話容易喘吼？ ಠ_ಠ」

鱈魚：「才不是，你才肥！你全家都肥！(╬ಠдಠ)」

助教：「希望本書能夠為大家帶來收穫，請大家多多指教！(´▽`)ノ」

鱈魚：「你別搶我台詞還擅自結束啊！Σ(´д`;)」

本書主要規劃結構如下為：

1. 開發環境：介紹在此專案中使用的各類套件與工具。

2. Vue 基本概念：介紹 Vue 基礎概念。

3. NestJS 基本概念：介紹何謂 NestJS 與伺服器開發概念。

4. 即時通訊：實作前後端即時通訊，奠定後續派對遊戲基礎。

5. 3D 網頁開發：利用 babylon.js 開發 3D 網頁並製作遊戲。

目錄

1 沒時間解釋了，快上車！

2 Vue 基本概念

3 NestJS 基礎概念

4 打造派對門面

11　小雞快飛

12　狐狸與老鼠

13　結束是另一個開始

14　後記

第**1**章

沒時間解釋了，
快上車！

首先讓我們依序介紹本書會用到的前後端框架與套件與專案內容。

1.1 原理與架構

第一步當然是先來了解一下原理與架構。

預期建立一個可以做為主機與搖桿的 Web 與提供 WebSocket 即時連線的 Server，畫個架構圖，協助思考並釐清結構吧。

為了讓架構更加簡單，開發階段會將 Web 與 Server 專案分離，Web 透過 Vite 之 Dev Server 提供網頁資源，Server 則純粹提供 WebSocket 連線。

▲ 圖 1-1 整體架構

整體架構基本上如上圖，基本流程為：

1. 瀏覽器從 Vite Dev Server 取得網頁

2. 成功取得網頁後，網頁會向 WebSocket Server 請求連線

3. 連線成功後 WebSocket Server 會提供網頁一個 Socket ID 表示已連線之 client

4. 最後網頁會不斷地向 WebSocket Server 傳遞資料，WebSocket Server 則會在所有 client 之間同步資料

以上是網頁即時通訊架構，接下來是網頁呈現的部分。

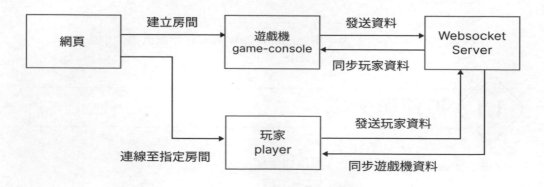

▲ 圖 1-2 網頁架構

首頁會提供兩個按鈕，分別為：

- 建立派對（建立房間）：

    點選後此 Web 會作為 game-console 開始連線，提供房間號碼與遊戲畫面。

- 加入遊戲（連線至指定房間）：

    點選後會開啟對話框，輸入房間號碼後做為玩家加入房間。

以上便是本書專案的基本架構說明，接著讓我們進入開發工具與套件。

1.2　開發工具

1.2.1 Visual Studio Code

▲ 圖 1-3　Visual Studio Code（圖片來源：https://code.visualstudio.com）

常被簡稱為「VS Code」，基本上前端開發一定聽過這個工具，因為其強大的智慧提示（IntelliSense）和豐富的外掛，根據 Stack Overflow2021 年開發者報告，目前 VS Code 的使用率佔了 71.06%，幾乎和鄉下路邊野狗一樣常見了。(๑′ω`๑)

Tips

雖然都是微軟出品，但是 Visual Studio Code 和 Visual Studio 基本上沒有甚麼關係，一個是文字編輯器一個是 IDE，定位不同。

在官網（https://code.visualstudio.com）下載並安裝，打開程式應該會出現下面這個畫面。

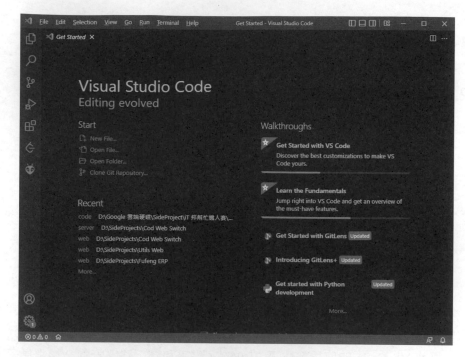

▲ 圖 1-4　VS Code 初始畫面

　　接下來讓我們安裝本書會用到的外掛，在左側的選單點選「4 個方塊」的圖示，即可開啟外掛商店，讓我們先安裝中文語系套件吧。

▲ 圖 1-5　安裝套件

按下紅框中的 Install 即可開始安裝。

安裝完成後，重啟 VS Code 應該就可以看到 VS Code 變成中文了！

 Tips

如果沒有變成中文，也可以依照以下步驟切換為中文：

1. 按下鍵盤 F1（或 Ctrl + Shift + P），開啟命令輸入框

2. 輸入 display language，選擇「Configure Display Language」

3. 選擇 zh-tw 後，VS Code 會提示需要重新開啟。

4. 重啟後畫面就會變成中文版本了。

接著為了讓開發更順利，請大家安裝以下外掛。

- Vue Language Features（Volar）

  用於提升 Vue.js 開發效率，提供例如：自動完成（Autocompletion）、智慧語法突顯（Intelligent Syntax Highlighting）等等方便功能。

  （沒有這個我都不會寫 Vue 啦。ᕕ(ﾟ ∀ ﾟ)ᕗ）

- Tailwind CSS IntelliSense

  用於方便使用 Tailwind CSS 開發，可以自動提示 Tailwind 支援之 Class，不用自己背名稱，若 Class 重複也會提供警告等等。

  （沒有這個我都不會寫 Tailwind 啦。ᕕ(ﾟ ∀ ﾟ)ᕗ）

- Code Spell Checker

  幫忙檢查拼寫錯誤，不然一堆錯字就糗大了。

- Sass

  可以簡化 CSS 寫法，寫起來更快樂。♪('∇')

以上外掛在 Web 專案下皆可用，但在 Server 專案下建議關閉以下外掛：

- Vue Language Features (Volar)

- Tailwind CSS IntelliSense

- Sass

以免造成影響。（伺服器專案也完全用不到就是了 (· ω ·)）

1.2.2　Google Chrome

▲ 圖 1-6　Chrome

瀏覽器則是使用知名的記憶體怪獸 Chrome。

 Vue.js devtools

✓ vuejs.org

★★★★★ 1,869 ⓘ ｜ 開發人員工具 ｜ 2,000,000+ 位使用者

▲ 圖 1-7　Vue.js devtools

請讀者前往應用程式商店，安裝「Vue.js devtools」這個外掛，我們在開發的過程中會使用此外掛 debug。

Tips

可以前往商店搜尋「Vue.js devtools」或是前往以下連結：

◆ https://chrome.google.com/webstore/detail/vuejs-devtools/nhdogjmejiglipccpnn
nanhbledajbpd

1.3 使用套件

1.3.1 TypeScript

▲ 圖 1-8 TypeScript（圖片來源：https://www.typescriptlang.org）

甚麼是 TypeScript？可以看看官網的說法：

TypeScript is a strongly typed programming language that builds on JavaScript, giving you better tooling at any scale.

簡單來說就是包含「強型別」的 JavaScript，這時候大家可能會問「為甚麼要加入型別？」。

TypeScript 的主要優點如下：

■ 編譯階段即可發現大部分錯誤，不會等到執行時才發現。

■ 增加可讀性與可維護性，藉由型別系統提供型別，配合編輯器可以提供程式碼自動完成、介面提示等等非常方便的功能。

　　但是導入 TypeScript 相對就會增加學習成本與短期開發成本，所以若是屬於創意展示、美術設計為主的網頁，我個人就不會非要使用 TypeScript，但若是屬於應用程式類型的網頁，就非常適合導入 TypeScript，所以本書這類會不停傳遞各類資料的主題就很適合使用 TypeScript。

　　過程中本書會循序漸進的帶領大家慢慢入門，不用擔心一堆火星文，看了就想睡！

1.3.2　Vue 3、Vue Router、Pinia

▲　圖 1-9　Vue（圖片來源：https://vuejs.org）

　　就是經典的 Vue 全家桶，涵蓋前端框架、路由管理、狀態管理功能。

　　選擇 Vue 3 則是因為其 TypeScript 支援度較佳，且有 Composition API，相較於 Vue 2 來說，更適合用於開發這類應用程式型的網頁。

　　後續章節會依序介紹 Vue 全家桶入門，請大家放心閱讀。

1.3.3　VueUse

▲　圖 1-10　VueUse（圖片來源：https://vueuse.org）

基於 Vue 3 之 Composition API 而生，內含 200 多個開發網頁常用功能，在本書中，我們會大量使用 VueUse 提供的功能，可以有效簡化程式，很推薦讀者們可以看看 VueUse 的原始碼，保證獲益良多。

1.3.4 Quasar

▲ 圖 1-11　Quasar（圖片來源：https://quasar.dev）

基於 Vue 之 UI Framework，依照 Material Design 2 設計規範，包含大量開發常用情境與功能並提供 CLI 命令，可以同時涵蓋 SPA、SSR、PWA、Mobile App、Electron、Browser Extensions 等等需求。

1.3.5 babylon.js

▲ 圖 1-12　babylon.js（圖片來源：https://www.babylonjs.com）

網頁 3D 引擎，大家可能最常聽過的網頁 3D 套件是 Three.js，不過 Three.js 主要專注於繪圖的部分，而 babylon.js 則包含了物理模擬、材質管理工具等等完整的功能，更在 v5 版本加入了 WebGPU，讓性能可以更上一階。

加上因為和 TypeScript 同一個老爸（都是微軟），所以對於 TypeScript 支援度極佳，因此本次專案選擇 babylon.js。

截至本書出版前（2023/09/04），babylon 已經推出 v6 版本，除了更強的性能還加入 3A 遊戲等級的物理引擎！ー=≡Σ(((つ•ω•)つ

不過本書目前維持使用 v5 版本，以免新版本 API 不穩定、功能異動等等。

1.3.6 Tailwind CSS

▲ 圖 1-13　Taildwind CSS（圖片來源：https://tailwindcss.com）

如同官網所述：

A utility-first CSS framework packed with classes like flex, pt-4, text-center and rotate-90 that can be composed to build any design, directly in your markup.

Taildwind CSS 是實用優先型的 CSS 框架，可以透過加入 class 就馬上加上指定樣式，可以簡單快速的完成介面。

Tips

Taildwind CSS 實務上其實有需多要注意的地方，如果放飛自我隨便使用，容易寫出一坨很難維護的程式，有興趣的讀者們可以參考以下連結：

◆ 用 Tailwind 來幫你實現真正的高效整潔

https://5xruby.tw/posts/tailwind-css-plugin

◆ Reusing Styles

https://tailwindcss.com/docs/reusing-styles

◆ 客觀評價 Tailwind CSS

  https://medium.com/@nightspirit622/%E5%AE%A2%E8%A7%80%E8%A9%9

  5%E5%83%B9-tailwindcss-af27581f6d9

也可以 Google 一下查詢更多資訊。

1.3.7 NestJS

▲ 圖 1-14 NestJS（圖片來源：https://nestjs.com）

　　基於 Node.js 之後端框架，結合了 OOP、FP 與 FRP 設計，並完全支援 TypeScript 開發，熟悉 Angular 的讀者應該會覺得很親切，因為 NestJS 許多設計就是啟發於 Angular。

　　本書專案中，NestJS 建立的 Server 之主要任務為，提供 WebSocket Server 連接所有 Web，同步所有 Web 狀態與房間配對等等功能。

Tips

如果對 JS 生態系有了解的讀者可能會想：「用 Express 或 Koa 不好嗎？」

說來很有趣，其實 NestJS 底層預設就是使用 Express，差別在於 NestJS 提供了一個完整的架構，可以更容易打造一個測試性、拓展性較高且更容易維護的專案。

雖然在小專案中會顯得笨重些，但是熟悉架構可以幫助增進設計能力，所以本書就讓我們來練習練習 NestJS 吧！

開發環境

接著讓我們介紹前端與後端開發環境，本書前端主要使用 Vue，後端則是
NestJS。

最重要的第一步是先安裝 Node.js：

1. 下載 Node.js：https://nodejs.org/zh-tw/download

2. 下載完成後只要一路按「下一步」就可以完成安裝了，相當簡單。

Tips

JavaScript 與 Node.js 是甚麼關係呢？ JavaScript 是程式語言，而 Node.js 是
runtime，也就是執行 JavaScript 這個語言的環境，截至目前為止（2023/04/01）
JavaScript runtime 除了最知名的 Node.js 以外，還有 Deno 與 Bun，有興趣的讀
者可以自行搜尋資料。

1.4.1 Vue 前端專案

首先開啟以下網址

- https://gitlab.com/deepmind1/animal-party/web/-/tree/init

會開啟 GitLab 網頁，如下圖點開下載按鈕，選擇喜歡的壓縮檔類型，這裡
我們選擇將原始碼下載為 zip 檔。

▲ 圖 1-15　下載前端專案

接著將壓縮檔解壓縮後用 VS Code 開啟，有以下兩種開啟方式：

1. 對著資料夾按右鍵，點選「以 Code 開啟」。

▲ 圖 1-16　右鍵快速開啟

先打開 VS Code，在「檔案」->「開啟資料夾」，找到並選擇剛剛解壓縮後的資料夾。

▲ 圖 1-17　開啟指定資料夾

完成以上步驟後，就會在 VS Code 的左方看到此資料夾下所有的檔案。

▲ 圖 1-18　前端專案目錄

接下來讓我們介紹目錄內重要的資料夾與檔案吧！

- public

  靜態資源目錄，此目錄內的檔案不會被打包工具額外處理

- src

  主程式資料夾，所有的程式內容皆集中至此。

- index.html

  打包程式會將所有的 CSS、JS 打包完成後，插入此 HTML 中。

- package.json

  記錄此專案詳細內容，例如專案名稱、相依套件、執行腳本等等。

- package-lock.json

  package.json 中預設只會紀錄套件主要版本號，而 lock 檔案中會記錄當下安裝的完整版本號，用於統一套件版本，以免在不同裝置中，大家的套件版本都有落差。

- vite.config.ts

  Vite 設定檔案。Vite 用於提供快速方便的開發環境，例如即使預覽、打包網頁等等功能。

接著讓我們詳細說明 src 內容吧。

- assets

  存放待打包的靜態資源，被打包的資源可以依照設定被自動處理，例如圖片壓縮等等。

 Tips

此專案主要使用 Vite，所以相關資源可以參考 Vite 文檔：

- 靜態資源處理：https://cn.vitejs.dev/guide/assets.html

- common

  存放專案中共用的程式碼，如常用函式、共用邏輯等。

- components

  存放 Vue 元件，這些元件通常是可重複使用的 UI 元件，例如按鈕、輸入框等。

- composables

  存放 Vue 3 中的 Composition API 相關的函式，這些函式可以在多個元件之間共享邏輯。

Tips

Composition API 是 Vue 3 之後新的 API，可以參考官方文檔說明：

◆ Composition API 問答：https://cn.vuejs.org/guide/extras/composition-api-faq.
html

- router

  此資料夾包含 Vue Router 的相關設定，用於定義專案的路由結構。後續
  會介紹何謂 Vue Router。

- stores

  存放狀態管理套件相關檔案，用於管理應用的狀態與資料，此專案我們
  使用 Pinia。後續會介紹何謂 Pinia。

- style

  存放專案的全域樣式檔案，例如 CSS、Sass、Less 等。

- types

  存放 TypeScript 的型別定義檔案，用於提供靜態型別檢查。

- views

  存放各個頁面元件，通常是由多個 components 組成的大型元件。

- App.vue

  Vue 專案中的根元件，用於組織其他元件，並在此元件中引入全域樣式
  檔案或是初始化各種全域物件或事件。

- env.d.ts

  定義額外環境變數的 TypeScript 型別，以便在專案中使用環境變數時獲
  得正確的型別提示。

- index.css

  Tailwind CSS 的入口點，用於宣告 Tailwind 指令。

- main.ts

  專案入口點，負責初始化 Vue、導入各種外掛（如 Vue Router、Pinia 等）
  並將根元件掛載到 DOM 上。

以上為 Vue 專案結構說明，後續章節我們會走一遍 Vue 的基礎教學，讓讀
者們更熟悉專案結構內容。

1.4.2　NestJS 後端專案

與前端專案相同，開啟以下網址

- https://gitlab.com/deepmind1/animal-party/server/-/tree/init

下載並使用 VS Code 開啟後，一樣先來認識目錄結構，由於根目錄下檔案
基本上組成與前端專案相同，所以在此只說說 src 中的檔案：

- app.controller.ts

  控制器負責處理 HTTP 請求，並且通常是根據用戶端輸入來操作伺服器
  端資料的主要入口點。

- app.module.ts

  應用程式的主要模組，並註冊了所有相關的控制器、提供者和其他模組
  等。還用於設定應用程式的全域中介軟體和全域資料庫連接等。

- app.service.ts

  主要服務，服務是用來處理應用程式業務邏輯的主要工具。通常會被控
  制器或其他服務使用，而且可以通過模組進行注入。

- main.ts

伺服器主進入點，用於啟動整個應用程式。可以設定伺服器的監聽端口和 IP 位置等，也可以啟用其他功能，例如全域中介軟體和全域例外處理器等等。

用具體一點的比喻來形容的話，main 就像一間餐廳，進入餐廳後由服務生（controller）接待並點餐，再由內場人員（service）接單並出菜，而經理（module）則負責指揮與調度。

以上我們便完成前後端專案內容介紹了，接下來讓我們進入下一章。

Vue 基本概念

已經了解 Vue 的讀者，可以跳過此章節。ー＝≡Σ(((つ•ω•)つ

Vue 是一種前端框架，最主要的特點是：聲明式渲染與響應性。

若想要將資料綁定至 DOM 中，以往 jQuery 要在資料更新時，自行更新相對應的 DOM，若資料大量重複使用時，此工作會變得繁重且容易出錯。

若是使用 Vue，只要在 DOM 中使用模板語法，Vue 就會自動綁定、更新資料。

只要在 HTML 中使用 Mustache 語法（也就是所謂的雙大括號），表示資料綁定位置。

```
<div class="flex flex-col gap-4 p-4">
  <input v-model="data">
  <div class=" text-red">
    {{ data }}
  </div>
</div>
```

接著在程式中給定資料。

```
import { ref } from 'vue';

const data = ref('123');
```

現在不管 data 怎麼變動，Vue 都會自動更新內容，心智負擔是不是大幅下降許多啊？(・ω・)y

最後是 Vue 的單檔案元件：Single-File Components（縮寫為 SFC），在擁有 Vue 完整建構環境的專案中，我們都會使用 SFC，其特點是將模板（HTML）、邏輯（JavaScript）與樣式（CSS）封裝在單一個檔案中。

以下程式便是一個標準的 SFC。

```html
<template>
  <div class="card">
    <input v-model="data">
    <div>
      {{ data }}
    </div>
  </div>
</template>

<script setup lang="ts">
import { ref } from 'vue';

const data = ref('123');
</script>

<style scoped lang="sass">
.card
  width: 100%
  height: 100%
  padding: 0
  margin: 0
</style>
```

可以看到 SFC 基本上由 <template>、<script> 與 <style>，共三個標籤組成。

這麼做有一個非常大的好處，可以根據元件的功能進行單一封裝，讓此元件高內聚、低耦合，能更簡單的復用在各個地方。

接下來讓我們循序漸進的認識 Vue 吧！♪(ˊωˋ)

想跟著實作的讀者可以下載以下連結專案，一起實作看看。

- https://gitlab.com/deepmind1/animal-party/tutorial-vue/-/tree/main

一樣下載、解壓縮專案並用 VS Code 開啟，接著我們需要開始安裝專案內需要使用的套件。

1. 首先在「檢視」中找到「終端」選項並點擊開啟終端機。

▲ 圖 2-1 開啟終端機

2. 接著在畫面下方的終端機中輸入「npm i」（等同於 npm install）後，等
   待安裝完成。

▲ 圖 2-2　輸入安裝命令

3. 安裝完成後，輸入「npm run dev」則會啟動 Vite 的開發模式。

4. 開發模式啟動後會出現預覽用網址，打開 Chrome 瀏覽器並輸入網址。

▲ 圖 2-3　開啟 dev server

5. 最後會在網頁看到以下畫面。

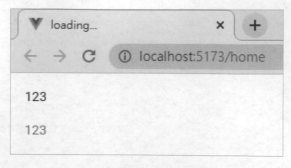

▲ 圖 2-4　輸入安裝命令

6. 現在可以開始開發網頁了！✧*｡٩(´ ꒳ ` *)و✧*｡。

基礎概念

此章節程式碼可以在以下連結取得。

- https://gitlab.com/deepmind1/animal-party/tutorial-vue/-/tree/vue-fundamentals

　　Vue 3 的響應式系統建立於 JavaScript 的 Proxy 物件之上，可以使用 ref 或 reactive 建立一個具有響應性的資料。

2.1.1　響應式與資料綁定

　　現在在 App.vue 應該可以看到以下內容。

→ src\App.vue

```
<template>
  <div class="input-box">
    <input
      v-model="data"
      class="input"
    >
    <input
      v-model="data"
      class="input"
    >
    <div>
      輸入框文字：{{ data }}
    </div>
  </div>
```

```
  </div>
</template>

<script setup lang="ts">
import { ref } from 'vue';

const data = ref('123');
</script>

<style lang="sass">
html, body, #app
  width: 100%
  height: 100%
  padding: 0
  margin: 0

#app
  display: flex
  flex-direction: column

.input-box
  padding: 1rem
  .input
    padding: 1rem
    border: 1px solid #AAA
    margin-bottom: 1rem
    margin-right: 0.4rem
</style>
```

Tips

大家可能會注意到 \<style> 標籤中有 lang=sass，這表示我們的 style 使用 Sass
進行撰寫，Sass 是 CSS Preprocessor（CSS 預處理器）的一種，也就是 Sass 的
內容最後都會編譯為 CSS，但是 Sass 提供了更為彈性、結構化的開發方式。

◆ https://sass-lang.com/guide/

目前應該會在網頁看到如下圖內容。

▲ 圖 2-5 基本內容

可以看到這裡使用 ref 建立了 data 變數，並使用 v-model 將 data 雙向綁定至 input 中。

雙向綁定表示 data 的資料會綁定於 input，而 input 內容增減時，也會即時改變 data，讀者可以嘗試輸入左右輸入框的內容，會發現數值不只會在兩個 input 之間同步，也會即時變更下方「輸入框文字」的內容。

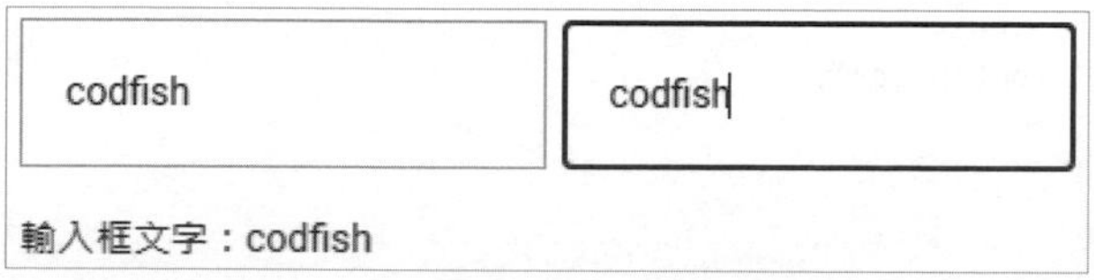

▲ 圖 2-6 雙向綁定

Vue 也可以將 HTML 的 style 與 class 進行動態綁定，新增以下內容。

➜ src\App.vue

```
<template>
  ...
  <div class="class-box">
    <input
      v-model="checked"
      type="checkbox"
    >
    <span :class="classObject">
      文字顏色
    </span>
    <span :class="classArray">
```

```
      文字粗細
    </span>
  </div>
</template>

<script setup lang="ts">
import { computed, ref } from 'vue';
...
const checked = ref(false);

const classObject = computed(() => ({
  'text-red': checked.value,
}));

const classArray = computed(() => {
  const result: string[] = [];

  if (checked.value) {
    result.push('font-black');
  }

  return result;
});
</script>

<style lang="sass">
...
.class-box
  width: 14rem
  padding: 1rem
  .text-red
    color: red
  .font-black
    font-weight: 900
</style>
```

　　class 可以使用物件或矩陣，這裡使用 computed 進行計算，computed 會自動收集內部依賴的響應式變數，變數發生變動時，computed 會立即重新計算結果，現在文字會在勾選 checkbox 加上指定 class 了。

☐ 文字顏色 文字粗細   ☑ 文字顏色 **文字粗細**

▲ 圖 2-7 綁定 class

2.1.2 條件渲染與事件

Vue 也可以根據不同條件渲染不同的 HTML 內容,而且處理 DOM 事件也變得相當容易。

我們新增一個可以任意新增文字的列表。

➜ src\App.vue

```
<template>
  ...
  <div class="list-box">
    <div class="tool">
      <input
        v-model="indexVisible"
        type="checkbox"
        class="input"
      >
      <input
        v-model="text"
        class="input"
        placeholder=" 輸入欲新增文字 "
      >
      <button class="button">
        新增
      </button>
    </div>

    <div class="list">
      <div
        v-for="item, i in list"
        :key="i"
        class="item"
      >
```

```
    <div
      v-if="indexVisible"
      class="index"
    >
      {{ i }}
    </div>

    <div class="text">
      {{ item }}
    </div>

    <div class="remove-btn">
      X
    </div>
  </div>
</div>
</div>
</template>

<script setup lang="ts">
...
const indexVisible = ref(true);
const text = ref('');
const list = ref(['codfish']);

</script>

<style lang="sass">
...

.list-box
  width: 24rem
  display: flex
  flex-direction: column
  padding: 1rem
  .tool
    display: flex
    gap: 1rem
    border: 1px solid #AAA
```

```
    padding: 1rem
    .input
      border: 1px solid #DDD
      padding: 0.4rem
      border-radius: 0.25rem
    .button
      background: #444
      padding: 0.4rem 1rem
      color: white
      border-radius: 0.25rem

  .list
    display: flex
    flex-direction: column
    gap: 0.4rem
  .item
    display: flex
    gap: 0.4rem
    padding: 1rem
    background: rgba(#000, 0.05)
    .index
      color: white
      background: #666
      padding: 0.1rem 0.6rem
      border-radius: 999px
    .text
      width: 100%
    .remove-btn
      padding: 0rem 0.4rem
      border: 1px solid #444
      border-radius: 0.25rem
      cursor: pointer
</style>
```

現在外觀如下。

▲ 圖 2-8　條件與列表渲染

v-for 可以用來將矩陣資料渲染為重複出現的 HTML 內容，v-if 則是在條件成立時，才會顯示 HTML 內容。

現在讀者們可以點擊 checkbox，變更狀態試試看，會發現列表的 index 會消失。

最後讓我們將事件綁定至 DOM 中吧，首先新增 function，用於新增列表文字與刪除列表項目。

➜ src\App.vue

```
...
<script setup lang="ts">
...
function addItem() {
  list.value.push(text.value);
  text.value = '';
}
function removeItem(index: number) {
  list.value.splice(index, 1);
}
</script>
...
```

現在只要使用 @ 符號並指定事件名稱，就可以綁定事件。

➜ src\App.vue

```
<template>
  ...
  <div class="list-box">
    <div class="tool">
      ...
      <button
        class="button"
        @click="addItem()"
      >
        新增
      </button>
    </div>

    <div class="list">
      <div ... >
        ...
        <div
          class="remove-btn"
          @click="removeItem(i)"
        >
          X
        </div>
      </div>
    </div>
  </div>
</template>
...
```

就是這麼簡單！ ﹑(≧∀≦)﹑

可以在輸入框輸入文字，按下新增後自動清空內容並在下方列表產生對應項目；點擊刪除按鈕則可刪除指定項目。

▲　圖 2-9　新增、刪除列表內容

除了點擊事件外，我們也可以綁定鍵盤事件。

➜　src\App.vue

```
<template>
  ...
  <div class="list-box">
    <div class="tool">
      ...
      <input
        ...
        placeholder=" 輸入欲新增文字 "
        @keydown.enter="addItem()"
      >
      ...
    </div>
    ...
  </div>
</template>
...
```

在輸入框內按下 Enter，可以直接新增項目了！ˋ(*´∽`*)ˊ

2.2　元件

此章節程式碼可以在以下連結取得。

- https://gitlab.com/deepmind1/animal-party/tutorial-vue/-/tree/components-basics

當網頁變得越來越複雜時，不難發現有很多區塊不只 HTML 重複，連 CSS 或 JS 邏輯也有大量重複，例如按鈕、輸入框等等，這時候我們就可以將其抽離，獨立為單一元件重複使用。

在 Vue 中推薦使用 Single-File Component（SFC）來定義元件，使用上相當簡單，就是一個包含 template、style、script 三個區塊，副檔名為 vue 檔案。

我們將剛才列表渲染中的 item 元件化試試看吧，新增 list-item.vue 檔案。

➜ src\components\list-item.vue

```
<template>
</template>

<script setup lang="ts">
</script>

<style scoped lang="sass">
</style>
```

首先來定義一下此元件可用的參數與事件，原本的 item 內包含 index 與列表文文字，由於內部有刪除按鈕，所以需要發出刪除事件。

➜ src\components\list-item.vue

```
...
<script setup lang="ts">
interface Props {
  indexVisible?: boolean;
  index: number;
  text?: string;
}
const props = withDefaults(defineProps<Props>(), {
  indexVisible: true,
  text: '',
});

const emit = defineEmits<{
  (e: 'remove', index: number): void;
}>();
</script>
...
```

接著讓我們加入 template 與 style。

➜ src\components\list-item.vue

```
<template>
  <div class="item">
    <div
      v-if="props.indexVisible"
      class="index"
    >
      {{ props.index }}
    </div>

    <div class="text">
      {{ props.text }}
    </div>
```

```
    <div
      class="remove-btn"
    >
      X
    </div>
  </div>
</template>

<script setup lang="ts">
...
</script>

<style scoped lang="sass">
.item
  display: flex
  gap: 0.4rem
  padding: 1rem
  background: rgba(#000, 0.05)
  .index
    color: white
    background: #666
    padding: 0.1rem 0.6rem
    border-radius: 999px
  .text
    width: 100%
  .remove-btn
    padding: 0rem 0.4rem
    border: 1px solid #444
    border-radius: 0.25rem
    cursor: pointer
</style>
```

最後加上點擊刪除按鈕發出刪除事件功能。

➜ src\components\list-item.vue

```
<template>
  <div class="item">
    ...
```

```
    <div
      class="remove-btn"
      @click="handleRemove()"
    >
      X
    </div>
  </div>
</template>

<script setup lang="ts">
...

function handleRemove() {
  emit('remove', props.index);
}
</script>
...
```

現在讓我們回到 App.vue 中，引入 list-item 元件吧。

➜ src\App.vue

```
<template>
  ...

  <div class="list-box">
    ...

    <div class="list">
      <list-item
        v-for="item, i in list"
        :key="i"
        :index="i"
        :text="item"
        @remove="removeItem"
      />
    </div>
  </div>
</template>
```

```
<script setup lang="ts">
...

import ListItem from './components/list-item.vue';
...
</script>

<style lang="sass">
...
</style>
```

App.vue 內的程式是不是減少了許多呢？元件化除了可以重複使用元素外，也可以用於封裝，讓父元件內的邏輯更加純粹，更不容易耦合。

同理，每個 box 內的內容完全獨立，我們也可以把每個 box 都封裝成單一元件。

➜ src\components\input-box.vue

```
<template>
  <div class="input-box">
    <input
      v-model="data"
      class="input"
    >
    <input
      v-model="data"
      class="input"
    >
    <div>
      輸入框文字：{{ data }}
    </div>
  </div>
</template>

<script setup lang="ts">
import { ref } from 'vue';
```

```
const data = ref('123');
</script>

<style scoped lang="sass">
.input-box
  padding: 1rem
  .input
    padding: 1rem
    border: 1px solid #AAA
    margin-bottom: 1rem
    margin-right: 0.4rem
</style>
```

➜ src\components\class-box.vue

```
<template>
  <div class="class-box">
    <input
      v-model="checked"
      type="checkbox"
    >
    <span :class="classObject">
      文字顏色
    </span>
    <span :class="classArray">
      文字粗細
    </span>
  </div>
</template>

<script setup lang="ts">
import { computed, ref } from 'vue';

const checked = ref(false);

const classObject = computed(() => ({
  'text-red': checked.value,
}));

const classArray = computed(() => {
```

```
  const result: string[] = [];

  if (checked.value) {
    result.push('font-black');
  }

  return result;
});
</script>

<style scoped lang="sass">
.class-box
  width: 14rem
  padding: 1rem
  .text-red
    color: red
  .font-black

    font-weight: 900
</style>
```

➜ src\components\list-box.vue

```
<template>
  <div class="list-box">
    <div class="tool">
      <input
        v-model="indexVisible"
        type="checkbox"
        class="input"
      >
      <input
        v-model="text"
        class="input"
        placeholder=" 輸入欲新增文字 "
        @keydown.enter="addItem()"
      >
      <button
        class="button"
        @click="addItem()"
      >
```

```
      新增
    </button>
  </div>

  <div class="list">
    <list-item
      v-for="item, i in list"
      :key="i"
      :index="i"
      :text="item"
      @remove="removeItem"
    />
  </div>
</div>
</template>

<script setup lang="ts">
import { ref } from 'vue';

import ListItem from './list-item.vue';

const indexVisible = ref(true);
const text = ref('');
const list = ref(['codfish']);

function addItem() {
  list.value.push(text.value);
  text.value = '';
}
function removeItem(index: number) {
  list.value.splice(index, 1);
}

interface Props {
  label?: string;
}
const props = withDefaults(defineProps<Props>(), {
  label: '',
});
```

```
const emit = defineEmits<{
  (e: 'update:modelValue', value: string): void;
}>();
</script>

<style scoped lang="sass">
.list-box
  width: 24rem
  display: flex
  flex-direction: column
  padding: 1rem
  .tool
    display: flex
    gap: 1rem
    border: 1px solid #AAA
    padding: 1rem
    .input
      border: 1px solid #DDD
      padding: 0.4rem
      border-radius: 0.25rem
    .button
      background: #444
      padding: 0.4rem 1rem
      color: white
      border-radius: 0.25rem

  .list
    display: flex
    flex-direction: column
    gap: 0.4rem
</style>
```

現在只要在 App.vue 引入以上元件。

➜ src\components\list-box.vue

```
<template>
  <input-box />
  <class-box />
```

```
  <list-box />
</template>

<script setup lang="ts">
import InputBox from './components/input-box.vue';
import ClassBox from './components/class-box.vue';
import ListBox from './components/list-box.vue';
</script>

<style lang="sass">
html, body, #app
  width: 100%
  height: 100%
  padding: 0
  margin: 0

#app
  display: flex
  flex-direction: column
</style>
```

最後會得到一模一樣的結果，但是每個 box 內的模板、樣式與程式邏輯變為完全獨立，不會互相影響，程式碼少了許多，這樣可以有效增加程式碼可讀性與專案維護性，至於到底怎麼拆分元件才好？那又是另一個學問了。乀(˚ ⱳ˚)「

Tips

想要深入了解的讀者可以參考以下連結：

◆ https://vuejs.org/guide/essentials/component-basics.html

以上就是 Vue 最基本的基礎知識了，剩下的部分讓我們邊做邊學吧。乀(•ω•)乁

NestJS 基礎概念

已經了解 NestJS 的讀者，可以跳過此章節。ー=≡Σ(((つ•ω•)つ

接下來 NestJS 入門的練習中，我們會假設要製作一個簡易的魚缸管理伺服器，提供 REST API 介面，可以增減魚兒。

想跟著實作的讀者可以下載以下連結專案，一起實作看看。

- https://gitlab.com/deepmind1/animal-party/tutorial-nest/-/tree/main

步驟與 Vue 專案大同小異（都是 JS 嘛 (´▽`)/）），一樣下載、解壓縮專案並用 VS Code 開啟，接著我們需要開始安裝專案內需要使用的套件。

1. 首先在「檢視」中找到「終端」選項並點擊開啟終端機。

▲ 圖 3-1 開啟終端機

2. 接著在畫面下方的終端機中輸入「npm i」（等同於 npm install）後，等待安裝完成。

▲ 圖 3-2 輸入安裝命令

3. 安裝完成後，輸入「npm run start:dev」則會啟動 NestJS 的開發模式。

```
[下午7:48:30] Starting compilation in watch mode...

[下午7:48:38] Found 0 errors. Watching for file changes.

[Nest] 12776  - 2023/09/07 下午7:48:44     LOG [NestFactory] Starting Nest application...
[Nest] 12776  - 2023/09/07 下午7:48:44     LOG [InstanceLoader] AppModule dependencies initialized +13ms
[Nest] 12776  - 2023/09/07 下午7:48:45     LOG [RoutesResolver] AppController {/}: +82ms
[Nest] 12776  - 2023/09/07 下午7:48:45     LOG [RouterExplorer] Mapped {/, GET} route +3ms
[Nest] 12776  - 2023/09/07 下午7:48:45     LOG [NestApplication] Nest application successfully started +2ms
```

▲ 圖 3-3 啟動開發模式

如此表示成功開啟 NestJS 開發模式了！✧*。٩(´ ロ ` *)و✧*。

3.1 建立資源

此章節程式碼可以在以下連結取得。

- https://gitlab.com/deepmind1/animal-party/tutorial-nest/-/tree/aquarium-module

每個資源一定都會有一個 module，負責包裝所有相關 provider 或 controller，還有一個 service 提供商業邏輯。

首先新增 aquarium.service。

➜ src\device\aquarium.service.ts

```
import { Injectable } from '@nestjs/common';

@Injectable()
export class AquariumService {
  //
}
```

接著新增 aquarium.module 並引入 aquarium.service。

➜ src\aquarium\aquarium.module.ts

```
import { Module } from '@nestjs/common';
import { AquariumService } from './aquarium.service';

@Module({
```

```
  imports: [],
  controllers: [],
  providers: [AquariumService],
})
export class AquariumModule {
  //
}
```

　　最後記得在 app.module 引入 aquarium.module。

➜ src\app.module.ts

```
...
import { AquariumModule } from './aquarium/aquarium.module';

@Module({
  imports: [AquariumModule],
  ...
})
export class AppModule {
  //
}
```

　　鱈魚：「如此 aquarium.module 就可以有作用了。(´▽`)/」

　　助教：「最好啦，一個 method 沒有啊。Σ(´д`;)」

　　鱈魚：「別急慢慢來嘛。(´˙ω˙`)」

　　現在我們需要儲存魚缸狀態還有魚的數量，首先讓我們新增型別定義。

➜ src\aquarium\type\index.ts

```
export interface Aquarium {
  fishList: Fish[];
}

export interface Fish {
  name: string;
}
```

接著讓我們在 service 新增變數。

➜ src\aquarium\aquarium.service.ts

```
import { Injectable } from '@nestjs/common';
import { Aquarium } from './type';

@Injectable()
export class AquariumService {
  private aquarium: Aquarium = {
    fishList: [],
  };
}
```

現在我們有魚缸，可以來養魚了！ㄟ(•ω•)ㄏ

新增增加、取得與刪除魚的 method。

➜ src\aquarium\aquarium.service.ts

```
...
export class AquariumService {
  ...

  get() {
    return this.aquarium;
  }

  addFish(name: string) {
    this.aquarium.fishList.push({
      name,
    });
  }

  removeFish(name: string) {
    const index = this.aquarium.fishList.findIndex((fish) => {
      return fish.name === name;
    });

    this.aquarium.fishList.splice(index, 1);
```

```
    }
}
```

現在可以養魚了！(·ω·)ノ ヾ(·ω·)

來開 REST API 吧，建立 API 的方式很簡單，NestJS 主要使用裝飾器（Decorator）實現各類功能，要建立一個 Get 方法，就使用 Get 裝飾器。

➜ src\aquarium\aquarium.controller.ts

```ts
import { Controller, Get } from '@nestjs/common';
import { AquariumService } from './aquarium.service';

@Controller('/')
export class AquariumController {
  constructor(private readonly aquariumService: AquariumService) {
    //
  }

  @Get('aquarium')
  get() {
    return this.aquariumService.get();
  }
}
```

記得要在 aquarium.module 引入 controller。

➜ src\aquarium\aquarium.controller.ts

```ts
...
import { AquariumController } from './aquarium.controller';

@Module({
  ...
  controllers: [AquariumController],
  ...
})
...
```

現在我們有一個 GET /aquarium API，可以用了。

我們使用 Postman 來測試 API 吧，前往官網（https://www.postman.com/ downloads）下載後安裝即可，過程我們就不贅述了，打開 Postman 後會看到以下畫面。

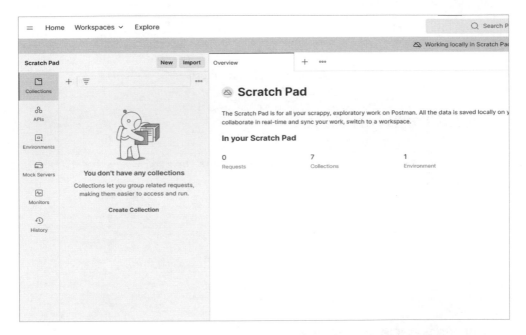

▲ 圖 3-4　Postman

按下 +，新增一個 Request。

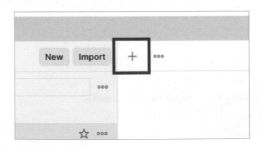

▲ 圖 3-5　新增 Get Request

輸入 http://localhost/aquarium，按下 Send。

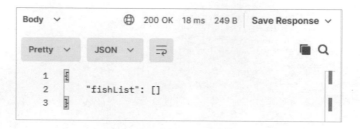

▲　圖 3-6　發送 Request

會在最下方看到 API 回應結果。

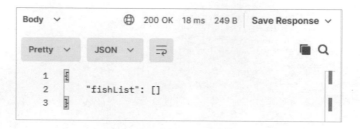

▲　圖 3-7　API 回應

現在讓我們新增加入魚的 API 吧。

➔　src\aquarium\aquarium.controller.ts

```
...
export class AquariumController {
  ...
  @Post('aquarium/fish')
  addFish(@Body() data: { name: string }) {
    console.log(`🐟 ~ data:`, data);
    return this.aquariumService.addFish(data.name);
  }
}
```

- @Post 裝飾器表示此 API 之 HTTP Method 需要使用 Post。

- @Body 裝飾器就可以取得 body 中的資料

在 Postman 新增此 API 吧。

▲ 圖 3-8　新增 Post Request

- 新增分頁後 URL 輸入：http://localhost/aquarium/fish

- Method 記得要選 POST

- 我們要透過 body 發送資料，我們將資料格式選擇 raw，類型選擇
  JSON，並在輸入框中輸入

```
{
    "name":"codfish"
}
```

現在按下 Send 後，會發現回應是 201，server 的 console 也會出現如下圖資
訊。

▲ 圖 3-9　Post 資料

現在我們再發送一次剛剛的 GET /aquarium API，會發現回應資料變成下圖。

▲ 圖 3-10 使用 Get API 取得新資料

鱈魚：「我們踏出建立 API 的第一步了！◇*。٩(´ロ｀*)ﻭ◇*。」

助教：「如果有人隨便 Post 資料會怎麼樣？(*´･д･)?」

鱈魚：「請他不要這麼做。ᐤ(ﾟ∀･)ᐣ」

助教：「最好是可以這樣啦。Σ(´д｀;)」

說的有點道理，讓我們加入資料驗證吧。

3.2 驗證資料

此章節程式碼可以在以下連結取得。

- https://gitlab.com/deepmind1/animal-party/tutorial-nest/-/tree/validation

驗證資料最直覺的方式就是檢查資料是否存在，有問題時回應 HttpCode 400 錯誤。

→　src\aquarium\aquarium.controller.ts

```
import {
  Body,
  Controller,
  Get,
  HttpException,
  HttpStatus,
  Post,
} from '@nestjs/common';
...
export class AquariumController {
  ...
  @Post('aquarium/fish')
  addFish(@Body() data: { name: string }) {
    const { name } = data;
    if (!name) {
      throw new HttpException('name 必填', HttpStatus.BAD_REQUEST);
    }

    return this.aquariumService.addFish(name);
  }
}
```

在 NestJS 中，可以直接使用 HttpException 拋出錯誤狀態，現在我們刻意發出缺少 name 資料的請求看看。

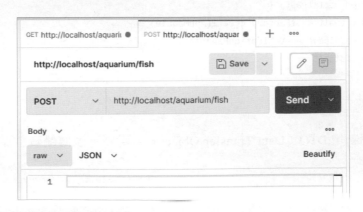

▲ 圖 3-11 Post 缺少必填資料

現在應該會看到回應變成 400，內容變成錯誤訊息了。

▲ 圖 3-12　400 錯誤訊息

不過每個 API 都要這樣驗證資料好麻煩，讓我們直接使用 ValidationPipe，可以同時完成資料定義、驗證和轉換，超方便呦。(ㅍﾉㅍ)✧

首先安裝必要套件。

```
npm i class-validator class-transformer
```

開啟全域驗證，這樣所有的 controller 都會啟用驗證。

➜ src\main.ts

```
...
import { ValidationPipe } from '@nestjs/common';

async function bootstrap() {
  const app = await NestFactory.create(AppModule);
  app.useGlobalPipes(new ValidationPipe());
  await app.listen(80);
}
bootstrap();
```

讓我們新增 DTO（Data Transfer Object），用於定義傳輸資料結構、驗證與轉換。

➡ src\aquarium\dto\index.ts

```ts
import { IsNotEmpty, IsString } from 'class-validator';

export class AddFishDto {
  @IsNotEmpty()
  @IsString()
  name!: string;
}
```

現在只要在 controller 引用 DTO 就可以了，同時刪除原本檢查 name 是否存在的邏輯。

➡ src\aquarium\aquarium.controller.ts

```ts
...
import { AddFishDto } from './dto';
...
export class AquariumController {
  ...
  @Post('aquarium/fish')
  addFish(@Body() data: AddFishDto) {
    const { name } = data;
    return this.aquariumService.addFish(name);
  }
}
```

現在再發送一次 POST /aquarium/fish API，會發現即使刪除了驗證邏輯，一樣有驗證效果。

▲ 圖 3-13　使用 dto 驗證

　　由於我們有使用 @IsString 驗證裝飾器，如果 name 的型別不是字串，一樣
會驗證失敗。

▲ 圖 3-14　name 必須為字串

　　是不是很棒啊！最後讓我們完成刪除魚的 API 吧。(´▽`)/

　　新增刪除 DTO。

➜ src\aquarium\dto\index.ts

```
...
export class RemoveFishDto {
  @IsNotEmpty()
  @IsString()
  name!: string;
}
```

➜ src\aquarium\aquarium.controller.ts

```
...
import { AddFishDto, RemoveFishDto } from './dto';
...
export class AquariumController {
```

```
...
@Delete('aquarium/fish/:name')
removeFish(@Param() data: RemoveFishDto) {
  const { name } = data;
  return this.aquariumService.removeFish(name);
}
}
```

刪除 API 的名稱資料放在 url 中，所以要使用 @Param 裝飾器，才可以取得 url 中的資料。

最後在 Postman 新增刪除 API，HTTP Method 記得選擇 DELETE。

▲ 圖 3-15 新增刪除 API

現在我們可以自由新增、取得與刪除魚缸中的魚了！ヽ(*´∀`*)ﾉ

以上我們完成 NestJS 基礎入門了，實際上 NestJS 還有超多功能可用，只是本書專案著重於前端開發，我們就不在這裡展開討論了。ヽ(•ω•)ﾉ

第 **4** 章

打造派對門面

現在讓我們進入「1.4.1 章」建立的派對遊戲前端專案，老樣子用 VS Code 開啟資料夾，打開終端機，輸入 npm run dev 開啟開發伺服器後，開啟網頁瀏覽器，準備動工！

Tips

還沒安裝相依套件的讀者記得要先運行一次 npm i 喔！

4.1 打造背景

現在畫面只有一個寂寞的「我是 Home Page」，派對怎麼可以沒有背景勒，讓我們來打造背景吧！

此章節程式碼可以在以下連結取得。

- https://gitlab.com/deepmind1/animal-party/web/-/tree/feat/creating-the-entrance-background

4.1.1　建立背景元件

希望畫面有各種多彩的飄浮多邊形，讓畫面有熱鬧、繽紛的感覺。

預計有以下參數可用：

- 自訂主色：背景主色

- 色塊顏色：內部色塊會隨機選擇指定顏色

- 初始數量：畫面出現時，內部初始多邊形數量，以免畫面空蕩蕩

- 最大數量：色塊最大數量，超過此數量時，會暫停產生多邊形，以免畫面爆炸

- 產生間距：越短產生多邊形速度越快

首先建立 background-polygons-floating 元件。（複製 src\components\base-ex.vue 檔案，改個檔名即可。）

➡ src\components\background-polygons-floating.vue

```
<template></template>

<script setup lang="ts">
import { ref } from 'vue';

interface Props {
  label?: string;
}
const props = withDefaults(defineProps<Props>(), {
  label: '',
});

const emit = defineEmits<{
  (e: 'update:modelValue', value: string): void;
}>();
</script>

<style scoped lang="sass">
</style>
```

　　依據剛剛提到的參數定義 Props。

```
interface Props {
  /** 背景顏色 */
  backgroundColor?: string;
  /** 多邊形顏色 */
  polygonColors?: string[];
  /** 初始數量，畫面出現時，內部初始方塊數量 */
  initialQuantity?: number;
  /** 色塊最大數量，超過此數量時，會暫停產生多邊形 */
  maxQuantity?: number;
  /** 產生間距，越短生成速度越快，單位 ms */
  generationInterval?: number;
}
const props = withDefaults(defineProps<Props>(), {
  backgroundColor: '#FF9258',
  polygonColors: () => ['#FF5D05', '#CBE64E', '#40FFF8', '#B14DFF'],
  initialQuantity: 10,
  maxQuantity: 50,
  generationInterval: 500,
});
```

　　接著加入背景 div，並建立背景樣式 backgroundStyle，我希望背景有漸層效果，讓顏色更豐富一些，所以透過 JS 計算顏色。

➔ src\components\background-polygons-floating.vue

```
<template>
  <div
    class="overflow-hidden"
    :style="backgroundStyle"
  >
  </div>
</template>

<script setup lang="ts">
import { computed, ref } from 'vue';
import { colors } from 'quasar';
```

```
const { lighten, textToRgb, rgbToHsv, hsvToRgb, rgbToHex } = colors;

...

const backgroundStyle = computed(() => {
  // 變亮並偏移色相
  const lightenColor = lighten(props.backgroundColor, 10);
  const hsvColor = rgbToHsv(textToRgb(lightenColor));
  hsvColor.h += 10;

  const result = rgbToHex(hsvToRgb(hsvColor));

  return {
    background: `linear-gradient(-10deg, ${props.backgroundColor}, ${result})`
  }
});
</script>
```

Tips

藉由 Quasar 提供之 Color Utils，可以輕鬆調整顏色。

- https://quasar.dev/quasar-utils/color-utils

接著在 the-home 元件中引入 background-polygons-floating 元件，來實際看看效果。

把 the-home 原本的內容清空，引入 background-polygons-floating。

→ src\views\the-home.vue

```
<template>
  <background-polygons-floating class="absolute inset-0" />
</template>

<script setup lang="ts">
import BackgroundPolygonsFloating from '../components/background-polygons-floating.vue';
</script>
```

現在應該可以看到背景出現了！(oﾟ▽ﾟ)o

▲ 圖 4-1 背景底色

4.1.2 建立可複用的多邊形元件

接著是最重要的多邊形部份了，定義一個 base-polygon 元件，產生各種樣式的多邊形，老樣子先定義多邊形 Props。

➜ src\components\base-polygon.vue

```ts
<script lang="ts">
export enum ShapeType {
  ROUND = 'round',
  TRIANGLE = 'triangle',
  SQUARE = 'square',
  PENTAGON = 'pentagon'
}

export enum FillType {
  SOLID = 'solid',
  FENCE = 'fence',
  SPOT = 'spot',
}
</script>

<script setup lang="ts">
import { ref } from 'vue';

interface Props {
```

```
  size?: string;
  color?: string;
  rotate?: string;
  opacity?: string | number;
  shape?: `${ShapeType}`,
  fill?: `${FillType}`
}
const props = withDefaults(defineProps<Props>(), {
  size: '10rem',
  color: 'white',
  rotate: '0deg',
  opacity: 1,
  shape: 'round',
  fill: 'fence',
});
</script>
```

 Tips

在 setup script 中 export 會發生異常，所以在此建立另一個一般的 script，用來 export 型別資料。

◆ https://cn.vuejs.org/api/sfc-script-setup.html#usage-alongside-normal-script

接著提供 div，準備切割成多邊形外觀

➜ src\components\base-polygon.vue

```
<template>
  <div class="frame">
    <div class="polygon" />
  </div>
</template>
```

切割形狀部分，使用 CSS 之 clipPath 實現，所以根據定義的 ShapeType，列舉一下對應的種類。

➔ src\components\base-polygon.vue

```
...

<script setup lang="ts">
import { computed, ref } from 'vue';

...

const clipPathMap = {
  [ShapeType.ROUND]: `circle(50% at 50% 50%)`,
  [ShapeType.SQUARE]: `polygon(50% 0%, 100% 50%, 50% 100%, 0% 50%)`,
  [ShapeType.TRIANGLE]: `polygon(50% 0%, 0% 100%, 100% 100%)`,
  [ShapeType.PENTAGON]: `polygon(50% 0%, 100% 38%, 82% 100%, 18% 100%, 0% 38%)`,
}
const clipPath = computed(() => clipPathMap?.[props.shape] ?? clipPathMap[ShapeType.ROUND]);
</script>
```

　　接著是填滿樣式（fill）的部分，實心很容易，但是網格狀或斑點狀的填滿效果要怎麼實現呢？有一個辦法，一樣透過萬能的 CSS 實現，也就是 maskImage！（CSS 萬歲♪(´▽｀)）

　　所以我們從百寶袋（大家可以直接前往 GitLab 專案中下載）變出 mask 要用的 svg 檔案：

- line.svg

- round.svg

放置於 public\images 目錄中，並列舉 fill 內容。

➔ src\components\base-polygon.vue

```
...

<script setup lang="ts">
...

const fillMap = {
```

```
  [FillType.SOLID]: ``,
  [FillType.FENCE]: `url(/images/line.svg)`,
  [FillType.SPOT]: `url(/images/round.svg)`,
}
const fill = computed(() => fillMap?.[props.fill] ?? fillMap[FillType.SPOT]);
</script>
```

最後把剛剛定義的所有內容集結成 style。

➜ src\components\base-polygon.vue

```
...

<script setup lang="ts">
...

const style = computed(() => ({
  width: props.size,
  height: props.size,
  backgroundColor: props.color,
  maskImage: fill.value,
  maskSize: `15%`,
  opacity: props.opacity,
  clipPath: clipPath.value,
  transform: `rotate(${props.rotate})`,
}));
</script>
```

並加入 div 中。

➜ src\components\base-polygon.vue

```
<template>
  <div class="frame">
    <div
      class="polygon"
      :style="style"
    />
  </div>
```

```
</template>

...
```

最後我們在 background-polygons-floating 中引入 base-polygon 看看效果吧！

➜ src\components\background-polygons-floating.vue

```
<template>
  <div
    class="overflow-hidden"
    :style="backgroundStyle"
  >
    <base-polygon
      shape="round"
      fill="solid"
    />
    <base-polygon
      shape="round"
      fill="spot"
    />
    <base-polygon
      shape="triangle"
      fill="fence"
    />
    <base-polygon
      shape="square"
      fill="spot"
    />
    <base-polygon
      shape="pentagon"
      fill="solid"
    />
  </div>
</template>

...
```

可以看到多樣的多邊形們出現了！ゝ(≧∀≦)ノ

▲ 圖 4-2 成功建立多邊形

4.1.3　漂浮吧！多邊形

現在讓多邊型不斷冒出來並漂起來吧！

要讓很多個多邊形出現，第一步就是建立一個產生多邊形的 function（這一句好像廢話(๑•́ω•̀๑)），先來定義一下多邊形需要的參數。

➜ src\components\background-polygons-floating.vue

```
...

<script setup lang="ts">
...

import BasePolygon, { FillType, ShapeType } from '../components/base-polygon.vue';

interface PolygonParams {
  left: string;
  top: string;
  size: string;
  rotate: string;
  opacity: string | number;
```

```
  shape: `${ShapeType}`;
  fill: `${FillType}`;
  color: string;
  animationDuration: string;
}
...
</script>
```

可以注意到有一個參數是 animationDuration，這是因為我們預期使用 CSS
之 animation 讓多邊形動起來，透過讓每個多邊形的動畫時間長度都不同，就可
以有交錯漂浮的效果，讓畫面更多元。

首先新增名為 createPolygonParams 的 function，負責產生多邊形參數。

➜ src\components\background-polygons-floating.vue

```
import { random, sample } from 'lodash-es';

...

function createPolygonParams() {
  const params: PolygonParams = {
    left: `${random(0, 100)}%`,
    top: `${random(0, 100)}%`,
    size: `${random(5, 20)}rem`,
    rotate: `${random(0, 90)}deg`,
    opacity: random(0.1, 0.2, true),
    color: sample(props.polygonColors) ?? 'white',
    shape: sample(Object.values(ShapeType)) ?? 'round',
    fill: sample(Object.values(FillType)) ?? 'solid',
    animationDuration: `${random(5, 20)}s`,
  }
  return params;
}
```

這裡引入了 lodash 的 random 與 sample，功能分別為：

- random：產生指定的隨機數值

- sample：隨機取出 Array 內某元素

接著就是實際產生多邊形的時候了，一般來說最常見的方式就是用一個矩陣將 PolygonParams 裝起來，並在 template 透過 v-for 產生即可，不過這樣有個小問題。

當多邊形動畫結束時，要移除矩陣中指定的多邊形會稍微有點麻煩，這裡用一個更簡單的方式實作，也就是 JavaScript 的 Map 物件。

 Tips

想要深入了解的讀者可以參考以下連結：

- https://developer.mozilla.org/zh-TW/docs/Web/JavaScript/Reference/Global_Objects/Map

其實使用 object 也可以，只是 Map 提供更簡便直覺的 Method，用來新增、刪除內容。

定義一個儲存多邊形用的 Map 物件。

➔ src\components\background-polygons-floating.vue

```
import { random, sample } from 'lodash-es';

...

function createPolygonParams() {
  const params: PolygonParams = {
    left: `${random(0, 100)}%`,
    top: `${random(0, 100)}%`,
    size: `${random(5, 20)}rem`,
    rotate: `${random(0, 90)}deg`,
    opacity: random(0.6, 0.8, true),
    color: sample(props.polygonColors) ?? 'white',
    shape: sample(Object.values(ShapeType)) ?? 'round',
    fill: sample(Object.values(FillType)) ?? 'solid',
```

```
    animationDuration: `${random(5, 20)}s`,
  }
  return params;
}
```

並新增用來新增、刪除多邊形的 function。

➜ src\components\background-polygons-floating.vue

```
import { nanoid } from 'nanoid';

...

function addPolygon(params: PolygonParams) {
  polygonsMap.value.set(nanoid(), params);
}
function removePolygon(id: string) {
  polygonsMap.value.delete(id);
}
```

這裡透過 nanoid 產生唯一 ID，方便後續直接刪除。

以上我們完成產生多邊形的程式了，接下來實作持續產生多邊形效果，有以下需求：

- 依據 generationInterval 之時間間距產生多邊形。

- 依據 initialQuantity 產生初始數量多邊形。

- 元件被解除時，需自動銷毀計時器，否則即使元件消失後也會在背景持續產生多邊形，會讓記憶體爆炸。

定時的部分，我們直接使用 VueUse 提供的 useIntervalFn。

➜ src\components\background-polygons-floating.vue

```
import { promiseTimeout, useIntervalFn } from '@vueuse/core';

...
```

```
useIntervalFn(async () => {
  // 到達最大數量後，停止生成
  if (polygonsMap.value.size >= props.maxQuantity) return;

  // 刻意延遲隨機時間，讓多邊形生成時機不會過度固定而顯得死板
  await promiseTimeout(random(300, 1000));
  addPolygon(createPolygonParams());
}, props.generationInterval);

function init() {
  // 預先建立多邊形
  for (let i = 0; i < props.initialQuantity; i++) {
    addPolygon(createPolygonParams());
  }
}
init();
```

現在讓我們打開 Vue DevTools。

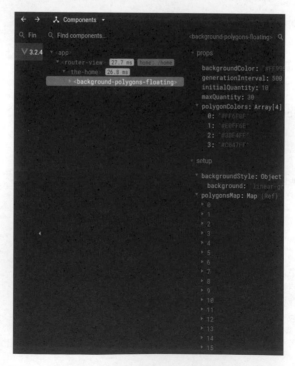

▲ 圖 4-3　持續產生多邊形

可以看到 polygonsMap 內應該會有持續冒出多邊形了。

讓我們將 polygonsMap 加到 template 裡面吧！將原本測試用的多邊形刪除，改為以下程式。

➡ src\components\background-polygons-floating.vue

```
<template>
  <div
    class="overflow-hidden"
    :style="backgroundStyle"
  >
    <base-polygon
      v-for="[key, polygon] in polygonsMap"
      :key="key"
      class="absolute"
      :style="`left: ${polygon.left}; top: ${polygon.top}; animation-duration:
${polygon.animationDuration}`"
      :color="polygon.color"
      :rotate="polygon.rotate"
      :shape="polygon.shape"
      :fill="polygon.fill"
      :size="polygon.size"
      :opacity="polygon.opacity"
      @animationend="removePolygon(key)"
    />
  </div>
</template>

...
```

讀者可能會注意到在此註冊了 **@**animationend 事件，這個事件用於 CSS 之 animation 播放結束時觸發，很適合用來刪除元素，所以將 removePolygon 綁定在此事件。

現在可以看到多邊形出現了。(‧∀‧)╱

▲ 圖 4-4 成功建立多邊形

最後我們讓多邊形動起來吧，實作相當簡單，建立動畫用的 class，並將
class 加入 template 就行惹。

➜ src\components\background-polygons-floating.vue

```
<template>
  <div ... >
    <base-polygon
      v-for="[key, polygon] in polygonsMap"
      :key="key"
      class="absolute polygon-floating"
      ...
    />
  </div>
</template>

<script setup lang="ts">
...
</script>

<style scoped lang="sass">
.polygon-floating
  animation: polygon-floating 10s forwards linear
@keyframes polygon-floating
  0%
```

```
    transform: translate(0px, 0px) rotate(0deg)
    opacity: 0
  10%
    opacity: 1
  90%
    opacity: 1
  100%
    transform: translate(-10rem, 10rem) rotate(6deg)
    opacity: 0
</style>
```

可以看到多邊形飄起來了！(。^▽^)

▲ 圖 4-5 多邊形漂浮

助教：「這個顏色看起來有點美感炸裂啊(#°Д°)」

鱈魚：「那就來調整一下透明度、顏色和尺寸吧。ᕕ('∀'ᕗ)」

→ src\components\background-polygons-floating.vue

```
...

const props = withDefaults(defineProps<Props>(), {
  ...
  polygonColors: () => ['#FFCF4F', '#58AC18', '#C88140', '#FF744F'],
  ...
```

```
});

...

function createPolygonParams() {
  const params: PolygonParams = {
    ...
    size: `${random(2, 15)}rem`,
    ...
    opacity: random(0.4, 0.6, true),
    ...
  }
  return params;
}

...
```

▲ 圖 4-6 多邊形漂浮

這樣看起來協調多了，到目前為止，我們完成背景了！✧*。٩(´ω｀*)و✧*。

大家也可以依照自己的喜好調整參數喔！

4.2 打造選單按鈕

此章節程式碼可以在以下連結取得。

- https://gitlab.com/deepmind1/animal-party/web/-/tree/feat/ creating-the-entrance-background

有背景後,接著打造選單按鈕吧,首先分析一下期望的按鈕功能:

- 可加入文字與文字外框並自訂顏色。

- 除了滑鼠互動外,也可以透過程式觸發 hover、active 等等效果,這樣就可以透過滑鼠以外的方式觸發按鈕,可用於讓玩家透過虛擬搖桿控制按鈕。

建立按鈕元件檔案。

➡ src\components\base-btn.vue

```
<template></template>

<script setup lang="ts">
import { ref } from 'vue';

interface Props {
  label?: string;
  labelColor?: string;
  labelHoverColor?: string;
  strokeColor?: string;
  strokeHoverColor?: string;
```

```
    strokeSize?: string;
}
const props = withDefaults(defineProps<Props>(), {
  label: '',
  labelColor: 'white',
  labelHoverColor: undefined,
  strokeColor: '#888',
  strokeHoverColor: undefined,
  strokeSize: '2'
});
</script>
```

　　新增一個變數，用來儲存目前按鈕狀態。

➔　src\components\base-btn.vue

```
...
<script lang="ts">
export interface State {
  active: boolean,
  hover: boolean,
}
</script>

<script setup lang="ts">
...
const state = ref<State>({
  active: false,
  hover: false,
});
</script>
...
```

　　這裡將 State 型別定義透過 export 匯出，讓其他需要定義此元件狀態時使用。

　　先讓我們把按鈕加到 the-home 中，讓我們一步一步地完成按鈕樣式。

➜ src\views\the-home.vue

```
<template>
  <background-polygons-floating class="absolute inset-0" />

  <div class="absolute inset-0 flex flex-col flex-center gap-20">
    <base-btn />
  </div>
</template>


<script setup lang="ts">
import BackgroundPolygonsFloating from '../components/background-polygons-floating.
vue';
import BaseBtn from '../components/base-btn.vue';
</script>
```

現在回到按鈕元件，首先是最外層的部分。

- 加入按鈕 CSS 並加入 active class 產生 active 時的動畫效果。

- 同時加入 slot 增加彈性並透過 scoped-slots 對外提供 state。

➜ src\components\base-btn.vue

```
<template>
  <div
    class="btn flex flex-center text-3xl p-12 rounded-full"
    :class="btnClass"
  >
    <slot v-bind="state" />
  </div>
</template>

<script lang="ts">
...
</script>

<script setup lang="ts">
...
```

```
const btnClass = computed(() => ({
  active: state.value.active,
}));
</script>


<style scoped lang="sass">
.btn
  backdrop-filter: blur(6px)
  background: rgba(white, 0.2)
  box-shadow: 2.8px 2.8px 2.2px rgba(0, 0, 0, 0.006), 6.7px 6.7px 5.3px rgba(0, 0,
0, 0.008), 12.5px 12.5px 10px rgba(0, 0, 0, 0.01), 22.3px 22.3px 17.9px rgba(0, 0,
0, 0.012), 41.8px 41.8px 33.4px rgba(0, 0, 0, 0.014), 100px 100px 80px rgba(0, 0, 0,
 0.02)
  user-select: none
  overflow: hidden
  cursor: pointer
  transition-timing-function: cubic-bezier(0.000, 1.650, 1.000, 1.650)
  transition-duration: 0.2s
  &.active
    transform: scale(0.98) rotate(-1deg)
    transition-timing-function: cubic-bezier(0.34, 1.56, 0.64, 1)
</style>
```

接著加入 label 文字的部份。

➜ src\components\base-btn.vue

```
<template>
  <div
    class="btn flex flex-center text-3xl p-12 rounded-full"
    :class="btnClass"
  >
    <slot v-bind="state" />

    <!-- label -->
    <div class="label relative font-black tracking-widest">
      {{ props.label }}
    </div>
  </div>
</template>
```

...

現在回到 the-home 元件中，在按鈕加入 label 參數。

➜ src\views\the-home.vue

```
<template>
  ...

  <div class="absolute inset-0 flex flex-col flex-center gap-20">
    <base-btn label=" 加入派對 " />
  </div>
</template>
...
```

目前看起來應該長這樣。

▲ 圖 4-7 按鈕雛型

看起來真不是普通的醜呢，別擔心，還沒結束 (˙ω˙)✧

把 Props 中的參數轉換成 label 樣式，新增 labelStyle。

➜ src\components\base-btn.vue

```
const labelStyle = computed(() => {
  let color = props.labelColor;
```

```
  if (props.labelHoverColor) {
    color = state.value.hover ? props.labelHoverColor : props.labelColor;
  }

  return {
    color,
  }
});
```

　　綁定至對應的標籤並加上 class，讓顏色變化有動畫。

➜ src\components\base-btn.vue

```
<template>
  <div ... >
    <slot v-bind="state" />

    <!-- label -->
    <div
      ...
      :style="labelStyle"
    >
      {{ props.label }}
    </div>
  </div>
</template>

...

<style scoped lang="sass">
...

.label
  transition-duration: 0.4s
</style>
```

接著實現文字外框的效果，這裡使用 SVG 實現文字外框效果，若要將 SVG 效果綁定至 class 中，需要指定 id，為了避免 id 重複，這裡使用 nanoid 產生唯一 ID

- 建立 strokeStyle，產生外框樣式。

- 將產生文字外框的元素設為絕對定位，和原本的 label 疊在一起，產生文字外框效果。

新增程式邏輯。

➜ src\components\base-btn.vue

```
const svgFilterId = `svg-filter-${nanoid()}`;
const strokeStyle = computed(() => {
  let color = props.strokeColor;

  if (props.strokeHoverColor) {
    color = state.value.hover ? props.strokeHoverColor : props.strokeColor;
  }

  return {
    color,
    filter: `url(#${svgFilterId})`
  }
});
```

接著新增 template 與 style

➜ src\components\base-btn.vue

```
<template>
  <div ... >
    ...
    <!-- label -->
    <div ... >
      ...
      <!-- stroke -->
      <div
```

```
        class="label-stroke absolute"
        :style="strokeStyle"
      >
        {{ props.label }}
      </div>
    </div>

    <svg
      version="1.1"
      style="display: none;"
    >
      <defs>
        <filter :id="svgFilterId">
          <feMorphology
            operator="dilate"
            :radius="props.strokeSize"
          />
          <feComposite
            operator="xor"
            in="SourceGraphic"
          />
        </filter>
      </defs>
    </svg>
  </div>
</template>
...

<style scoped lang="sass">
...
.label-stroke
  top: 0px
  transition-duration: 0.4s
</style>
```

這裡使用了 SVG 的 filter 功能，其步驟如下：

1. 使用 feMorphology 讓邊界膨脹，並設定 radius 指定膨脹幅度，達成設定 border 寬度效果。

2. 再透過 feComposite 設為 xor，僅保留膨脹部分，以免影響文字。

3. 最後使用 CSS filter 功能，將 SVG 效果透過 ID 綁定至 DOM 中。

外框就這樣出現了！∠(ᐛ 」∠)_

▲ 圖 4-8 加入文字外框

現在讓我們指定一下顏色。

➜ src\views\the-home.vue

```
<template>
  ...
  <div ...>
    <base-btn
      label=" 加入派對 "
      label-hover-color="#ff9a1f"
      stroke-color="#856639"
      stroke-hover-color="white"
    />
  </div>
</template>
...
```

看起來比較像樣一點了。(￣▽￣)

▲　圖 4-9　加入外框顏色

最後也是最重要的部分，綁定各類事件，讓按鈕產生互動效果吧。

➜ src\components\base-btn.vue

```
<template>
  <div
    class="btn flex flex-center text-3xl p-12 rounded-full"
    :class="btnClass"
    @click="handleClick()"
    @mouseenter="handleMouseenter"
    @mouseleave="handleMouseleave"
    @mousedown="handleMousedown"
    @mouseup="handleMouseup"
  >
   ...
  </div>
</template>
...

<script setup lang="ts">
...

const emit = defineEmits<{
  (e: 'click'): void;
}>();
```

```
...

function handleClick() {
  emit('click');
}
function handleMouseenter() {
  state.value.hover = true;
}
function handleMouseleave() {
  state.value.hover = false;
}
function handleMousedown() {
  state.value.active = true;
}
function handleMouseup() {
  state.value.active = false;
}
</script>
...
```

現在 hover 和 click 時，按鈕都會有反應了！

▲ 圖 4-10 綁定按鈕事件

助教：「可是看起來有夠陽春，很遜捏。(°ㅁ°) 」

鱈魚：「沒問題，讓我們加點裝飾吧！(. ❛ ᴗ ❛.) 」

這時候 slot 就派上用場了，首先把按鈕拉寬一點。

➔ src\views\the-home.vue

```
<template>
  ...
  <div class="absolute inset-0 flex flex-col flex-center gap-20">
    <base-btn
      label=" 加入派對 "
      ...
      class=" w-[30rem]"
    />
  </div>
</template>
```

接著使用 slot 插入裝飾用元素並加入 CSS 效果。

➔ src\views\the-home.vue

```
<template>
  ...
  <div class="absolute inset-0 flex flex-col flex-center gap-20">
    <base-btn
      v-slot="{ hover }"
      label=" 加入派對 "
      ...
    >
      <div
        class="btn-content absolute inset-0"
        :class="{ ‹hover›: hover }"
      >
        <base-polygon
          class="absolute btn-polygon-lt"
          size="13rem"
          shape="round"
          fill="spot"
          opacity="0.6"
        />

        <q-icon
```

```
        name="sports_esports"
        color="white"
        size="8rem"
        class="absolute game-icon"
      />
    </div>
  </base-btn>
 </div>
</template>

<script setup lang="ts">
...
import BasePolygon from '../components/base-polygon.vue';
</script>

<style scoped lang="sass">
.btn-polygon-lt
  left: 0
  top: 0
  transform: translate(-50%, -60%)


.game-icon
  right: 0
  bottom: 0
  transform: translate(12%, 24%) rotate(-10deg)
  opacity: 0.6

.btn-content
  transform: scale(1)
  transition-duration: 0.4s
  transition-timing-function: cubic-bezier(0.545, 1.650, 0.520, 1.305)
  &.hover
    transform: scale(0.96) rotate(-2deg)
    transition-timing-function: cubic-bezier(0.34, 1.56, 0.64, 1)
</style>
```

▲ 圖 4-11　加上按鈕裝飾

看起來不會那麼單調了吧！(´▽`)ノ

讀者們如果想要更複雜的效果，可以自行魔改喔！

現在把「加入遊戲」按鈕也加上去吧。

➡ src\views\the-home.vue

```
<template>
  ...
  <div class="absolute inset-0 flex flex-col flex-center gap-20">
    ...
    <base-btn
      v-slot="{ hover }"
      label=" 加入遊戲 "
      label-hover-color="#ff9a1f"
      stroke-color="#856639"
      stroke-hover-color="white"
      class=" w-[30rem]"
    >
      <div
        class="btn-content absolute inset-0"
        :class="{ ‹hover›: hover }"
      >
        <base-polygon
          class="absolute btn-polygon-lt"
          size="13rem"
          rotate="144deg"
          shape="pentagon"
```

```
        opacity="0.6"
      />
      <q-icon
        name="person_add"
        color="white"
        size="7.8rem"
        class="absolute join-icon"
      />
    </div>
  </base-btn>
</div>
</template>
...
<style scoped lang="sass">
...

.join-icon
  right: 0
  bottom: 0
  transform: translate(6%, 20%) rotate(-10deg)
  opacity: 0.6

</style>
```

最後換個與遊戲風格更協調的字體。

➜ src\views\the-home.vue

```
...

<style lang="sass">
@import url('https://fonts.googleapis.com/css2?family=Noto+Sans+TC:wght@100;300;400;50
0;700;900&display=swap')

html, body, #app
  width: 100%
  height: 100%
  padding: 0
  margin: 0
  font-family: 'Noto Sans TC', sans-serif
```

```
...
</style>
```

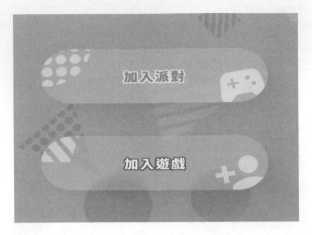

▲ 圖 4-12 加上按鈕裝飾

主選單按鈕完成！(/≧▽≦)/

4.3 打造標題 Logo

最後讓我們打造標題 Logo 吧！

此章節程式碼可以在以下連結取得。

■ https://gitlab.com/deepmind1/animal-party/web/-/tree/feat/create-title-logo

建立 title-logo 元件。

➜ src\components\title-logo.vue

```
<template>
</template>

<script setup lang="ts">
</script>

<style scoped lang="sass">
</style>
```

將 title-logo 加入 the-home，讓我們邊做邊預覽樣式。

➜ src\views\the-home.vue

```
<template>
  ...
  <div class="absolute inset-0 flex flex-center gap-20">
    <title-logo />

    <div class="flex flex-col flex-center gap-20">
      <base-btn
        ...
        label=" 加入派對 "
        ...
      >
        ...
      </base-btn>

      <base-btn
        ...
        label=" 加入遊戲 "
        ...
      >
        ...
      </base-btn>
    </div>
  </div>
```

```
</template>

<script setup lang="ts">
...
import TitleLogo from '../components/title-logo.vue';
</script>
```

第一步先讓我們把標題文字打出來。

➜　src\components\title-logo.vue

```
<template>
  <div class="flex flex-col flex-center">
    <div>
      Animals
    </div>
    <div>
      Party
    </div>
    <div>
      動物派對嗨起來！
    </div>
  </div>
</template>
...
```

▲ 圖 4-13 加入 title-logo

看起來遜到炸裂。(´・ω・`)

讓我們繼續加點魔法,加點字體和樣式。

先在 App.vue 引入字體。

➜ src\App.vue

```
...
<style lang="sass">
...
@import url('https://fonts.googleapis.com/css2?family=Luckiest+Guy&display=swap')


...
</style>
```

接著調整字體和字級。

➜ src\components\title-logo.vue

```
<template>
  <div class=" relative leading-none">
    <div
      class="flex flex-col flex-center text-white pb-6"
    >
      <div class="text-[8rem] font-game">
        ANIMALS
      </div>
      <div class="text-[12.5rem] font-game">
        PARTY
      </div>
      <div class="text-[4.2rem] font-black mt-3">
        動物派對嗨起來!
      </div>
    </div>
  </div>
</template>
...
<style scoped lang="sass">
.font-game
```

```
  font-family: 'Luckiest Guy'
</style>
```

▲　圖 4-14　調整文字

看起來有模有樣了！、(≧∀≦)、

接著和選單按鈕的原理一樣，利用 SVG 加入文字外框。

➜　src\components\title-logo.vue

```
<template>
  <div class=" relative leading-none">
    <div
      ref="titleDiv"
      class="flex flex-col flex-center text-white pb-6"
    >
      ...
    </div>

    <!-- stroke -->
    <div
      class=" absolute top-0 flex flex-col flex-center "
      :style="strokeStyle"
      v-html="titleDiv?.innerHTML"
    />

    <svg
      version="1.1"
      style="display: none;"
```

```
    >
      <defs>
        <filter :id="svgFilterId">
          <feMorphology
            operator="dilate"
            radius="10"
          />
          <feComposite
            operator="xor"
            in="SourceGraphic"
          />
        </filter>
      </defs>
    </svg>
  </div>
</template>

<script setup lang="ts">
import { nanoid } from 'nanoid';
import { ref, computed } from 'vue';

const titleDiv = ref<HTMLDivElement>();

const svgFilterId = `svg-filter-${nanoid()}`;
const strokeStyle = computed(() => {
  return {
    color: ‹#222›,
    filter: `url(#${svgFilterId})`
  }
});
</script>
...
```

　　這裡我們使用一個技巧，直接取得 titleDiv 的 innerHTML，並用 v-html 直接將 HTML 內容放到 stroke 的 div 中，這樣就可以自動複製一模一樣的 HTML 了，不用自己手動重複 HTML。

▲　圖 4-15　加入外框

接著加入動態吧，我們讓英文字的部分會像果凍那樣 QQ 彈吧！ㄟ(•∀•)ㄏ

將英文部分獨立區塊並加入 class。

➜　src\components\title-logo.vue

```
<template>
  <div class=" relative leading-none">
    <div ... >
      <div class="flex flex-col flex-center jelly-bounce">
        <div class="text-[8rem] font-game">
          ANIMALS
        </div>
        <div class="text-[12.5rem] font-game">
          PARTY
        </div>
      </div>
      ...
    </div>
    ...
  </div>
</template>
...
<style scoped lang="sass">
...

.jelly-bounce
  animation: jelly-bounce 3.2s infinite
```

```
    transform-origin: 50% 100%

@keyframes jelly-bounce
  0%, 70%, 100%
    transform: scale( 1 )
  75%
    transform: scale( 1.2, 0.8 )
  80%
    transform: scale( 0.85, 1.15 )
  85%
    transform: scale( 1.05, 0.95 )
  90%
    transform: scale( 0.98, 1.02 )
  95%
    transform: scale( 1.01, 0.99 )

</style>
```

▲ 圖 4-16　彈跳動畫

現在文字會很 Q 彈的跳動了！ᕙ(˚∀˚)ᕗ

最後我們在中文字的部份加上色彩和動態吧。

➜ src\components\title-logo.vue

```
<template>
  <div class=" relative leading-none">
    <div
      ref="titleDiv"
```

```
      class=" flex flex-col flex-center text-white pb-6"
    >
      ...
      <div class=" text-[4rem] font-black mt-3 flex gap-2 joy-bounce">
        <span class="text-red-400"> 動 </span>
        <span class="text-lime-400"> 物 </span>
        <span class="text-sky-400"> 派 </span>
        <span class="text-purple-400"> 對 </span>
        <span class="text-amber-400"> 嗨 </span>
        <span class="text-teal-400"> 起 </span>
        <span class="text-pink-400"> 來 </span>
        <span class="text-fuchsia-400"> ！ </span>
      </div>
    </div>
    ...
  </div>
</template>
...
<style scoped lang="sass">
...

// 避免顏色覆蓋外框顏色
.stroke-color
  color: #170e01
  span
    color: #170e01


.joy-bounce
  animation: joy-bounce 1s infinite
  transform-origin: 50% 100%


@keyframes joy-bounce
  0%, 100%
    transform: scale( 1 )
    animation-timing-function: cubic-bezier(0.895, 0.030, 0.685, 0.220)
  50%
    transform: scale( 1.05, 0.9 )
    animation-timing-function: cubic-bezier(0.165, 0.840, 0.440, 1.000)
</style>
```

▲ 圖 4-17 標題 Logo 完成

門面完成！大家可以依照喜好自行改造喔！✧*｡٩(´ㅇ`*)ﻭ✧*｡

第 **5** 章

開趴前先 loading 一下

再來先來實現「建立派對」的功能，讓我們準備開趴！

第一步先來建立遊戲機頁面，並在 router 新增 RouteName 定義。

➜ src\views\the-game-console.vue

```
<template>
  <div class="flex">
    我是 game-console
  </div>
</template>

<script setup lang="ts">
import { ref } from 'vue';
</script>

<style scoped lang="sass">
</style>
```

➜ src\router\router.ts

```
import { createRouter, createWebHistory, RouteRecordRaw } from 'vue-router'
...
export enum RouteName {
  HOME = 'home',
  GAME_CONSOLE = 'game-console',
```

```
}

const routes: Array<RouteRecordRaw> = [
  ...
  {
    path: `/home`,
    name: RouteName.HOME,
    component: () => import('../views/the-home.vue')
  },
  {
    path: `/game-console`,
    name: RouteName.GAME_CONSOLE,
    component: () => import('../views/the-game-console.vue'),
  },
...
]
...
```

Tips

Vue Router 可以讓我們利用 URL 切換頁面，想要深入了解的讀者可以參考以下
連結：

- https://router.vuejs.org/zh/introduction.html

鱈魚：「頁面建立完成後，可以直接控制 route，切換頁面至 game-console
了」

助教：「等等，這樣轉場不給力啊，遊戲應該要有遊戲專用的轉場才對不
是嗎？Σ(´Д`;)」

鱈魚：「偷懶被發現了，那就來做自定義的 loading 效果吧。ヽ('~'̑)ﾉ」

助教：「不要正大光明的偷懶啊！ ㅇ(·ロ`·ㅇ)」

建立 loading 相關元件

此章節程式碼可以在以下連結取得。

- https://gitlab.com/deepmind1/animal-party/web/-/tree/feat/create-loading-screen

由於我們可以在任意元件中切換畫面，不難想像 loading 相關的狀態必須是全域狀態，所以首先新增 loading.store，用來管理 loading 相關狀態。

➜ src\stores\loading.store.ts

```
import { defineStore } from 'pinia';
import { ref } from 'vue';

export const useLoadingStore = defineStore('loading', () => {
  return {
    isLoading: ref(false),
    /** 表示 loading screen 正在進入畫面 */
    isEntering: ref(false),
    /** 表示 loading screen 正在離開畫面 */
    isLeaving: ref(false),
    /** loading 樣式，預留未來可以切換多種樣式 */
    type: ref(''),
  }
})
```

Tips

Pinia 是 Vue 3 推薦使用的狀態管理套件，可以讓使用者輕鬆管理全域範圍的狀態，想要深入了解的讀者可以參考以下連結：

◆ https://pinia.vuejs.org/zh/introduction.html

接下來預計需要 3 個元件：

- transition-mask：提供過場效果並偵測過場狀態。

- 讀取畫面：提供全版面的讀取背景。

- loading-overlay：包裝以上 2 個元件。

首先，在 Vue 中要建立過場效果，大家一定都會想到內建元件 transition，這裡將 transition 再次封裝成 transition-mask，簡化事件處理並提供過場效果，方便過場動畫相關元件使用。

新增 transition-mask 元件並定義參數。

→ src\components\transition-mask.vue

```ts
...
<script setup lang="ts">
import { computed, reactive, ref, watch } from 'vue';

interface Props {
  modelValue: boolean;
  type?: `${AnimationType}`;
}
const props = withDefaults(defineProps<Props>(), {
  type: AnimationType.ROUND,
});
</script>
...
```

接著新增狀態變數與各類轉場 hook，並將資料與事件綁定至 template 中。

➔ src\components\transition-mask.vue

```
<template>
  <transition
    :name="props.type"
    @before-enter="handleBeforeEnter"
    @after-enter="handleAfterEnter"
    @before-leave="handleBeforeLeave"
    @after-leave="handleAfterLeave"
  >
    <div
      v-if="props.modelValue"
      class="mask"
    >
      <slot />
    </div>
  </transition>
</template>

<script lang="ts">
...

export interface State {
  isEntering: boolean,
  isLeaving: boolean,
}
</script>

<script setup lang="ts">
...

const emit = defineEmits<{
  (e: 'update', state: State): void;
}>();

const state = reactive<State>({
  isEntering: false,
  isLeaving: false,
```

```
});

watch(state, () => emit('update', state.value), {
  deep: true
});

function handleBeforeEnter() {
  state.isEntering = true;
  state.isLeaving = false;
}
function handleAfterEnter() {
  state.isEntering = false;
}

function handleBeforeLeave() {
  state.isEntering = false;
  state.isLeaving = true;
}
function handleAfterLeave() {
  state.isLeaving = false;
}

/** 定義元件可對外提供之資料 */
defineExpose({
  state
});
</script>
...
```

　　還差設計轉場用的 CSS，但是先讓我們把總體結構都完成，等等測試時再設計動畫。

Tips

　想要深入了解 transition 的讀者可以參考以下連結：

- https://cn.vuejs.org/guide/built-ins/transition.html#css-based-transitions

最後我們預期會有各種不同的讀取畫面，所以需要一個容器將不同的讀取畫面與過場效果包裹在一起。

新增 loading-overlay 元件，用來包裝所有的 loading 相關元件。

→ src\components\loading-overlay.vue

```
<template>
  <transition-mask class="absolute inset-0">
    <!-- 先放一個全畫面的白底，用來檢視是否有效果 -->
    <div class="absolute inset-0 bg-white" />
  </transition-mask>
</template>

<script setup lang="ts">
import { } from 'vue';

import TransitionMask from './transition-mask.vue';
</script>
```

到目前為止提供過場畫面、轉場效果的元件的基本功能已經完成。讀取畫面還沒製作，目前我們先用全畫面白底，用來測試看看 loading 是否有效果

測試功能的第一步是建立 use-loading，用來提供所有元件 loading 功能。

→ src\composables\use-loading.ts

```
import { promiseTimeout, until, watchOnce } from '@vueuse/core';
import { defaults } from 'lodash-es';
import { storeToRefs } from 'pinia';
import { ref, watch } from 'vue'
import { useLoadingStore } from '../stores/loading.store'

export interface State {
  isEntering: boolean,
  isLeaving: boolean,
}

interface UseLoadingParams {
```

```
  /** 最小持續時間 (ms，預設 1000)
   *
   * 讀取頁面進入完成至開始離開之間的最小時間，可用來展示動畫。
   */
  minDuration?: number;
}
const defaultParams: Required<UseLoadingParams> = {
  minDuration: 1000,
}

export function useLoading(paramsIn?: UseLoadingParams) {
  const params = defaults(paramsIn, defaultParams);

  const store = useLoadingStore();
  const { isLoading, isEntering, isLeaving } = storeToRefs(store);

  /** 讀取畫面顯示狀態 */
  const visible = ref(false);

  watch(isLoading, (value) => {
    visible.value = value;
  }, {
    immediate: true
  });

  watch(visible, async (value) => {
    // show() 要立即顯示
    if (value) {
      store.$patch({
        isLoading: true,
      });
      return;
    }

    await promiseTimeout(params.minDuration);

    // hide() entering 為 false，直接隱藏
    if (!isEntering.value) {
      store.$patch({
```

```
    isLoading: false,
  });
  return;
}

// entering 結束之後才可隱藏
watchOnce(isEntering, () => {
  store.$patch({
    isLoading: false,
  });
})
}, {
  deep: true
});

/**
 * 等到過場動畫進入完成後，才會完成 Promise
 * 可以避免動畫還沒完成就跳頁的問題
 */
async function show() {
  visible.value = true;

  await until(isEntering).toBe(true);
  await until(isEntering).toBe(false);
}

/** 等到過場動畫離開完成後，才會完成 Promise */
async function hide() {
  visible.value = false;

  await until(isLeaving).toBe(true);
  await until(isLeaving).toBe(false);
}

/** 處理狀態更新 */
function handleUpdate({ isEntering, isLeaving }: State) {
  store.$patch({
    isEntering,
    isLeaving
```

```
    });
  }

  return {
    isLoading,
    isEntering,
    isLeaving,

    show,
    hide,
    handleUpdate
  }
}
```

在 loading-overlay 中導入 use-loading，綁定狀態吧！

➜ src\components\loading-overlay.vue

```html
<template>
  <transition-mask
    v-model="isLoading"
    class="absolute inset-0"
    @update="handleUpdate"
  >
    <div class="absolute inset-0 bg-white" />
  </transition-mask>
</template>

<script setup lang="ts">
import { } from 'vue';
import { useLoading } from '../composables/use-loading';

import TransitionMask from './transition-mask.vue';

const { isLoading, handleUpdate } = useLoading();
</script>
```

現在讓我們把 loading-overlay 加到 App.vue 並使用 use-loading，實測看看過場效果。

➔ src\App.vue

```
<template>
  <router-view />
  <loading-overlay />
</template>

<script setup lang="ts">
import LoadingOverlay from './components/loading-overlay.vue';

import { useLoading } from './composables/use-loading';

const loading = useLoading();

setTimeout(() => {
  loading.show();
  setTimeout(() => {
    loading.hide();
  }, 2000);
}, 2000);
</script>
...
```

　　助教：「... 怎麼會是 2 秒後忽然全白，再 2 秒後白色消失？(⊙_⊙)」

　　鱈魚：「別擔心，那是因為我們還沒在 transition-mask 中加入轉場用 class，所以只會瞬間進入、瞬間消失。」

　　助教：「你確定不是 Bug ？你這個 Bug 魚。」

　　鱈魚：「真的啦，不要幫我取奇怪的綽號啊！(っ °Д °;)っ」

　　現在讓我們回到 transition-mask 加上轉場用 class。

➔ src\components\transition-mask.vue

```
...
<style scoped lang="sass">
.round-enter-active, .round-leave-active
  transition-duration: 0.4s
```

```
  .round-enter-from, .round-leave-to
    opacity: 0 !important
</style>
```

　　現在是不是變成 2 秒後漸入全白，再 2 秒後全白漸出呢？這就表示 loading 功能正常運作了！

　　現在讓我們在 transition-mask 加入有趣的轉場 class 吧！ ˇ(≧▽≦*)o

➜ src\components\transition-mask.vue

```
...
<style scoped lang="sass">
.round-enter-active
  animation-duration: 1.4s
.round-leave-active
  transition-duration: 0.4s
  transition-timing-function: ease-in-out
.round-enter-from, .round-enter-to
  animation-name: round-in
  animation-fill-mode: forwards
@keyframes round-in
  0%
    clip-path: circle(3% at 46% -50%)
    animation-timing-function: cubic-bezier(0.005, 0.920, 0.060, 0.99)
  40%
    clip-path: circle(3% at 50% 50%)
    animation-timing-function: cubic-bezier(0.630, -0.170, 0.140, 0.980)
  100%
    clip-path: circle(70.7% at 50% 50%)
.round-leave-from
  clip-path: circle(70.7% at 50% 50%)
.round-leave-to
  clip-path: circle(40% at 140% 140%)
</style>
```

　　目前過場效果應該會從畫面上方跑出一個白球到畫面中央，變大、蓋住全畫面後從右下方離開。

▲ 圖 5-1　過場效果

鱈魚：「大成功！ヽヽ(✿ﾟ▽ﾟ)ノ」

助教：「是吼，一片空白也算成功吼(ﾟﾛﾟ)」

鱈魚：「才不是只有這樣（•`ω•´）✧，接下來讓我們打造有趣的讀取畫面吧！」

5.2　載入就應該要有載入的樣子

此章節程式碼可以在以下連結取得。

- https://gitlab.com/deepmind1/animal-party/web/-/tree/feat/create-loading-background

讓我們把原本的全白畫面換成「載入畫面」吧！

預期希望畫面中會有三個跳躍的多邊形，讓畫面看起來有正在努力讀取的感覺。

新增元件並定義一下參數。

➜　src\components\background-loading-jumping-polygon.vue

```
<template></template>

<script setup lang="ts">
import { ref } from 'vue';

interface Props {
  /** 背景顏色 */
  backgroundColor?: string;
  /** 方塊顏色 */
  polygonColors?: string[];
}
const props = withDefaults(defineProps<Props>(), {
  backgroundColor: '#c8e6b1',
  polygonColors: () => ['#AEF2BD', '#BCFCB6', '#EDFCB6'],
});
</script>
...
```

再來新增最外層放置多邊形用的容器，並將自動計算出來的 style 綁上去，產生背景顏色。

➜　src\components\background-loading-jumping-polygon.vue

```
<template>
  <div
    class="flex flex-center overflow-hidden"
    :style="backgroundStyle"
  >
  </div>
</template>
```

```ts
<script setup lang="ts">
import { computed, ref } from 'vue';
import { colors } from 'quasar';
const { lighten, textToRgb, rgbToHsv, hsvToRgb, rgbToHex } = colors;
...
const backgroundStyle = computed(() => {
  // 變亮
  const lightenColor = lighten(props.backgroundColor, 20);

  // 變暗並偏移色相
  const darkColor = lighten(props.backgroundColor, -14);

  const hsvColor = rgbToHsv(textToRgb(darkColor));
  hsvColor.h -= 15;

  const offsetColor = rgbToHex(hsvToRgb(hsvColor));

  return {
    background: `linear-gradient(-30deg, ${offsetColor}, ${props.backgroundColor},
${lightenColor}, ${props.backgroundColor}, ${offsetColor})`
  }
});
</script>
```

接著暫時在 App.vue 中引入元件，方便讓我們看看目前的模樣，同時刪除 loading.show() 的程式，以免產生干擾。

➜ src\App.vue

```ts
<template>
  <router-view />
  <loading-overlay />
  <background-loading-jumping-polygon class=" absolute w-full h-full" />
</template>

<script setup lang="ts">
import BackgroundLoadingJumpingPolygon from './components/background-loading-jumping-
polygon.vue';
```

```
import LoadingOverlay from './components/loading-overlay.vue';

import { useLoading } from './composables/use-loading';
</script>
...
```

目前看起來應該長這樣。

▲ 圖 5-2　讀取畫面背景顏色

助教：「嗯…... 一點都不像讀取畫面呢。(ﾟﾍﾟ)（準備打包走人）」

鱈魚：「先別急著走啊，多邊形還沒登場啊！ ꒰(･´ᴗ･ ꒱」

讓我們引入多邊形並定義一下多邊形的樣式。

➔ src\components\background-loading-jumping-polygon.vue

```
...
<script setup lang="ts">
...
import BasePolygon, { ShapeType } from './base-polygon.vue';
...

const polygons = computed(() => [
  {
    shape: ShapeType.SQUARE,
    color: props.polygonColors[0],
```

```
  },
  {
    shape: ShapeType.ROUND,
    color: props.polygonColors[1],
  },
  {
    shape: ShapeType.TRIANGLE,
    color: props.polygonColors[2],
  },
]);
</script>
```

接著在 template 中，用 v-for 產生多邊形。

➔ src\components\background-loading-jumping-polygon.vue

```
<template>
  <div ... >
    <div class="flex gap-20">
      <div
        v-for="(poly, i) in polygons"
        :key="i"
      >
        <base-polygon
          size="9rem"
          :shape="poly.shape"
          fill="solid"
          :color="poly.color"
        />
      </div>
    </div>
  </div>
</template>
...
```

多邊形外層還包一個 div，是為了提升加入 CSS 動畫的彈性。

現在多邊形出現了！φ(　▽` *)♪

▲ 圖 5-3　3 個多邊形

助教：「這顏色 ... ๑_๒」。

鱈魚：「...... 讓我們換個顏色吧。(´。＿。`)」

調整以下預設顏色

➜ src\components\background-loading-jumping-polygon.vue

```
<script setup lang="ts">
...
const props = withDefaults(defineProps<Props>(), {
  ...
 polygonColors: () => ['#7c8f6d60', '#B1CC9D60', '#90A68060'],
});
...
</script>
```

▲ 圖 5-4　調整多邊形顏色

現在看起來好多了吧 ⋯(´,,•ω•,,)

現在讓這三個多邊形動起來吧，設計邏輯為：

- 多邊形套上果凍動畫

- 多邊形容器套上跳動動畫

這兩種動畫合併後就可以產生 Q 彈的果凍跳躍的效果了！(˙•ω•˙)y

➜ src\components\background-loading-jumping-polygon.vue

```
<template>
  <div ... >
    <div class="flex gap-20">
      <div
        v-for="(poly, i) in polygons"
        :key="i"
        class="box"
      >
        <base-polygon
          ...
          class="jelly-bounce"
        />
      </div>
    </div>
  </div>
</template>
...
<style scoped lang="sass">
.box
  animation: jump 1.4s infinite ease-in-out

.jelly-bounce
  animation: jelly-bounce 1.4s infinite ease-in-out
  transform-origin: 50% 100%

@keyframes jump
  0%
    transform: translateY(-30%)
```

```
  45%
    transform: translateY(0%)
  100%
    transform: translateY(-30%)

@keyframes jelly-bounce
  0%
    transform: scale( 1 )
  30%
    transform: scale( 1 )
  50%
    transform: scale( 1.2, 0.8 )
  70%
    transform: scale( 0.85, 1.15 )

  80%
    transform: scale( 1.05, 0.95 )
  90%
    transform: scale( 0.98, 1.02 )
  100%
    transform: scale( 1 )
</style>
```

▲ 圖 5-5 跳躍多邊形

看起來真不錯，真想揉揉看。 ㄘ (´∀`ㄘ)

　　最後我們讓三個多邊形的動畫交錯，這樣看起來比較有趣，直接用 animation-delay 簡單解決！

➜ src\components\background-loading-jumping-polygon.vue

```
<template>
  <div ... >
    <div class="flex gap-20">
      <div
        ...
        class="box"
        :style="`animation-delay: ${i * 0.1}s`"
      >
        <base-polygon
          ...
          :style="`animation-delay: ${i * 0.1}s`"
        />
      </div>
    </div>
  </div>
</template>
```

▲ 圖 5-6 沒默契三兄弟

成功產生三個很沒默契的多邊形！ ᐕ(˚ ∀ ˚)ᐛ

最後我們把這個背景加到載入畫面使用吧！

首先刪除 App.vue 中的 background-loading-jumping-polygon。

➜ src\App.vue

```
<template>
  <router-view />
  <loading-overlay />
</template>

<script setup lang="ts">
import LoadingOverlay from './components/loading-overlay.vue';
</script>

...
```

　　在 loading-overlay 引入 background-loading-jumping-polygon，替換原先純白的 div。

➜ src\components\loading-overlay.vue

```
<template>
  <transition-mask ...  >
    <background-loading-jumping-polygon class="absolute inset-0" />
  </transition-mask>
</template>

<script setup lang="ts">
...
import BackgroundLoadingJumpingPolygon from './background-loading-jumping-polygon.
vue';
...
</script>
```

　　讀取背景加入完成，接下來就是來實際試試看效果了。

　　在 the-home 新增 startParty()，實際使用 use-loading 功能並跳轉 route 看看。

➜ src\views\the-home.vue

```
<template>
  ...
  <div ... >
    ...
```

```
    <div class="flex flex-col flex-center gap-20">
      <base-btn
        v-slot="{ hover }"
        label=" 加入派對 "
        ...
        @click="startParty"
      >
        ...
      </base-btn>
      ...
    </div>
  </div>
</template>

<script setup lang="ts">
import { RouteName } from '../router/router';
...
import { useLoading } from '../composables/use-loading';
import { useRouter } from 'vue-router';

const loading = useLoading();
const router = useRouter();

async function startParty() {
  await loading.show();
  router.push({
    name: RouteName.GAME_CONSOLE
  });
}
</script>
...
```

最後在 the-game-console 中呼叫 loading.hide()，讓讀取畫面消失。

src\views\the-game-console.vue

```
...
<script setup lang="ts">
```

```
import { ref } from 'vue';
import { useLoading } from '../composables/use-loading';

const loading = useLoading();
loading.hide();
</script>
```

現在按下「加入派對」應該會出現讀取畫面後，進入 game-console 頁面。

▲ 圖 5-7　讀取並跳轉

成功讀取並跳轉畫面了！ ◇*｡٩(´ ❍ ｀*)ﻭ◇*｡

接下來讓我們啟動伺服器，準備開始連線！

第 **6** 章

讓前後端接上線

　　主要是為了盡可能多練習 NestJS，所以在此專案中，後端伺服器決定採用 NestJS 進行開發，而本次專案也不會使用到資料庫，所以規模更為單純，更適合和大家一起練習練習。

　　記得先同「1.4.2 NestJS 後端專案」章節步驟把專案下載下來，接著在開始之前先記得安裝 Nest CLI。（已經安裝過的讀者就不用再次安裝）

　　打開終端機並輸入以下指定安裝：

```
npm install -g @nestjs/cli
```

　　那就讓我們開始吧！ヾ(◍'◡`◍)ノﾞ

　　此專案中的網頁與伺服器需要即時雙向通訊，所以我們採用 WebSocket 進行通訊，具體實作我們使用 Socket.IO 這個套件。

　　Socket.IO 套件能夠提供客戶端與伺服器端實現低延遲的雙向通訊，主要建構在 WebSocket 協定上，並能夠在不支援 WebSocket 時，自動退回 HTTP long-polling，使用者可以簡單且一致的使用其 API 實現雙向通訊。

 Tips

想要深入了解的讀者可以參考以下連結：

● https://socket.io/zh-CN/docs/v4/

Socket.IO 連線時，會配給 client 一個 socket id，但是重新連線時 socket id 都不一樣，為了實現「就算 socket id 不同，也要能夠辨識出是誰」的功能需求。

就先讓我們新增負責處理 Socket.IO 連線的資源吧。

我們使用 Nest CLI 指令替我們自動新增資源，打開終端機後輸入以下指令：

```
nest g resource ws-client
```

Nest CLI 會詢問一系列設定，協助我們快速建立並自動引用資源模組。

- What transport layer do you use?

  選擇「WebSockets」

- Would you like to generate CRUD entry points?

  因為用不到 CRUD，也沒有資料庫，所以這裡請輸入「n」

接著就會看到終端機跑出一堆訊息。

```
? What transport layer do you use? WebSockets
? Would you like to generate CRUD entry points? No
CREATE src/ws-client/ws-client.gateway.spec.ts (547 bytes)
CREATE src/ws-client/ws-client.gateway.ts (232 bytes)
CREATE src/ws-client/ws-client.module.ts (244 bytes)
CREATE src/ws-client/ws-client.service.spec.ts (475 bytes)
CREATE src/ws-client/ws-client.service.ts (92 bytes)
UPDATE src/app.module.ts (332 bytes)
```

而且專案資料夾也發生變化了。

▲ 圖 6-1　新增 ws-client 資源

　　如上圖，Nest CLI 自動幫我們產生 ws-client 所需要的所有檔案，還自動在 app.module.ts 中引入檔案，是不是非常的快速且貼心啊！(′„•ω•„)

　　接下來就準備讓我們開始實作連線功能，後續章節中，我們會不斷的在 web 與 server 專案之間切換。

　　大家坐穩了，讓我們出發吧！ ─═≡Σ(((つ•̀ω•́)つ

6.1 後端提供連線

　　此章節程式碼可以在以下連結取得。

- https://gitlab.com/deepmind1/animal-party/server/-/tree/feat/create-ws-client

　　目前我們已經成功新增 ws-client 模組，還差最重要的一步，那就是安裝相依套件，請在終端機輸入以下命令開始安裝。

```
npm i
```

　　接著安裝 WebSocket 所需套件。

```
npm i @nestjs/websockets @nestjs/platform-socket.io
```

　　安裝完成後讓我們進入主題吧。 φ(゜▽゜*)♪

　　ws-client 模組的主要功能是「管理 socketId 與 clientId 的映射關係」。

目的是「只要 client 傳輸的 id 相同，即使 socket 改變，都可以取得先前的連線資料。」

基本概念為：

1. client 發起連線並傳送 id。

2. server 取得 id 後，檢查是否有此 id 連線資料。

3. 已存在則替換新的 socket id；不存在則建立新資料。

接著來介紹一下，目前 ws-client 資料夾內的檔案分別功能為何：

- ws-client.gateway.ts

  負責處理請求，概念與 MVC 模型中的 controller 功能類似。

- ws-client.gateway.spec.ts：

  gateway 測試檔案，負責測試 ws-client.gateway 功能，本次專案不會用到。

- ws-client.service.ts

  負責提供 ws-client 功能邏輯。

- ws-client.service.spec.ts

  service 測試檔案，負責測試 ws-client.service 功能，本次專案不會用到。

- ws-client.module.ts

  負責包裝以上檔案成為一個獨立模組。

首先讓我們完成 ws-client.service 功能邏輯，設計一下預期使用的型別定義。

➜ src\ws-client\ws-client.service.ts

```
import { Injectable } from '@nestjs/common';

export enum ClientType {
```

```
  /** 遊戲機，負責建立派對房間 */
  GAME_CONSOLE = 'game-console',
  /** 玩家，通常是手機端網頁 */
  PLAYER = 'player',
}

export type ClientId = string;

export interface Client {
  id: ClientId;
  socketId: string;
  type: `${ClientType}`;
}

@Injectable()
export class WsClientService {}
```

新增 Map 物件儲存目前已連線之 client。

➜ src\ws-client\ws-client.service.ts

```
...
export class WsClientService {
  clientsMap = new Map<ClientId, Client>();
}
```

接著建立當使用者連線時，用來新增或替換 Client 的 method。

➜ src\ws-client\ws-client.service.ts

```
...

export interface PutClientParams {
  socketId: string;
  clientId: ClientId;
  type: `${ClientType}`;
}

@Injectable()
```

```
export class WsClientService {
  ...
  /** 不存在則新增，存在則更新 */
  putClient(params: PutClientParams) {
    const { clientId, socketId, type } = params;

    const client = this.clientsMap.get(clientId);

    // 新增
    if (!client) {
      const newClient = { id: clientId, socketId, type };
      this.clientsMap.set(clientId, newClient);
      return newClient;
    }

    // 更新
    client.socketId = socketId;
    client.type = type;

    this.clientsMap.set(clientId, client);

    return client;
  }
}
```

儲存 client 之後，當然還要可以取出才行，最後我們新增取得 client 用的 method。

→ src\ws-client\ws-client.service.ts

```
...
/** 允許使用 socketId 或 clientId 取得 */
export type GetClientParams = { socketId: string } | { clientId: string };

@Injectable()
export class WsClientService {
  ...
  getClient(params: GetClientParams) {
    // 存在 clientId，使用 clientId 取得
```

```
  if ('clientId' in params) {
    return this.clientsMap.get(params.clientId);
  }

  // 否則用 socketId 查詢
  const clients = [...this.clientsMap.values()];
  const target = clients.find(({ socketId }) => socketId === params.socketId);
  return target;
  }
}
```

目前我們可以使用 ws-client.service，儲存、更新並取得使用者了，service 基本上這樣就完成了，接下來完成 gateway 的功能吧。(´▽`)ノ

ws-client.gateway 的工作其實非常單純，只專注於 client 連線事件。並於連線時，取得 clientId 並呼叫 service。

根據 NestJS 官網的介紹，只要新增名為 handleConnection 的 method，NestJS 就會在 client 連線時，自動呼叫此名稱的 method。

→ src\ws-client\ws-client.gateway.ts

```
...
import { Socket } from 'socket.io';

@WebSocketGateway()
export class WsClientGateway {
  constructor(private readonly wsClientService: WsClientService) { }

  handleConnection(socket: Socket) {
  }
}
```

method 好了，所以我們要怎麼取得 client 連線時傳輸的訊息？

根據 Socket.IO 文件的說法，伺服器可以透過 socket.handshake.query 的方式取得資料。

Tips

想要深入了解的讀者可以參考以下連結：

◆ https://socket.io/docs/v4/server-socket-instance/#sockethandshake

➜ src\ws-client\ws-client.gateway.ts

```
...
export class WsClientGateway {
  ...
  handleConnection(socket: Socket) {
    const queryData = socket.handshake.query;
  }
}
```

　　像這樣就可以取得 client 連線時傳輸的資料，只是現在有另一個問題：「我們要怎麼確定 queryData 的資料定義正確？」

　　在此需要一個可以判斷物件型別是否正確的 function，所以讓我們新增新的模組，用來處理各類資料吧。

　　輸入以下指令，建立 utils 模組。

```
nest g resource utils*
```

- What transport layer do you use?

  選哪一個都可以，因為用不到 controller

- Would you like to generate CRUD entry points?

  輸入「n」

之後就可以看到目錄下新增了 utils 模組。

▲ 圖 6-2　新增 utils 資源

接著讓我們刪除 controller（因為這個模組不處理請求）與 spec 檔案，並調整 utils.module 內容。

➜ src\utils\utils.module.ts

```
import { Global, Module } from '@nestjs/common';
import { UtilsService } from './utils.service';

@Global()
@Module({
  providers: [UtilsService],
  exports: [UtilsService],
})
export class UtilsModule { }
```

- @Global() 裝飾器會讓此模組變成全域模組，也就是在任意地方都可以存取。

- exports 則表示此模組要對外提供那些內容，此模組對外提供 utils.service 功能。

接著讓我們完成 utils.service 功能，我們希望 queryData 要包含 clientId 與 type 資料，新增名為 isSocketQueryData 的 method 進行判斷。

➜ src\utils\utils.service.ts

```
...
import { ClientType } from 'src/ws-client/ws-client.service';
```

```
interface SocketQueryData {
  clientId: string;
  type: `${ClientType}`;
}

@Injectable()
export class UtilsService {
  isSocketQueryData(data: any): data is SocketQueryData {
    // 沒有必要屬性
    if (!('clientId' in data) || !('type' in data)) {
      return false;
    }

    // type 不屬於列舉類型
    if (!Object.values(ClientType).includes(data['type'])) {
      return false;
    }

    return true;
  }
}
```

接下來讓我們回到 ws-client.gateway，來使用 utils 模組功能吧！

➜ src\ws-client\ws-client.gateway.ts

```
...
import { UtilsService } from 'src/utils/utils.service';

@WebSocketGateway()
export class WsClientGateway {
  constructor(
    private readonly wsClientService: WsClientService,
    private readonly utilsService: UtilsService,
  ) {
    //
  }

  handleConnection(socket: Socket) {
```

```
    const queryData = socket.handshake.query as unknown;

    // 若資料無效，則中斷連線
    if (!this.utilsService.isSocketQueryData(queryData)) {
      socket.disconnect();
      return;
    }

    const { clientId, type } = queryData;
  }
}
```

現在我們可以確定 queryData 內容一定是我們希望的資料了，接續完成此 method 吧。

其實只要呼叫 this.wsClientService.putClient 就完成了。(´„•ω•„)

不過還是加個 logger，方便查看訊息吧。

➜ src\ws-client\ws-client.gateway.ts

```
...
import { Logger } from '@nestjs/common';

@WebSocketGateway()
export class WsClientGateway {
  private logger: Logger = new Logger(WsClientGateway.name);
  ...
  handleConnection(socket: Socket) {
    this.logger.log(`client connected : ${socket.id}`);

    const queryData = socket.handshake.query;
    this.logger.log(`queryData : `, queryData);

    // 若資料無效，則中斷連線
    if (!this.utilsService.isSocketQueryData(queryData)) {
      socket.disconnect();
      return;
    }
```

```
    const { clientId, type } = queryData;
    this.wsClientService.putClient({
      socketId: socket.id,
      clientId,
      type,
    });
  }
}
```

 Tips

以此專案規模而言，使用 console.log 就足夠了，只是在持續增長的專案中，
用 logger 可以保留未來拓展彈性，所以養成好習慣，這裡就使用 logger 吧。
(´▽`)/

　　最終在 ws-client.module 加上 exports，以便未來有需求時，別的模組也能
使用 service。

➜ src\ws-client\ws-client.module.ts

```
...

@Module({
  providers: [WsClientGateway, WsClientService],
  exports: [WsClientService],
})
export class WsClientModule {}
```

　　以上我們已經準備好可以連線的伺服器了，現在讓我們啟動伺服器吧！

　　在終端機輸入以下命令即可啟動伺服器：

```
npm run start:dev
```

此命令表示使用 dev 模式啟動伺服器，終端機應該會出現以下訊息。

```
[下午10:15:52] Starting compilation in watch mode...

[下午10:15:59] Found 0 errors. Watching for file changes.

[Nest] 1064  - 2023/04/03 下午10:16:06     LOG [NestFactory] Starting Nest application...
[Nest] 1064  - 2023/04/03 下午10:16:06     LOG [InstanceLoader] UtilsModule dependencies initialized +16ms
[Nest] 1064  - 2023/04/03 下午10:16:06     LOG [InstanceLoader] AppModule dependencies initialized +1ms
[Nest] 1064  - 2023/04/03 下午10:16:06     LOG [InstanceLoader] WsClientModule dependencies initialized +0ms
[Nest] 1064  - 2023/04/03 下午10:16:07     LOG [RoutesResolver] AppController {/}: +832ms
[Nest] 1064  - 2023/04/03 下午10:16:07     LOG [RouterExplorer] Mapped {/, GET} route +3ms
[Nest] 1064  - 2023/04/03 下午10:16:07     LOG [NestApplication] Nest application successfully started +3ms
```

▲ 圖 6-3 dev 模式啟動伺服器

啟動成功！✧*｡٩(´ ꓓ ｀*)و✧*｡現在讓我們回到 Web 專案！

6.2　前端開始連線

此章節程式碼可以在以下連結取得。

- https://gitlab.com/deepmind1/animal-party/web/-/tree/feat/connect-to-server

第一步老樣子，來安裝相依套件吧。(´▽`)/

輸入以下命令安裝 socket.io client 套件。

```
npm i socket.io-client
```

首先新增 main.type 定義 client type 基本資料。

➜ src\types\main.type.ts

```
export enum ClientType {
  /** 遊戲主機 */
  GAME_CONSOLE = 'game-console',
  /** 玩家 */
  PLAYER = 'player',
}
```

再新增 main.store 用來儲存 socket.io 之 client。

➜ src\stores\main.store.ts

```
import { defineStore } from 'pinia';
import { ref } from 'vue';
import { ClientType } from '../types';
import { nanoid } from 'nanoid';
import { Socket } from 'socket.io-client';

export const useMainStore = defineStore('main', () => {
  /** 檢查 localStorage 是否有儲存 client id，沒有則產生 id */
  const savedId = localStorage.getItem(`animals-party:clientId`);
  const clientId = savedId ?? nanoid();
  localStorage.setItem(`animals-party:clientId`, clientId);

  const client = ref<Socket>();
  const type = ref<`${ClientType}`>();

  return {
    clientId,
    client,
    type,
  }
})
```

可以注意到每次初始化時取得 localStorage 已儲存 id，若沒有 id，則使用 nanoid 建立同時儲存至 localStorage。

Tips

client id 更好的做法還可以改為使用裝置的 fingerprint，不過這裡我們就簡單一
點，直接產生 id 字串即可。

接著讓我們加入連線和斷線的功能。

→ src\stores\main.store.ts

```
...
import { Socket, io } from 'socket.io-client';

export const useMainStore = defineStore('main', () => {
  ...
  function connect(clientType: `${ClientType}`) {
    // 已經存在
    if (client.value) {
      client.value.connect();
      return client.value;
    }

    // 建立連線，傳送 query data
    client.value = io({
      query: {
        clientId,
        type: clientType,
      }
    });
    type.value = clientType;

    return client.value;
  }

  function disconnect() {
    client.value?.disconnect();
  }

  return {
```

```
    clientId,
    client,
    type,

    connect,
    disconnect,
  }
})
```

　　最後回到 the-home，讓我們實際在 startParty() 中，試試看啟動連線吧！
(≧∀≦)

➜ src\views\the-home.vue

```
...

<script setup lang="ts">
...
import { useMainStore } from '../stores/main.store';

const mainStore = useMainStore();
const loading = useLoading();
const router = useRouter();

async function startParty() {
  await loading.show();

  mainStore.connect('game-console');

  router.push({
    name: RouteName.GAME_CONSOLE
  });
}
</script>

...
```

　　這裡還有一件非常非常重要的事情，就是要設定 Vite 的 proxy 功能，請
Vite 的 Dev Server 協助代理傳輸資料，否則會怎麼樣都無法連線喔！

讓我們前往 vite.config 設定。

➔ vite.config.ts

```ts
import { defineConfig } from 'vite';
import vue from '@vitejs/plugin-vue';
import { quasar, transformAssetUrls } from '@quasar/vite-plugin';

// https://vitejs.dev/config/
export default defineConfig(({ command, mode }) => {
  return {
    plugins: [
      ...
    ],
    server: {
      proxy: {
        '/socket.io': {
          target: 'ws://localhost/socket.io',
          ws: true
        }
      }
    }
  }
})
```

現在按下首頁的「建立派對」後，應該會在伺服器的終端機中看到以下
訊息。

```
[Nest] 27804   - 2023/04/03 下午11:48:45     LOG [WsClientGateway] Object:
{
  "clientId": "WrKCcRW8xILu6fNpiQ1Ty",
  "type": "game-console",
  "EIO": "4",
  "transport": "polling",
  "t": "OT7pdLS"
}
```

▲ 圖 6-4 前後端連線成功

我們成功讓前後端連線了！✧*。٩(ˊᗜˋ*)و✧*。

6.3 開房間！開派對！♪(´ω`)و

連線成功後的下一步，就讓我們來建立房間吧！

房間可以用來儲存連線代號和玩家，並與作為遊戲機的網頁端進行資料同步，可以讓遊戲機網頁取得目前玩家數量與其 ID。

6.3.1 伺服器建立房間資源

此章節程式碼可以在以下連結取得。

- https://gitlab.com/deepmind1/animal-party/server/-/tree/feat/build-a-game-room

首先到伺服器專案，讓我們建立 room 模組，輸入以下命令：

```
nest g resource room
```

選擇與 ws-client 模組相同，選擇「WebSockets」後輸入「n」。

首先在 room.module 中加入 exports。

➡ src\room\room.module.ts

```
...
@Module({
  ...
  exports: [RoomService],
})
export class RoomModule {}
```

接著來完成 service 功能，與 ws-client 概念相同，先來定義資料型別。

➡ src\room\room.service.ts

```
import { Injectable } from '@nestjs/common';
import { ClientId } from 'src/ws-client/ws-client.service';

/** 房間 ID，6 位數字組成 */
export type RoomId = string;

export interface Room {
  /** 房間 ID，6 位數字組成 */
  id: RoomId;
  founderId: ClientId;
  playerIds: ClientId[];
}

@Injectable()
export class RoomService { }
```

加入 logger 與 roomsMap。

➡ src\room\room.service.ts

```
import { Injectable, Logger } from '@nestjs/common';
...
@Injectable()
export class RoomService {
  private logger: Logger = new Logger(RoomService.name);
  /** 紀錄已建立的房間 */
  roomsMap = new Map<RoomId, Room>();
}
```

加入新增房間 method，這裡我們需要 nanoid 建立房間號碼。

首先輸入命令安裝套件：

```
npm i nanoid@^3
```

Tips

由於 nanoid 在 v4 版本移除了 CommonJS 支援，為了避免在 NestJS 環境發生引入異常，所以這裡指定安裝 v3 版本。

→ src\room\room.service.ts

```
...
import { customAlphabet } from 'nanoid';
const createRoomId = customAlphabet('1234567890', 6);

/** 房間 ID，6 位數字組成 */
export type RoomId = string;

export interface Room {
  /** 房間 ID，6 位數字組成 */
  id: RoomId;
  founderId: ClientId;
  playerIds: ClientId[];
}

@Injectable()
export class RoomService {
  ...
  addRoom(clientId: string) {
    let roomId = createRoomId();
    while (this.roomsMap.has(roomId)) {
      roomId = createRoomId();
    }

    const newRoom = {
      id: roomId,
      founderId: clientId,
      playerIds: [],
    };
    this.roomsMap.set(roomId, newRoom);
    this.logger.log(`created room : `, newRoom);
```

```
      return newRoom;
  }
}
```

使用 nanoid 提供之 customAlphabet 簡單產生房間 ID。

有房間之後，讓我們接著新增一系列與房間操作有關的 method。

➜ src\room\room.service.ts

```
...
export type GetRoomParams = { founderId: string } | { playerId: string };

@Injectable()
export class RoomService {
  ...
  getRoom(params: GetRoomParams) {
    const result = [...this.roomsMap.values()].find((room) => {
      if ('founderId' in params) {
        return room.founderId === params.founderId;
      }

      return room.playerIds.includes(params.playerId);
    });

    return result;
  }

  hasRoom(roomId: string) {
    return this.roomsMap.has(roomId);
  }

  deleteRooms(founderId: string) {
    const rooms = [...this.roomsMap.values()].filter(
      (room) => room.founderId === founderId,
    );

    rooms.forEach(({ id }) => {
      this.roomsMap.delete(id);
```

```
  });
}


deletePlayer(clientId: string) {
  this.roomsMap.forEach((room, key) => {
    const index = room.playerIds.indexOf(clientId);
    if (index < 0) return;

    room.playerIds.splice(index, 1);
    this.roomsMap.set(key, room);
  });
}
}
```

最後讓我們前往 room.gateway，處理連線與斷線事件，基本邏輯為：

- 連線時，若 client type 類型為 game-console，則建立房間。

- 斷線時，若 client type 類型為 game-console，則自動刪除房間。

載入 UtilsService 後，新增 handleConnection、handleDisconnect 之 method。

➜ src\room\room.gateway.ts

```
import { WebSocketGateway } from '@nestjs/websockets';
import { RoomService } from './room.service';
import { Socket } from 'socket.io';
import { UtilsService } from 'src/utils/utils.service';

@WebSocketGateway()
export class RoomGateway {
  constructor(
    private readonly roomService: RoomService,
    private readonly utilsService: UtilsService,
  ) {
    //
  }

  handleConnection(socket: Socket) {
  }
```

```
  handleDisconnect(socket: Socket) {
  }
}
```

實作連線邏輯，與 ws-client 概念相同，從 handshake.query 取得連線資訊後，若 client type 為 game-console 則建立房間。

→ src\room\room.gateway.ts

```
...
@WebSocketGateway()
export class RoomGateway {
  ...
  handleConnection(socket: Socket) {
    const queryData = socket.handshake.query;

    if (!this.utilsService.isSocketQueryData(queryData)) return;

    const { clientId, type } = queryData;

    /** 只有 game-console 才建立房間 */
    if (type !== 'game-console') return;

    const room = this.roomService.addRoom(clientId);

    /** 加入 Socket.IO 提供的 room 功能,
     * 這樣可以簡單輕鬆的對所有成員廣播資料
     *
     * https://socket.io/docs/v4/rooms/#default-room
     */
    socket.join(room.id);

    /** 發送房間建立成功事件 */
    socket.emit('game-console:room-created', room);
  }
  ...
}
```

　　最後讓我們完成斷線邏輯，由於我們需要依 socketId 取得對應 client，這裡需要用到 ws-client.service 功能，所以必須在 room.module 引入 ws-client.module 才行。

➜ src\room\room.module.ts

```
import { Module } from '@nestjs/common';
import { RoomService } from './room.service';
import { RoomGateway } from './room.gateway';
import { WsClientModule } from 'src/ws-client/ws-client.module';

@Module({
  imports: [WsClientModule],
  providers: [RoomGateway, RoomService],
  exports: [RoomService],
})
export class RoomModule {}
```

　　然後在 room.gateway 之 constructor 引用 ws-client.service

➜ src\room\room.gateway.ts

```
...
import { WsClientService } from 'src/ws-client/ws-client.service';

@WebSocketGateway()
export class RoomGateway {
  constructor(
    private readonly roomService: RoomService,
    private readonly utilsService: UtilsService,
    private readonly wsClientService: WsClientService,
  ) {
    //
  }
  ...
}
```

現在讓我們接著完成 handleDisconnect 邏輯吧。

➜ src\room\room.gateway.ts

```
...
export class RoomGateway {
  ...
  handleDisconnect(socket: Socket) {
    const client = this.wsClientService.getClient({
      socketId: socket.id,
    });
    if (!client) return;

    if (client.type === 'game-console') {
      this.roomService.deleteRooms(client.id);
      return;
    }

    if (client.type === 'player') {
      this.roomService.deletePlayer(client.id);
      return;
    }
  }
}
```

完成！但是這裡其實有個地方不太好，聰明的讀者們有發現是哪裡嗎？
(^∀^●)

答案是 handleConnection 中的 socket.emit('game-console:room-created')
部分！

為甚麼說這樣不好呢？因為這個 emit 事件的名稱沒有被列舉，這樣會很難
保證未來在功能增減的過程中保證事件名稱統一，所以讓我們改進這個部分吧！

根據 Socket.IO 文檔描述，我們可以自行列舉所有的 on 或 emit 事件，所以
讓我們新增 socket.type 檔案。

➜ types\socket.type.ts

```ts
import { Socket } from 'socket.io';
import { Room } from 'src/room/room.service';

interface OnEvents {
  '': () => void;
}

interface EmitEvents {
  'game-console:room-created': (data: Room) => void;
}

export type ClientSocket = Socket<OnEvents, EmitEvents>;
```

並新增 index 統一匯出。

➜ types\index.ts

```ts
export * from './socket.type';
```

回到 room.gateway 替換一下型別定義。

→ src\room\room.gateway.ts

```
...
import { ClientSocket } from 'types';
...
export class RoomGateway {
  ...

  handleConnection(socket: ClientSocket) {...}
  handleDisconnect(socket: ClientSocket) {...}
}
```

然後神奇的事情發生了！

▲ 圖 6-5 引入 ClientSocket 型別定義

當你輸入 emit 內容時，編譯器就會自動提示有那些事件可以輸入了！

改進大成功！、(●`∀´●)/

現在讓我們切回網頁專案，來實際接收一下房間資訊吧。

6.3.2 網頁建立派對

此章節程式碼可以在以下連結取得。

- https://gitlab.com/deepmind1/animal-party/web/-/tree/feat/build-a-game-room

　　把伺服器定義事件的 socket.type 複製過來，然後交換一下 OnEvents 與 EmitEvents 內容並補充內容。

➜ src\types\socket.type.ts

```
import { Socket } from 'socket.io-client';

export interface Room {
  /** 房間 ID，6 位數字組成 */
  id: string;
  founderId: string;
  playerIds: string[];
}

interface OnEvents {
  'game-console:room-created': (data: Room) => void;
}

interface EmitEvents {
  '': () => void;
}

export type ClientSocket = Socket<OnEvents, EmitEvents>;
```

　　記得一樣在 index.ts 統一匯出。

➜ src\types\index.ts

```
export * from './main.type';
export * from './socket.type';
```

　　接著把 main.store 的 Socket 型別換掉，這樣網頁端也有 socket 事件提示了。

➜ src\stores\main.store.ts

```
...
import { ClientSocket, ClientType } from '../types';
...
```

```
export const useMainStore = defineStore('main', () => {
  ...

  const client = ref<ClientSocket>();
  ...
})
```

目前伺服器已經可以提供網頁端建立房間，網頁端也可以開啟連線了！

網頁端主要可以分為「game-console」與「player」兩種身分，先來製作 game-console 功能吧！♪(´▽｀)

然而「game-console」與「player」兩種身分的功能與事件雖然部分重複，但還是會有獨有邏輯，而且在網頁中必定有多個元件都會需要使用這兩種身分的功能，所以先讓我們新增 use-client-game-console 模組，用來封裝並提供 game-console 與伺服器事件與功能邏輯。

➜ src\composables\use-client-game-console.ts

```
import { useMainStore } from "../stores/main.store";
import { Room } from "../types";

export function useClientGameConsole() {
  const { connect, disconnect } = useMainStore();

  async function startParty() {
    disconnect();

    // 開始連線
    const client = connect('game-console');

    return new Promise<Room>((resolve, reject) => {
      // 5 秒後超時
      const timer = setTimeout(() => {
        disconnect();
        client.removeAllListeners();

        reject(' 連線 timeout');
```

```
    }, 3000);

    // 發生連線異常
    client.once('connect_error', (error) => {
      client.removeAllListeners();
      reject(error);
    });

    // 房間建立成功
    client.once('game-console:room-created', async (room) => {
      client.removeAllListeners();
      clearTimeout(timer);
      resolve(room);
    });
  });
}

return {
  /** 開始派對
   *
   * 建立連線，並回傳房間資料
   */
  startParty,
}
}
```

　　讓我們回到 the-home，使用 use-client-game-console 功能開啟並取得房間 ID 吧！(´∀`)/ ⌐

　　如果連線發生錯誤，希望直接使用 Quasar 提供之 notify 提示使用者，根據文檔描述，需要先在 main.ts 中安裝。

➜ src\main.ts

```
...
import { Quasar, Notify } from 'quasar';
...
createApp(App)
  .use(Quasar, {
```

```
  plugins: {
    Notify,
  },
  lang: quasarLang,
})
...
```

 Tips

想要深入了解的讀者可以參考以下連結：

◆ https://quasar.dev/quasar-plugins/notify#installation

如此便可以使用 Quasar 的 notify 功能了。

接著讓我們修改一下 the-home 中的 startParty() 內容。

➔ src\views\the-home.vue

```
...
<script setup lang="ts">
...
import to from 'await-to-js';
...
import { useQuasar } from 'quasar';
import { useClientGameConsole } from '../composables/use-client-game-console';

const gameConsole = useClientGameConsole();
const loading = useLoading();
const router = useRouter();
const $q = useQuasar();

async function startParty() {
  await loading.show();

  const [err, room] = await to(gameConsole.startParty());
  if (err) {
    console.error(`[ startParty ] err : `, err);
```

```
    $q.notify({
      type: 'negative',
      message: '建立派對失敗，請吸嗨後再度嘗試'
    });
    return;
  }

  console.log(`roomId : `, room.id);

  router.push({
    name: RouteName.GAME_CONSOLE
  });
}
</script>
...
```

 Tips

Promise 使用 await-to-js 包裝後配合 Return Early Pattern，可以有效減少程式縮排、提升可讀性。詳細說明可以參考以下連結：

- https://www.npmjs.com/package/await-to-js

至於甚麼是 Return Early Pattern，簡單來說就是提早中斷函數。詳細說明可以參考以下連結：

- https://gomakethings.com/the-early-return-pattern-in-javascript

現在按下「建立派對」，應該會在讀取畫面消失後，於 DevTool 的 console 看到以下訊息。

```
roomId : 518728
```

我們成功建立房間並取得房間 ID 了！✧*｡٩(´ ▽ ` *)ﻭ✧*｡

第 **7** 章

歡迎光臨遊戲大廳

既然已經可以建立房間了，那就讓我們來把遊戲大廳做出來吧！ヽ(•ω•)ﾉ

預期會有以下內容：

- 目前房間 ID。

  顯示此房間 ID，讓玩家加入。

- 主選單：開始遊戲、結束派對。

  可以開始或結束此遊戲室。

- 遊戲選單

  在 3D 場景中展示可選擇遊戲，切換遊戲時移動鏡頭至對應的遊戲選項。

- 所有玩家頭像

  顯示目前所有連線玩家，玩家所有操作會以對話泡泡方式出現。

目前可以看到建立遊戲室後 route 會導向至 /game-console，接著進入大廳與各個遊戲時，會往子路徑延伸。

例如：

- 大廳是 /game-console/lobby

- 某遊戲是 /game-console/game

以此類推。

現在讓我們新增大廳頁面元件吧！╰(*´︶`*)╯

7.1 大廳頁面

此章節程式碼可以在以下連結取得。

- https://gitlab.com/deepmind1/animal-party/web/-/tree/feat/create-game-console-lobby

首先新增頁面元件。

➜ src\views\the-game-console-lobby.vue

```
<template>
  <div class="flex">
    我是 game-console-lobby
  </div>
</template>

<script setup lang="ts">
import { ref } from 'vue';
import { useLoading } from '../composables/use-loading';

const loading = useLoading();
loading.hide();
</script>

<style scoped lang="sass">
</style>
```

將頁面加到 Router 中。

➜ src\router\router.ts

```
...
export enum RouteName {
  HOME = 'home',
  GAME_CONSOLE = 'game-console',
  GAME_CONSOLE_LOBBY = 'game-console-lobby',
}

const routes: Array<RouteRecordRaw> = [
  ...
  {
    path: `/game-console`,
    name: RouteName.GAME_CONSOLE,
    component: () => import('../views/the-game-console.vue'),
    children: [
      {
        path: `lobby`,
        name: RouteName.GAME_CONSOLE_LOBBY,
        component: () => import('../views/the-game-console-lobby.vue')
      },
    ]
  },
  ...
]
...
```

the-game-console 元件負責註冊 game-console 相關事件並依狀態跳轉頁面等等，但前提是要先有狀態資料才行，所以讓我們新增 game-console.store 儲存資料。

➜ src\stores\game-console.store.ts

```
import { defineStore } from 'pinia';
import { ref } from 'vue';

export const useGameConsoleStore = defineStore('game-console', () => {
```

```
    return {}
})
```

　　加入型別與資料，目前第一個遊戲預計是「第一隻企鵝（the first penguin）」。

→ src\stores\game-console.store.ts

```typescript
import { defineStore } from 'pinia';
import { ref } from 'vue';

/** 遊戲狀態 */
export enum GameConsoleStatus {
  /** 首頁 */
  HOME = 'home',
  /** 大廳 */
  LOBBY = 'lobby',
  /** 遊戲中 */
  PLAYING = 'playing',
}

/** 列舉遊戲名稱 */
export enum GameName {
  THE_FIRST_PENGUIN = 'the-first-penguin',
}

/** 玩家 */
export interface Player {
  clientId: string;
}

export const useGameConsoleStore = defineStore('game-console', () => {
  const status = ref(GameConsoleStatus.HOME);
  const gameName = ref(GameName.THE_FIRST_PENGUIN);
  const roomId = ref<string>();
  const players = ref<Player[]>([]);

  return {
    status,
```

```
    gameName,
    roomId,
    players,
  }
})
```

接著提供「更新狀態」與「設定 room ID」的函數。

➜ src\stores\game-console.store.ts

```
...
export interface UpdateParams {
  status?: `${GameConsoleStatus}`;
  gameName?: `${GameName}`;
  players?: Player[];
}

export const useGameConsoleStore = defineStore('game-console', () => {
  const roomId = ref<string>();

  function setRoomId(id: string) {
    roomId.value = id;
  }


  const status = ref<`${GameConsoleStatus}`>('home');
  const gameName = ref<`${GameName}`>('the-first-penguin');
  const players = ref<Player[]>([]);

  function updateState(state: UpdateParams) {
    status.value = state.status ?? status.value;
    gameName.value = state.gameName ?? gameName.value;
    players.value = state.players ?? players.value;
  }

  return {
    status,
    gameName,
    roomId,
```

```
    players,

    setRoomId,
    updateState,
  }
})
```

還記得我們在「6.3.2 網頁建立派對」中新增 use-client-game-console 用來封裝 game-console 相關操作嗎？現在有了 game-console.store 後，讓我們調整一下 use-client-game-console 內容。

引入 game-console.store 並儲存 roomId 吧。

➡ src\composables\use-client-game-console.ts

```
import { useGameConsoleStore } from "../stores/game-console.store";
...
export function useClientGameConsole() {
  const { connect, disconnect } = useMainStore();
  const gameConsoleStore = useGameConsoleStore();

  async function startParty() {
    ...
    return new Promise<Room>((resolve, reject) => {
      ...
      // 房間建立成功
      client.once('game-console:room-created', async (room) => {
        client.removeAllListeners();
        clearTimeout(timer);

        gameConsoleStore.setRoomId(room.id);

        resolve(room);
      });
    });
  }
...
}
```

現在開啟派對房間後，房間 ID 會被儲存起來了，最後讓我們回到 game-console 頁面元件，新增跳轉邏輯，讓房間建立完成後，遊戲機的畫面會跳轉至 game-console-lobby。

首先調整 template 內容，只要一個 router-view 負責顯示子頁面，並依據目前狀態跳轉頁面功能。

➔ src\views\the-game-console.vue

```
<template>
  <router-view />
</template>

<script setup lang="ts">
import { RouteName } from '../router/router';

import { useRouter } from 'vue-router';
import { useLoading } from '../composables/use-loading';
import { useGameConsoleStore } from '../stores/game-console.store';

const loading = useLoading();
const router = useRouter();
const gameConsoleStore = useGameConsoleStore();

function init() {
  // 房間 ID 不存在，跳回首頁
  if (!gameConsoleStore.roomId) {
    router.push({
      name: RouteName.HOME
    });
    loading.hide();
    return;
  }

  // 跳轉至遊戲大廳
  router.push({
    name: RouteName.GAME_CONSOLE_LOBBY
  });
}
```

```
init();
</script>
```

　　現在會在房間建立完成後，自動降落在 game-console-lobby 頁面了！φ(゜▽゜*)♪

7.2　建立遊戲選單場景

　　此章節程式碼可以在以下連結取得。

■ https://gitlab.com/deepmind1/animal-party/web/-/tree/feat/create-game-menu

　　一片空白的遊戲大廳，真是讓人覺得開了個寂寞派對，讓我們來打造一個背景吧！、(●`∀´●)/

　　讓我們回顧一下故事背景：

　　運送肥宅快樂粉的超大型航空母雞在高空中爆炸啦！

　　快樂粉隨風飄至全世界。

　　遠至南極，近至水溝，忽然間所有的動物們都嗨起來了！

　　基於上述背景，遊戲選單打算做成一個海平面，不同的遊戲屬於一個獨立的小島，切換遊戲時鏡頭會在小島之間互相切換。

最重要的第一步當然是安裝套件，輸入以下命令開始安裝：

```
npm i @babylonjs/core@^5 @babylonjs/loaders@^5
```

Tips

這裡指定 v5 版本，其中：

◆ @babylonjs/core

babylon 3D 核心功能。

◆ @babylonjs/loaders

用於載入 3D 模型，支援 OBJ、gLTF 格式。

接著讓我們利用 babylon.js 打造 3D 場景吧！ㄟ(•ω•)ㄏ babylon.js 產生遊戲畫面需要以下要素：

- 取得 canvas DOM

- 建立 Engine

- 建立 Scene

- 建立至少一個光源

- 建立至少一個 camera

這樣才能產生一個看的到的畫面。

首先建立遊戲選單元件 game-menu。

➔ src\components\game-menu.vue

```
<template>
  <canvas
    ref="canvas"
    class=" outline-none"
  />
```

```
</template>

<script setup lang="ts">
import { ref } from 'vue';

interface Props {
  label?: string;
}
const props = withDefaults(defineProps<Props>(), {
  label: '',
});

const emit = defineEmits<{
  (e: 'update:modelValue', value: string): void;
}>();
</script>
```

可以看到 template 中只要一個 canvas 即可，因為實際畫面都由 JS 產生。

接著先把元件暫時放到 App.vue 中，方便及時預覽目前狀態。

➜　src\App.vue

```
<template>
  <router-view />
  <loading-overlay />

  <game-menu class=" w-full h-full absolute" />
</template>

<script setup lang="ts">
import LoadingOverlay from './components/loading-overlay.vue';
import GameMenu from './components/game-menu.vue';
</script>
...
```

現在讓我們引入 babylon.js 並依序新增建立場景所需的 function 吧。

➡ src\components\game-menu.vue

```
...
<script setup lang="ts">
import { ref } from 'vue';
import { ArcRotateCamera, Engine, Scene, Vector3 } from '@babylonjs/core';
...
const canvas = ref<HTMLCanvasElement>();
let engine: Engine;
let scene: Scene;
let camera: ArcRotateCamera;

function createEngine(canvas: HTMLCanvasElement) {
  const engine = new Engine(canvas, true);
  return engine;
}
function createScene(engine: Engine) {
  const scene = new Scene(engine);
  /** 使用預設光源 */
  scene.createDefaultLight();
  return scene;
}
function createCamera(scene: Scene) {
  const camera = new ArcRotateCamera(
    'camera',
    Math.PI / 2,
    Math.PI / 4,
    34,
    new Vector3(0, 0, -2),
    scene
  );

  return camera;
}
</script>
```

- 建立引擎基本上就是傳入 canvas DOM 與開啟反鋸齒，沒有其他特別參數。

- 場景力求簡單，先直接使用預設光源

- 相機使用最簡單的 ArcRotateCamera

延伸閱讀：

Arc Rotate Camera

- https://doc.babylonjs.com/features/featuresDeepDive/cameras/camera_introduction#arc-rotate-camera

Create Default Light

- https://doc.babylonjs.com/features/featuresDeepDive/scene/fastBuildWorld#create-default-light

接著加入以下邏輯：

- 增加初始化用的函數 init，建立場景。

- 定義 'complete' emit 事件，可以對外發出初始化完成事件。

- 在 onMounted 事件中呼叫 init，

- 在 onBeforeUnmount，也就是元件銷毀前關閉引擎。

熟悉 Vue 的讀者們一定都知道會在 onMounted 中才初始化，因為這樣才取得到 canvas DOM。

Tips

onMounted 會在 Vue 實例的 DOM 元素被插入到真實 DOM 中後觸發,所以操作 DOM 元素時,需要於 onMounted 觸發後才可進行操作。否則 Vue 實例的 DOM 元素尚未被插入到真實的 DOM 中,嘗試操作 DOM 元素會因為尚未存在導致錯誤。

- https://cn.vuejs.org/api/composition-api-lifecycle.html#onmounted

➜ src\components\game-menu.vue

```ts
...
<script setup lang="ts">
import { onBeforeUnmount, onMounted, ref } from 'vue';
...
const emit = defineEmits<{
  (e: 'complete'): void;
}>();
...
function init() {
  if (!canvas.value) {
    console.error('無法取得 canvas DOM');
    return;
  }

  engine = createEngine(canvas.value);
  scene = createScene(engine);
  camera = createCamera(scene);

  /** 反覆渲染場景,這樣畫面才會持續變化 */
  engine.runRenderLoop(() => {
    scene.render();
  });

  /** 發出完成事件,表示畫面初始化完成 */
  emit('complete');
}
```

```
onMounted(() => {
  init();
});

onBeforeUnmount(() => {
  engine.dispose();
});
</script>
```

再加入考慮畫面隨視窗自動縮放的功能。

➔ src\components\game-menu.vue

```
...
<script setup lang="ts">
...
onMounted(() => {
  init();
  window.addEventListener('resize', handleResize);
});

onBeforeUnmount(() => {
  engine.dispose();
  window.removeEventListener('resize', handleResize);
});

function handleResize() {
  engine.resize();
}
</script>
```

▲ 圖 7-1 產生 3D 場景

鱈魚：「現在會發現 3D 場景出現了！ ◖(°∀°)◗」

助教：「出個毛線！這坨深藍色我用 CSS 都能產生！ \(°д°\)」

鱈魚：「那是背景預設的顏色啦，讓我們慢慢來嘛。('„•ω•„)」

組成一個完整 3D 物件最基本構成是 Mesh 與 Material，前者負責組成「骨架」，後者負責「皮膚」。

現在讓我們建立一片海洋（其實就是一片地板），過程如下：

1. 建立海洋要用的網格

   使用 MeshBuilder.CreateGround，可以產生一個不具厚度的網格，因為地板不需要厚度。

2. 建立材質

   將顏色設定為藍色後附加於網格上。

以上步驟便可以完成海洋地板了，現在建立新增海洋函數。

➜ src\components\game-menu.vue

```
...
<script setup lang="ts">
...
import { ArcRotateCamera, BackgroundMaterial, Color3, Engine, MeshBuilder, Scene,
Vector3 } from '@babylonjs/core';
...
function createCamera(scene: Scene) { ... }

function createSea(scene: Scene) {
  const sea = MeshBuilder.CreateGround('sea', { height: 1000, width: 1000 });

  const material = new BackgroundMaterial("seaMaterial", scene);
  material.useRGBColor = false;
  material.primaryColor = new Color3(71 / 256, 174 / 256, 250 / 256);

  sea.material = material;
```

```
  return sea;
}

function init() {
  ...
  camera = createCamera(scene);
  createSea(scene);

  ...
}
...
</script>
```

▲ 圖 7-2 產生海洋

鱈魚：「現在是一片藍了！ヽ(•ω•)ﾉ」

助教：「所以我說那個 3D 呢？(○´･д･)ﾉ」

鱈魚：「那就讓我們建立企鵝島吧！(`•ω•´)✧」

　　建立 penguin-island 檔案，用來產生立體企鵝島，讓我們先定義 interface 後再實作每個小島吧。

小島包含需要以下內容：

- 提供 active 狀態，讓使用者在挑選遊戲時，能夠看出目前選定的遊戲
- 提供模型原點，用於移動、旋轉或縮放模型。

➜ src\types\main.type.ts

```
import { TransformNode } from "@babylonjs/core";

export enum ClientType {...}

export interface ModelIsland {
  /** 模型原點 */
  rootNode: TransformNode;
  setActive(value: boolean): void;
}
```

　　接著來建立企鵝島，新增建立 PenguinIsland 物件之函數，其中 Penguin Island 必須實作 ModelIsland 介面。

➜ src\components\game-menu\the-first-penguin.ts

```
import { Scene } from "@babylonjs/core";
import { ModelIsland } from "../../types";
import { ref } from "vue";

export type PenguinIsland = ModelIsland;

export async function createPenguinIsland(scene: Scene): Promise<PenguinIsland> {
  const active = ref(false);
  function setActive(value: boolean) {
    active.value = value;
  }

  return {
    setActive,
  }
}
```

接著我們新增一個 TransformNode，用來表示整個模型原點，其他 3D 物件皆依附在此原點上。

➔ src\components\game-menu\the-first-penguin.ts

```ts
import { ModelIsland } from "../../types";
import {
  Scene, TransformNode,
} from "@babylonjs/core";
...


export async function createPenguinIsland(scene: Scene): Promise<PenguinIsland> {
  const rootNode = new TransformNode("penguinIsland");
  ...
}
```

Tips

想要深入了解的讀者可以參考以下連結：

- https://doc.babylonjs.com/features/featuresDeepDive/mesh/transforms/parent_pivot/transform_node

先讓我們在場景中引入企鵝島，以便查看目前外觀。

➔ src\components\game-menu.vue

```vue
...
<script setup lang="ts">
...
import { createPenguinIsland } from './game-menu/the-first-penguin';
...
function createIslands(scene: Scene) {
  createPenguinIsland(scene);
}

function init() {
 ...
```

```
  createIslands(scene);
  ...
}
...
</script>
```

因為只有一個看不見的原點，所以目前看起來還是甚麼都沒有。(. ＾ ᵕ ＾ .)

現在讓我們建立第一塊浮冰吧，新增一個名為 createIces 的 method。

➔ src\components\game-menu\the-first-penguin.ts

```
...
export async function createPenguinIsland(scene: Scene): Promise<PenguinIsland> {
  ...

  function createIces() {
    const material = new StandardMaterial('iceMaterial', scene);
    const ice = MeshBuilder.CreateBox(`ice`, {
      width: 4,
      depth: 4,
      height: 2,
    });
    ice.material = material;
    ice.position = Vector3.Zero();
    ice.setParent(rootNode);
  }

  createIces();

  return {
    setActive,
  }
}
```

可以看到第一塊浮冰出現了！

▲ 圖 7-3 第一塊浮冰

鱈魚：「接下來讓我們繼續產生模型吧！ ヾ(*°▽°*)ノ」

助教：「視角不能隨意調整，這樣好難開發喔」

鱈魚：「說的有道理，讓我們開啟鏡頭控制和 babylon Inspector 工具吧」

助教：「你是不會早一點變出來喔。 ℠(´口`℠)」

鱈魚：「這樣才能正大光明的增長內容啊。 ヽ(•ω•)ノ」

助教：「... ('◉ ᴗ ◉`)」

第一步先開啟鏡頭控制，只要一行程式即可。

➔ src\components\game-menu.vue

```
...
<script setup lang="ts">
...
function init() {
  ...
```

```
  camera = createCamera(scene);
  camera.attachControl(canvas.value, true);
  /** 調整滾輪縮放程度 */
  camera.wheelDeltaPercentage = 0.01;
  ...
}
...
</script>
```

現在可以使用滑鼠左鍵控制旋轉，右鍵控制平移，滾輪控制縮放了。

▲ 圖 7-4 鏡頭控制

接下來開啟 Inspector 工具前，需要先安裝相依套件，指定使用 v5 版本。

```
npm i @babylonjs/inspector@^5 @babylonjs/gui@^5 @babylonjs/gui-editor@^5 @babylonjs/
materials@^5 @babylonjs/serializers@^5
```

接著引入並啟用 Inspector。

➜ src\components\game-menu.vue

```
...
<script setup lang="ts">
...
```

```
import {...} from '@babylonjs/core';
import '@babylonjs/loaders';
import "@babylonjs/core/Debug/debugLayer";
import "@babylonjs/inspector";
...
function init() {
  ...
  window.addEventListener('keydown', (ev) => {
    // 按下 Shift+I 可以切換視窗
    if (ev.shiftKey && ev.keyCode === 73) {
      if (scene.debugLayer.isVisible()) {
        scene.debugLayer.hide();
      } else {
        scene.debugLayer.show();
      }
    }
  });

  /** 發出完成事件，表示畫面初始化完成 */
  emit('complete');
}
...
</script>
```

現在按下 Shift+I，開啟 Inspector 會看到 ..... 畫面跑掉啦！(＃°Д°)

▲ 圖 7-5 引入 Inspector

這是因為 Inspector 的 HTML 打亂原本的排版，要解決這個問題很容易，讓我們使用全域 CSS 覆蓋它的樣式吧。

➜ src\style\global.sass

```
...

#scene-explorer-host, #inspector-host
  position: fixed !important
  top: 0 !important
  z-index: 99999
  opacity: 0
  transition-duration: 0.4s
  &:hover
    opacity: 0.8
  input, select
    color: black
```

現在可以看到一左一右的工具欄，而且變成滑鼠觸碰後才出現了。、(≧∀≦)、

▲ 圖 7-6 調整 Inspector 版面

來簡介一下如何使用。

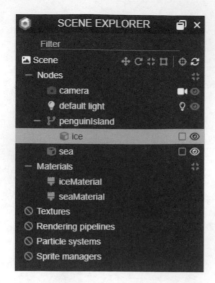

▲ 圖 7-7 scene exploter 與 inspector

　　scene exploter 呈現目前場景中所有的成員，而且可以隱藏、移動、旋轉、縮放。

　　inspector 則可顯示目前設定參數，或是在選擇物體後顯示對應物體的參數細節。

　　讓我們繼續調整浮冰，把冰塊周圍加上一點水波紋的效果，可以預期浮冰的內容會越來越多，所以讓我們把浮冰獨立成一個檔案。

➡ src\components\game-menu\the-first-penguin\ice.ts

```
import { StandardMaterial, MeshBuilder, Scene, Vector3 } from "@babylonjs/core";

interface Params {
  width: number;
  depth: number;
  height: number;
  position: Vector3;
  material?: StandardMaterial;
```

```
}

export function createIce(name: string, scene: Scene, params: Params) {
  const material = params.material ?? new StandardMaterial(`${name}-material`, scene);

  // 建立浮冰
  const mesh = MeshBuilder.CreateBox(name, params);
  mesh.material = material;
  mesh.position = params.position;

  return {
    mesh
  }
}
```

調整 createIces。

➜ src\components\game-menu\the-first-penguin.ts

```
...
export async function createPenguinIsland(scene: Scene): Promise<PenguinIsland> {
  ...

  function createIces() {
    const material = new StandardMaterial('ice-material', scene);
    const ice = createIce(`ice`, scene, {
      width: 4,
      depth: 4,
      height: 2,
      position: new Vector3(0, 0, 0),
      material,
    });
    ice.mesh.setParent(rootNode);
  }
...
}
```

雖然改寫了程式，但是畫面應該是一點都沒變。(´„•ω•„)

來幫浮冰加入水波紋吧，與建立海面類似，讓我們建立一個白色平面。

➜ src\components\game-menu\the-first-penguin\ice.ts

```
...
export function createIce(name: string, scene: Scene, params: Params) {
  ...
  // 建立浮冰
  ...

  // 建立水波紋
  function createRipple() {
    const ripple = MeshBuilder.CreateGround('ripple', {
      height: params.depth + 0.3,
      width: params.width + 0.3,
    });
    ripple.setParent(mesh);
    // 往上移一點，要比海面高一點才看的到
    ripple.position = new Vector3(0, 0.01, 0);

    // 設為半透明白色
    const material = new StandardMaterial('ripple-material', scene);
    material.diffuseColor = Color3.White();
    material.alpha = 0.5;

    ripple.material = material;
  }
  createRipple();

  return {
    mesh
  }
}
```

▲ 圖 7-8 加入水波紋

　　最後讓我們加入水波紋動畫，動畫的部分比較複雜一點，主要分為「Animation」與「Easing Function」兩個部分。

- Animation

  負責定義動畫細節，例如每幀的內容、播放模式等等。

- Easing Function

  負責定義每幀之前如何過渡。

　　水波紋動畫實現很單純就是在水平面方向（XZ 平面）縮放即可，來建立 Animation 部分。

```
/** 總幀數 */
const frameRate = 20;
const breatheAnimation = new Animation(
  /** 動畫名稱 */
  'breathe',
  /** 目標屬性 */
  'scaling',
  /** 每秒播放幀數 */
  frameRate / 2,
  /** 目標屬性類型 */
  Animation.ANIMATIONTYPE_VECTOR3,
  /** 循環方式 */
```

```
  Animation.ANIMATIONLOOPMODE_CYCLE
);
```

再來定義 keyframe，也就是所謂的關鍵幀，與 CSS 的 animation 中的 keyframe 概念相同，用來描述動畫關鍵點的狀態。

```
const keyFrames: IAnimationKey[] = [
  {
    frame: 0,
    value: new Vector3(1, 1, 1),
  },
  {
    frame: frameRate / 2,
    value: new Vector3(1.05, 1, 1.05),
  },
  {
    frame: frameRate,
    value: new Vector3(1, 1, 1)
  },
];
breatheAnimation.setKeys(keyFrames);
ripple.animations.push(breatheAnimation);
```

 Tips

想要深入了解的讀者可以參考以下連結：

◆ https://doc.babylonjs.com/features/featuresDeepDive/animation/animation_method

以上我們就完成水波紋動畫了，現在讓我們開始播放動畫。

→ src\components\game-menu\the-first-penguin\ice.ts

```
...
export function createIce(name: string, scene: Scene, params: Params) {
  ...
```

```
// 建立水波紋
function createRipple() {
  ...
  // 建立動畫
  /** 總幀數 */
  const frameRate = 20;
  const breatheAnimation = new Animation(...);

  const keyFrames: IAnimationKey[] = [...];
  breatheAnimation.setKeys(keyFrames);
  ripple.animations.push(breatheAnimation);

  /** 開始播放動畫 */
  scene.beginAnimation(ripple, 0, frameRate, true);
}
...
}
```

▲ 圖 7-9 水波紋動畫

現在應該會看到水波紋有縮放的動畫了！ヽ(*´∀`*)ノ

只是動畫看起來有點死板，這是每個關鍵幀之間為線性插幀，這時候讓我們引入 Easing Function 吧。

Easing Function 與 CSS 的 animation-timing-function 概念相同。

Tips

想要深入了解的讀者可以參考以下連結：

◆ https://ithelp.ithome.com.tw/articles/10200709

讓我們新增 Easing Function 吧。

➜ src\components\game-menu\the-first-penguin\ice.ts

```
import {
  StandardMaterial, MeshBuilder, Scene, Vector3, Color3,
  Animation, AnimationGroup, IAnimationKey, AnimationKeyInterpolation, CircleEase,
EasingFunction,
} from "@babylonjs/core";
...
export function createIce(name: string, scene: Scene, params: Params) {
  ...
  // 建立水波紋
  function createRipple() {
    ...
    // 建立動畫
    /** 總幀數 */
    const frameRate = 20;
    const breatheAnimation = new Animation(...);

    const keyFrames: IAnimationKey[] = [...];
    breatheAnimation.setKeys(keyFrames);
    ripple.animations.push(breatheAnimation);

    // 建立 easing function
    const easingFunction = new CircleEase();
    easingFunction.setEasingMode(EasingFunction.EASINGMODE_EASEINOUT);
    breatheAnimation.setEasingFunction(easingFunction);
    ripple.animations.push(breatheAnimation);

    /** 開始播放動畫 */
    scene.beginAnimation(ripple, 0, frameRate, true);
```

```
    }
    ...
}
```

順利的話可以發現水波紋動起來更有節奏感了！♪(´ω｀)♪

只有一個冰塊感覺真寂寞，讓我們回到 createPenguinIsland 調整一下 createIces，產生更多浮冰。

➡ src\components\game-menu\the-first-penguin.ts

```
...
export async function createPenguinIsland(scene: Scene): Promise<PenguinIsland> {
  ...
  function createIces() {
    const list = [
      {
        width: 2.5,
        depth: 2.5,
        height: 3,
        position: new Vector3(1, 0, 2),
      },
      {
        width: 4,
        depth: 4,
        height: 2,
        position: Vector3.Zero(),
      },
      {
        width: 2,
        depth: 2,
        height: 1.5,
        position: new Vector3(8, 0, 3.2),
      },
      {
        width: 2.5,
        depth: 2.5,
        height: 1,
        position: new Vector3(-3.5, 0, 1.5),
```

```
    },
    {
      width: 1,
      depth: 1,
      height: 1,
      position: new Vector3(3, 0, -2.5),
    },
  ];

  const material = new StandardMaterial('iceMaterial', scene);
  list.forEach((data, i) => {
    const ice = MeshBuilder.CreateBox(`ice${i}`, data);
    ice.material = material;
    ice.position = data.position;
    ice.setParent(rootNode);
  });
}

createIces();

return {
  setActive,
}
}
```

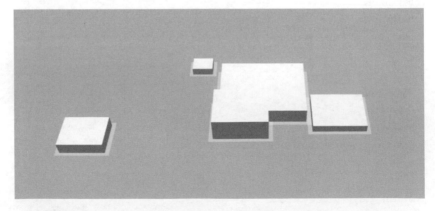

▲ 圖 7-10 更多浮冰

最後讓我們把企鵝放上去吧！(´▽`)ノ

先把鏡頭視角轉到一個好觀察的角度，並在 scene explorer 點選 Nodes 中的 camera，並把右方 inspector 中的 transforms 的數值記下來。

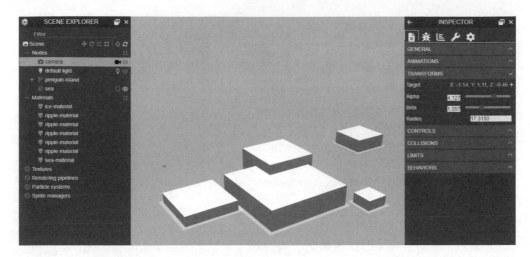

▲ 圖 7-11　更多浮冰

回到 game-menu 中，將數值填入初始化 camera 的地方，這樣就不會每次儲存的時候，都要重新調整鏡頭了。

➜ src/components/game-menu.vue

```
...
<script setup lang="ts">
...
function createCamera(scene: Scene) {
  const camera = new ArcRotateCamera(
    'camera',
    4.127,
    1.097,
    17.315,
    new Vector3(-1.14, 1.11, -0.46),
    scene
  )」
```

```
  return camera;
}
...
</script>
```

新增建立企鵝用的檔案。

➜ src\components\game-menu\the-first-penguin\penguin.ts

```
import {
  Scene, Vector3, AnimationGroup, SceneLoader,
} from "@babylonjs/core";
import { defaults } from "lodash-es";

export async function createPenguin(name: string, scene: Scene) {
}
```

讓我們載入企鵝的 3D 模型吧，首先把 glb 檔案放至 public 目錄中。

▲ 圖 7-12 新增企鵝 3D 模型檔案

讀者可以前往 GitLab 下載模型：

- https://gitlab.com/deepmind1/animal-party/web/-/blob/feat/create-game-menu/
  public/the-first-penguin/penguin.glb

接著載入模型。

➜ src\components\game-menu\the-first-penguin\penguin.ts

```
export async function createPenguin(name: string, scene: Scene) {
  const result = await SceneLoader.ImportMeshAsync('', '/the-first-penguin/', 'penguin.
glb', scene);
}
```

讓我們回到企鵝島引入企鵝模型，方便觀察狀態。

➜ src\components\game-menu\the-first-penguin.ts

```
...
import { createPenguin } from './the-first-penguin/penguin';

export type PenguinIsland = ModelIsland;

export async function createPenguinIsland(scene: Scene): Promise<PenguinIsland> {
  ...
  createIces();

  async function createPenguins() {
    createPenguin('penguin', scene);
  }
  createPenguins();
  ...
}
```

現在畫面應該會跑出一個一直轉的巨型企鵝。ᕕ(ﾟ∀ﾟ)ᕗ

▲ 圖 7-13　企鵝登場

現在讓我們定義設定參數，讓企鵝尺寸正常一點。

➜ src\components\game-menu\the-first-penguin.ts

```
interface Option {
  scaling?: number;
  position?: Vector3;
  rotation?: Vector3;
  /** 企鵝播放動畫 */
  playAnimation?: 'attack' | 'walk' | 'idle' | '';
}

const defaultOption: Required<Option> = {
  scaling: 1,
  position: Vector3.Zero(),
  rotation: Vector3.Zero(),
  playAnimation: '',
}
```

實際帶入設定參數並先暫停動畫,企鵝才不會一直轉。

➡ src\components\game-menu\the-first-penguin.ts

```
...
export async function createPenguin(name: string, scene: Scene, option?: Option) {
  const {
    scaling, position, rotation
  } = defaults(option, defaultOption);

  const result = await SceneLoader.ImportMeshAsync('', '/the-first-penguin/', 'penguin.
glb', scene);

  const mesh = result.meshes[0];
  mesh.name = name;
  mesh.position = position;
  mesh.rotation = rotation;
  mesh.scaling = new Vector3(scaling, scaling, scaling);

  function initAnimation(animationGroups: AnimationGroup[]) {
    animationGroups.forEach((animationGroup) => {
      animationGroup.stop();
    });
  }
  initAnimation(result.animationGroups);

  return { mesh }
}
```

 Tips

const mesh = result.meshes[0];

這行程式表示取一個 mesh 作為企鵝節點,至於為甚麼呢?因為在建立企鵝模型時,第一個 mesh 即為整個企鵝模型的根 mesh,其他部位皆依附於此,所以取根 mesh 等取到了整個企鵝模型了。

接下來調整 createPenguins 實際加入設定吧。

➜ src\components\game-menu\the-first-penguin.ts

```
...
  async function createPenguins() {
    /**
     * 從 createPenguin 函數獲取第三個參數的類型（索引為 2，因為索引從 0 開始）
     */
    const list: Parameters<typeof createPenguin>[2][] = [
      {
        scaling: 0.4,
        position: new Vector3(8, 0.8, 3),
        rotation: new Vector3(0, Math.PI, 0),
        playAnimation: 'idle',
      },
    ];

    /** 建立 createPenguin 的 Promise 矩陣 */
    const tasks = list.map((option, i) =>
      createPenguin(`penguin-${i}`, scene, option)
    );
    /** 當所有的 Promise 都完成時，回傳所有 Promise 的結果。 */
    const results = await Promise.allSettled(tasks);

    /** 將成功建立的企鵝綁定至 rootNode */
    results.forEach((result) => {
      if (result.status === 'rejected') return;
      result.value.mesh.setParent(rootNode);
    });
  }
...
```

現在可以看到第一隻企鵝就定位了。✧*.٩(´ ロ ` *)ﻭ✧*。

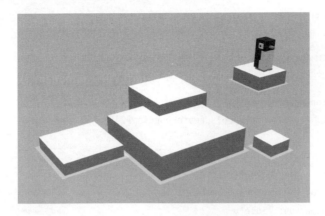

▲ 圖 7-14 第一隻企鵝就定位

哎呀，忘記加入動畫了，讓我們追加一下。('„•ω•„)

➜ src\components\game-menu\the-first-penguin\penguin.ts

```
...
export async function createPenguin(name: string, scene: Scene, option?: Option) {
  ...
  function initAnimation(animationGroups: AnimationGroup[]) {
    animationGroups.forEach((animationGroup) => {
      animationGroup.stop();
    });

    /** 播放指定動畫 */
    if (option?.playAnimation) {
      const animation = animationGroups.find(({ name }) => name === option.
playAnimation);
      animation?.play(true);
    }
  }
  initAnimation(result.animationGroups);

  return { mesh }
}
```

現在企鵝動起來了！那就讓我們加入更多企鵝吧！

➜ src\components\game-menu\the-first-penguin.ts

```
...
  async function createPenguins() {
    const list: Parameters<typeof createPenguin>[2][] = [
      {
        scaling: 0.4,
        position: new Vector3(8, 0.8, 3),
        rotation: new Vector3(0, Math.PI, 0),
        playAnimation: 'idle',
      },
      {
        scaling: 0.45,
        position: new Vector3(-0.5, 1, 0),
        rotation: new Vector3(0, Angle.FromDegrees(125).radians(), 0),
        playAnimation: 'walk',
      },
      {
        scaling: 0.45,
        position: new Vector3(0.5, 1, -0.5),
        rotation: new Vector3(0, Angle.FromDegrees(310).radians(), 0),
        playAnimation: 'walk',
      },
      {
        scaling: 0.35,
        position: new Vector3(-4.2, 0.8, 1.7),
        rotation: new Vector3(-Math.PI / 2, -Math.PI / 4, 0),
      },
    ];
    ...
  }
...
```

Tips

Angle.FromDegrees(125).radians()，表示將角度轉換成徑度。

▲　圖 7-15　企鵝到齊

　　可以看到預設光線讓模型的光影看起來有點單調，讓我們為企鵝島多加一盞燈吧。(´▽`)/ ₋

➜　src\components\game-menu\the-first-penguin.ts

```
import {
  Scene, TransformNode, MeshBuilder, StandardMaterial, Vector3, Angle, PointLight,
  Color3
} from '@babylonjs/core';
...
export async function createPenguinIsland(scene: Scene): Promise<PenguinIsland> {
  ...
  createPenguins();

  function createLight() {
    const light = new PointLight(
      'penguin-island-light',
      new Vector3(-7.8, 7.1, -3),
      scene,
    );
    light.diffuse = new Color3(186 / 255, 229 / 255, 1);
    light.intensity = 0.6;
    light.parent = rootNode;
```

```
  }
  createLight();

  return {
    rootNode,
    setActive,
  }
}
```

現在可以看到浮冰的陰影不會那麼單調了！ヽ(•ω•)ﾉ

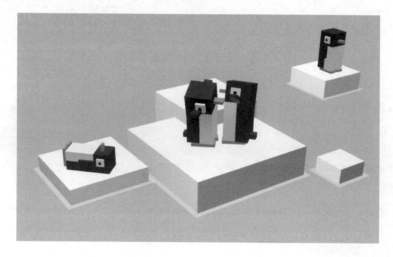

▲ 圖 7-16　增加企鵝島燈光

最後讓我們加個文字吧，與企鵝概念相同，讓我們建立檔案。

 Tips

在 babylon 中建立立體中文字相當麻煩，所以這裡直接引入建立好的立體中文字模型，讀者們可前往這裡下載：

- https://gitlab.com/deepmind1/animal-party/web/-/blob/feat/create-game-menu/public/the-first-penguin/penguin-title.glb

下載後將檔案放在 public\the-first-penguin\penguin-title.glb 路徑。

➜ src\components\game-menu\the-first-penguin\title.ts

```ts
import {
  Scene, Vector3, SceneLoader,
  Animation, BackEase, SineEase,
  StandardMaterial, Color3,
} from "@babylonjs/core";
import { defaults } from "lodash-es";

interface Option {
  scaling?: number;
  position?: Vector3;
  rotation?: Vector3;
}

const defaultOption: Required<Option> = {
  scaling: 1,
  position: Vector3.Zero(),
  rotation: Vector3.Zero(),
}

export async function createTitle(name: string, scene: Scene, option?: Option) {
  const {
    scaling, position, rotation
  } = defaults(option, defaultOption);

  const result = await SceneLoader.ImportMeshAsync('', '/the-first-penguin/', 'penguin-
title.glb', scene);

  const scalingVector = new Vector3(scaling, scaling, scaling);

  /** 這裡取第二個 mesh 是因為這裡需要加入 material
   * 但是第一個 mesh 是 root，root 無法使用材質，
   * 所以要取第二個 mesh，也就是文字本體。
   */
  const mesh = result.meshes[1];
  mesh.name = name;
  mesh.position = position;
  mesh.rotation = rotation;
  mesh.scaling = scalingVector;
```

```
  return { mesh }
}
```

在 Option 中提供外部可以調整文字的位置、旋轉和縮放。

現在引入 createTitle 實際建立文字吧。

➜ src\components\game-menu\the-first-penguin.ts

```
...
import { createTitle } from './the-first-penguin/title';
...
export async function createPenguinIsland(scene: Scene): Promise<PenguinIsland> {
  ...

  async function createText() {
    const title = await createTitle('penguin-title', scene, {
      position: new Vector3(-0.17, -0.08, -2.94),
      rotation: new Vector3(0, Math.PI, Math.PI),
      scaling: 0.9,
    });
    title.mesh.parent = rootNode;

    return title;
  }
  createText();

  return {
    rootNode,
    setActive,
  }
}
```

現在可以看到標題出現了！(ˊ„•ω•„)

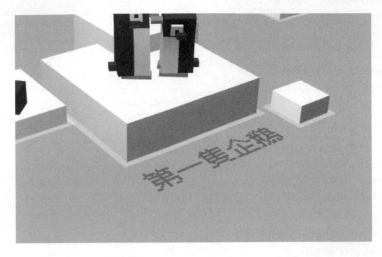

▲ 圖 7-17 企鵝島 title

接著與浮冰概念相同，讓我們加入材質與動畫吧。

➡ src\components\game-menu\the-first-penguin\title.ts

```
...
export async function createTitle(name: string, scene: Scene, option?: Option) {
  ...
  function initMaterial() {
    const material = new StandardMaterial('penguin-title-material', scene);
    material.diffuseColor = Color3.White();
    material.emissiveColor = Color3.FromHexString('#101519');
    material.specularColor = Color3.Black();
    mesh.material = material;
    return material;
  }
  initMaterial();

  // 建立動畫
  function initAnimation() {
    const frameRate = 20;
    const floatAnimation = new Animation(
```

```
  'float', 'position', frameRate / 3,
  Animation.ANIMATIONTYPE_VECTOR3,
  Animation.ANIMATIONLOOPMODE_CYCLE
);

floatAnimation.setKeys([
  {
    frame: 0,
    value: position.add(new Vector3(0, 0, 0)),
  },
  {
    frame: frameRate / 2,
    value: position.add(new Vector3(0, 0.05, 0)),
  },
  {
    frame: frameRate,
    value: position.add(new Vector3(0, 0, 0)),
  },
]);
mesh.animations.push(floatAnimation);
floatAnimation.setEasingFunction(new SineEase());

const scaleAnimation = new Animation(
  'scale', 'scaling', frameRate / 2,
  Animation.ANIMATIONTYPE_VECTOR3,

  Animation.ANIMATIONLOOPMODE_CYCLE
);
const scaleValue = 0.01;
scaleAnimation.setKeys([
  {
    frame: 0,
    value: scalingVector.add(new Vector3(-scaleValue, 0, scaleValue)),
  },
  {
    frame: frameRate / 2,
    value: scalingVector.add(new Vector3(scaleValue, 0, -scaleValue)),
  },
  {
    frame: frameRate,
```

```
      value: scalingVector.add(new Vector3(-scaleValue, 0, scaleValue)),
    },
  ]);
  mesh.animations.push(scaleAnimation);
  scaleAnimation.setEasingFunction(new BackEase());

  scene.beginAnimation(mesh, 0, frameRate, true);
}
initAnimation();

return { mesh }
}
```

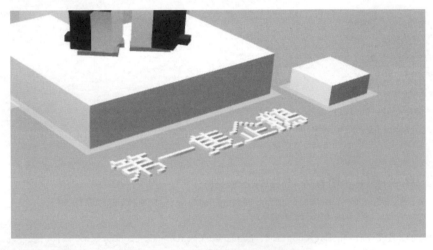

▲ 圖 7-18 企鵝島 title 材質與動畫

看起來好多了。(´▽`)╯

現在讓我們完成前面提到的企鵝島 active 的效果後，企鵝島基本上就大功告成了！✧*。٩(´ロ`*)۶✧*。

目前的想像是 title 文字平時在海面下方，active 後才會浮出海面，這裡的動畫和先前的浮冰有點差異，之前的 animation 都是預先指定好關鍵幀在播放，這個 active 的動畫則是在觸發時將狀態渦渡至指定目標。

　　換句話說，就是每次動畫的目標都不固定，只有呼叫的當下才會知道目標。

　　所以讓我們建立一個簡單產生動畫的函數。

➜ src\common\utils.ts

```typescript
import { Animation, Color3, CubicEase, EasingFunction, Vector3 } from "@babylonjs/
core";
import { defaults, get } from "lodash-es";

export interface CreateAnimationOption {
  /** 總幀數
   * @default 60
   */
  frameRate?: number;
  /** 播放倍數
   * @default 1
   */
  speedRatio?: number;
  /** @default CubicEase */
  easingFunction?: EasingFunction;
}

const defaultEasingFunction = new CubicEase();
defaultEasingFunction.setEasingMode(EasingFunction.EASINGMODE_EASEINOUT);

const defaultCreateAnimationOption: Required<CreateAnimationOption> = {
  frameRate: 60,
  speedRatio: 1,
  easingFunction: defaultEasingFunction,
}

/** 建立從目前狀態至目標狀態的動畫
 *
 * @param target 目標物件
 * @param property 過度的目標屬性
 * @param to 目標狀態
 * @param option 參數
 * @returns
 */
```

```
export function createAnimation(target: any, property: string, to: number | Vector3 |
Color3, option?: CreateAnimationOption) {
  const {
    frameRate, speedRatio, easingFunction,
  } = defaults(option, defaultCreateAnimationOption);

  const keys = [
    {
      frame: 0, value: get(target, property),
    },
    {
      frame: frameRate, value: to,
    },
  ];

  let animationType = Animation.ANIMATIONTYPE_FLOAT;
  if (typeof to === 'number') {
    animationType = Animation.ANIMATIONTYPE_FLOAT
  } else if (to instanceof Vector3) {
    animationType = Animation.ANIMATIONTYPE_VECTOR3
  } else if (to instanceof Color3) {
    animationType = Animation.ANIMATIONTYPE_COLOR3
  }

  const animation = Animation.CreateAnimation(
    property,
    animationType,
    frameRate * speedRatio,
    easingFunction
  );
  animation.setKeys(keys);
  return { animation, frameRate };
}
```

第一步先來調整一下 title 內的結構，與企鵝島 rootNode 的概念相同，新增 TransformNode 作為原點，這樣原點的移動動畫與文字模型本身的動畫就可以完全隔離，不會互相影響。

➔ src\components\game-menu\the-first-penguin\title.ts

```
...
export async function createTitle(name: string, scene: Scene, option?: Option) {
  ...

  const rootNode = new TransformNode('penguin-island-title');
  ...

  /** 這裡取第二個 mesh 是因為這裡需要加入 material
   * 但是第一個 mesh 是 root，root 無法使用材質，
   * 所以要取第二個 mesh，也就是文字本體。
   */
  ...
  mesh.parent = rootNode;
  ...
  return { node: rootNode }
}
```

接著回到企鵝島，讓我們新增處理 active 時產生動畫功能。

➔ src\components\game-menu\the-first-penguin.ts

```
...
export async function createPenguinIsland(scene: Scene): Promise<PenguinIsland> {
  ...

  async function createText() {
    ...
    title.node.parent = rootNode;

    return title;
  }
  const titleModel = await createText();

  const active = ref(false);
  function setActive(value: boolean) {
    active.value = value;
  }
  watch(active, (value) => {
```

```
      playActiveAnimation(value);
  }, { immediate: true });

  function playActiveAnimation(value: boolean) {
    /** 浮出水面或下降 */
    const offset = value ? 0 : -1;
    const rotation = value ? 0 : Angle.FromDegrees(10).radians();

    const results = [
      createAnimation(titleModel.node, 'position', new Vector3(0, offset, 0)),
      createAnimation(titleModel.node, 'rotation', new Vector3(rotation, 0,
rotation)),
    ];
    const animations = results.map(({ animation }) => animation);

    scene.beginDirectAnimation(titleModel.node, animations, 0, results[0].frameRate);
  }

  return {
    rootNode,
    setActive,
  }
}
```

　　現在企鵝島的標題文字會在 active 發生變化時，隱藏或浮出水面了。
ˋ(•ω•)ˊ

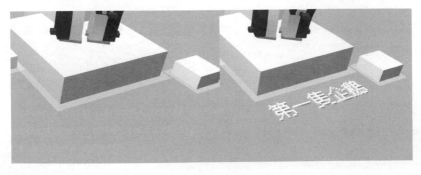

▲ 圖 7-19 標題動畫

最後讓我們完成遊戲選單畫面的鏡頭運鏡功能，預期鏡頭一開始會在很高的高空，接著移動並聚焦至企鵝島，讓我們一步一步加入程式吧。

首先新增 game.type 定義遊戲相關的型別。

➜　src\types\game.type.ts

```ts
import { RouteName } from "../router/router";

export enum GameName {
  THE_FIRST_PENGUIN = 'the-first-penguin',
}

export interface GameInfo {
  name: `${GameName}`;
  title: string;
  description: string;
  routeName: `${RouteName}`;
}
```

一樣由 index.ts 統一匯出。

➜　src\types\index.ts

```ts
export * from './main.type';
export * from './socket.type';
export * from './game.type';
```

讓我們回到 game-menu 元件，新增用來儲存已選擇的遊戲變數並定義各個小島的鏡頭位置參數。

➜　src\components\game-menu.vue

```vue
...
<script setup lang="ts">
...

interface Shot {
  name: `${GameName}`;
```

```
  camera: {
    target: Vector3;
    alpha: number;
    beta: number;
    radius: number;
  }
}
...
const canvas = ref<HTMLCanvasElement>();
let engine: Engine;
let scene: Scene;
let camera: ArcRotateCamera;

const selectedGame = ref<`${GameName}`>('the-first-penguin');
...
</script>
```

接著將鏡頭拉到高空並在場景中新增霧的效果。

➡ src\components\game-menu.vue

```
...
<script setup lang="ts">
...
function createScene(engine: Engine) {
  const scene = new Scene(engine);
  /** 使用預設光源 */
  scene.createDefaultLight();

  scene.fogMode = Scene.FOGMODE_LINEAR;
  scene.fogStart = 50;
  scene.fogEnd = 400;

  const color = Color3.FromHexString('#f0faff');
  scene.clearColor = color.toColor4(1);
  scene.fogColor = color;

  return scene;
}
```

```
function createCamera(scene: Scene) {
  const camera = new ArcRotateCamera(
    'camera',
    1.5, 0, 360,
    new Vector3(-28, 1.3, -0.5),
    scene
  );

  return camera;
}
...
</script>
```

可以看到企鵝島變成模糊的小點了。

▲ 圖 7-20 高空鏡頭

接著定義企鵝島的鏡位與移動鏡頭的功能。

➜ src\components\game-menu.vue

```ts
...
<script setup lang="ts">
...
function createCamera(scene: Scene) {...}
const shots: Shot[] = [
  {
    name: 'the-first-penguin',
    camera: {
      target: new Vector3(-4, 0.67, 1.31),
      alpha: 4.339,
      beta: 1.034,
      radius: 13.046,
    }
  }
];

const ease = new BackEase();
ease.setEasingMode(EasingFunction.EASINGMODE_EASEINOUT);
async function moveCamera(first = false) {
  const shot = shots.find(({ name }) => name === selectedGame.value);
  if (!shot) return;

  const results = [
    createAnimation(camera, 'target', shot.camera.target, { easingFunction: ease }),
    createAnimation(camera, 'alpha', shot.camera.alpha, { easingFunction: ease }),
    createAnimation(camera, 'beta', shot.camera.beta, { easingFunction: ease }),
    createAnimation(camera, 'radius', shot.camera.radius),
  ]

  const animations = results.map(({ animation }) => animation);

  const speed = first ? 0.15 : 0.5;
  scene.beginDirectAnimation(camera, animations, 0, results[0].frameRate, false,
speed);
}
...
</script>
```

最後我們希望在畫面載入完成後才移動鏡頭，所以要將 init 改為 async function。

➡ src\components\game-menu.vue

```
...
<script setup lang="ts">
...
async function init() {...}

onMounted(async () => {
  window.addEventListener('resize', handleResize);

  await init();
  moveCamera(true);
});
...
</script>
```

這樣還不夠，還得將 init 內部含有非同步讀取功能的 function 都改為 async function 才行。

目前只有企鵝島的部分要調整。

➡ src\components\game-menu.vue

```
...
<script setup lang="ts">
...
/** 將所有島嶼儲存至此 */
let islands: ModelIsland[] = [];
async function createIslands(scene: Scene) {
  const result = await createPenguinIsland(scene);
  islands.push(result);
}

async function init() {
  ...
  createSea(scene);
```

```
  await createIslands(scene);
  ...
  /** 發出完成事件，表示畫面初始化完成 */
  emit('complete');
}
...
</script>
```

現在應該會有鏡頭移動動畫了！ヾ(*´∇`*)ノ

 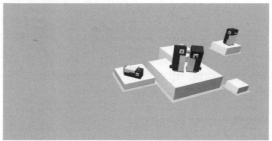

▲ 圖 7-21 鏡頭聚焦至企鵝島

現在只差讓企鵝島的標題浮出水面了！這個部分實現很單純，只要在鏡頭移動動畫結束後將島嶼的 active 設為 true 即可。

首先島嶼的型別定義追加 name 欄位。

➜ src\types\main.type.ts

```
...
export interface ModelIsland {
  name: `${GameName}`;
  /** 模型原點 */
  rootNode: TransformNode;
  setActive(value: boolean): void;
}
```

➜ src\components\game-menu\the-first-penguin.ts

```
...
export async function createPenguinIsland(scene: Scene): Promise<PenguinIsland> {
  ...
  return {
    name: 'the-first-penguin',
    rootNode,
    setActive,
  }
}
```

　　最後在 moveCamera 動畫結束後設定狀態。

➜ src\components\game-menu.vue

```
...
<script setup lang="ts">
...
async function moveCamera(first = false) {
  ...
  const animations = results.map(({ animation }) => animation);

  const speed = first ? 0.15 : 0.5;
  const ref = scene.beginDirectAnimation(camera, animations, 0, results[0].frameRate,
false, speed);
  await ref.waitAsync();

  const target = islands.find(({ name }) => name === selectedGame.value);
  target?.setActive(true);
}
...
</script>
```

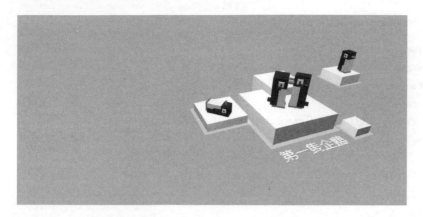

▲　圖 7-22　選單入場運鏡完成

目前我們成功完成選單背景畫面了！✧*。٩(ˊᗜˋ*)و✧*。

7.3　建立選單按鈕

此章節程式碼可以在以下連結取得。

- https://gitlab.com/deepmind1/animal-party/web/-/tree/feat/create-game-menu-btn

現在右側的企鵝島畫面完成了，讓我們完成左側的選單按鈕吧。

調整一下 template 內容，讓現在的 canvas 變為背景。

➜ src\components\game-menu.vue

```
<template>
  <div class="flex">
    <canvas
      ref="canvas"
      class="outline-none absolute w-full h-full"
    />

    <div class=" w-1/2">
    </div>
  </div>
</template>
...
```

接著引入 base-btn 元件。ʊ(´∀`ʊ)

➜ src\components\game-menu.vue

```
<template>
  <div class="flex">
    <canvas
      ref="canvas"
      class="outline-none absolute w-full h-full"
    />

    <div class=" w-1/2 h-full flex flex-col items-center justify-center gap-12 pr-24 pb-10">
      <base-btn
        label=" 開始遊戲 "
        class=" w-[28rem]"
        label-hover-color="#3676a3"
        stroke-color="#456b87"
        stroke-hover-color="white"
      />

      <base-btn
        label=" 結束派對 "
        class=" w-[28rem]"
        label-hover-color="#3676a3"
```

```
        stroke-color="#456b87"
        stroke-hover-color="white"
      />
    </div>
  </div>
</template>
...
```

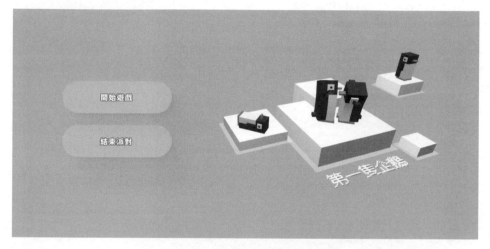

▲ 圖 7-23　遊戲選單按鈕基礎樣式

如同首頁的按鈕一樣，我們也把大廳內的主選單加上裝飾吧。

為了不要讓畫面太花，這裡的按鈕只有 hover 的時候才會出現裝飾效果。

引入 base-polygon 作為裝飾吧。(´„•ω•„)

➡ src\components\game-menu.vue

```
<template>
  <div class="flex">
    ...

    <div class=" ...">
      <base-btn
        v-slot="{ hover }"
```

```
    label=" 開始遊戲 "
  ...
>
  <transition name="opacity">
    <div
      v-if="hover"
      class="btn-content absolute inset-0"
    >
      <div class="polygon-lt">
        <base-polygon
          size="13rem"
          shape="round"
          fill="spot"
          opacity="0.3"
        />
      </div>

      <div class="polygon-rb">
        <base-polygon
          size="13rem"
          shape="round"
          fill="fence"
          opacity="0.2"
        />
      </div>
    </div>
  </transition>
</base-btn>

<base-btn
  v-slot="{ hover }"
  label=" 結束派對 "
  ...
>
  <transition name="opacity">
    <div
      v-if="hover"
      class="btn-content absolute inset-0"
    >
```

```html
            <div class="polygon-lt">
              <base-polygon
                size="13.4rem"
                shape="square"
                fill="spot"
                opacity="0.3"
              />
            </div>
            <div class="polygon-rb">
              <base-polygon
                size="13.3rem"
                shape="square"
                fill="fence"
                opacity="0.2"
              />
            </div>
          </div>
        </transition>
      </base-btn>
    </div>
  </div>
</template>
...
<style scoped lang="sass">
.btn-content
  background: #307da1

.polygon-lt
  position: absolute
  left: -6rem
  top: -6rem
.polygon-rb
  position: absolute
  right: -6rem
  bottom: -6rem
</style>
```

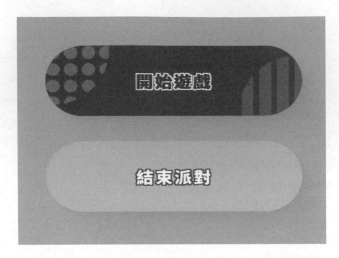

▲ 圖 7-24　選單按鈕 hover 效果

裝飾出現了，不過感覺有點死板，那就加個動畫吧！ヽ(≧∀≦)ﾉ

➜ src\components\game-menu.vue

```
<template>
  <div class="flex">
    ...
    <div ... >
      <base-btn
        v-slot="{ hover }"
        label=" 開始遊戲 "
        ...
      >
        <transition name="opacity">
          <div ... >
            <div class="polygon-lt">
              <base-polygon
                ...
                class="polygon-beat"
              />
            </div>

            <div class="polygon-rb">
              <base-polygon
```

```
          ...
          class="polygon-beat"
        />
      </div>
    </div>
  </transition>
</base-btn>

<base-btn
  v-slot="{ hover }"
  label=" 結束派對 "
  ...
>
  <transition name="opacity">
    <div ... >
      <div class="polygon-lt">
        <base-polygon
          ...
          class="polygon-swing"
        />
      </div>
      <div class="polygon-rb">
        <base-polygon
          ...
          class="polygon-swing"
        />
      </div>
    </div>
  </transition>
</base-btn>
    </div>
  </div>
</template>
...
<style scoped lang="sass">
...
.polygon-lt
  position: absolute
  left: -6rem
```

```
    top: -6rem
    animation: polygon-rotate 50s infinite linear
.polygon-rb
    position: absolute
    right: -6rem
    bottom: -6rem
    animation: polygon-rotate 40s infinite linear

@keyframes polygon-rotate
    0%
        transform: rotate(0deg)
    100%
        transform: rotate(360deg)

.polygon-beat
    animation: polygon-beat 1.4s infinite

.polygon-swing
    animation: polygon-swing 1.8s infinite

@keyframes polygon-beat
    0%
        transform: scale(1)
        animation-timing-function: cubic-bezier(0.000, 0.000, 1.000, 0.000)
    50%
        transform: scale(0.9)
        animation-timing-function: cubic-bezier(0.000, 1.000, 1.000, 1.000)
    100%
        transform: scale(1)

@keyframes polygon-swing
    0%
        transform: scale(1)
        animation-timing-function: cubic-bezier(0.870, 0.000, 0.180, 0.995)
    50%
        transform: scale(0.9)
        animation-timing-function: cubic-bezier(0.870, 0.000, 0.260, 1.375)
    100%
```

```
    transform: scale(1)
</style>
```

沒有意外的話按鈕內部的多邊形動起來了！(ㄒ_ㄒ)✧

 # 7.4 顯示房間 ID 與玩家

此章節程式碼可以在以下連結取得。

- https://gitlab.com/deepmind1/animal-party/web/-/tree/feat/create-id-chip-and-players

除了選單按鈕外，我們還需要顯示房間 ID 與目前玩家。

首先讓我們新增 room-id-chip 元件，用來顯示房間 ID。

➔ src\components\room-id-chip.vue

```
<template>
  <div class=" relative overflow-hidden">
    <div class="chip">
      {{ id }}
    </div>
  </div>
</template>

<script setup lang="ts">
import { computed, ref } from 'vue';
import { useGameConsoleStore } from '../stores/game-console.store';
```

```
const gameConsoleStore = useGameConsoleStore();

const id = computed(() => gameConsoleStore.roomId ?? '123456');
</script>

<style scoped lang="sass">
</style>
```

接著在 game-menu 引入 room-id-chip，方便觀察。

→ src\components\game-menu.vue

```
<template>
  <div class="flex">
    <canvas ... class="... z-0" />

    <div class=" ... z-0">
      <room-id-chip />
      ...
    </div>
  </div>
</template>

<script setup lang="ts">
...
import RoomIdChip from './room-id-chip.vue';
...
</script>
...
```

Tips

由於背景的 canvas 加入了 absolute，這樣可能會破壞原本的 DOM 的堆疊順序，導致某些元素消失（被覆蓋於下方），所以我們加入 z-0 確保堆疊順序。

▲ 圖 7-25 引入 room-id-chip

出現了！還真不是普通的醜。（⊙ω⊙）

讓我們加點裝飾吧，首先換個字體和底色。

➜ src\components\room-id-chip.vue

```
<template>
  <div class=" relative overflow-hidden">
    <div class="chip rounded-full flex items-center gap-6">
      {{ id }}
    </div>
  </div>
</template>
...
<style scoped lang="sass">
@import url('https://fonts.googleapis.com/css2?family=Noto+Sans+TC:wght@100;300;400;50
0;700;900&display=swap')
@import url('https://fonts.googleapis.com/css2?family=Chakra+Petch&family=Noto+Sans+TC
:wght@100;300;400;500;700;900&display=swap')

.chip
  color: white
  font-size: 1.8rem
  padding: 2rem 3rem
```

```
  font-weight: 700
  font-family: 'Chakra Petch', sans-serif
  background: rgba(#3676a3, 0.8)
  letter-spacing: 3px
  text-shadow: 0px 0px 10px rgba(#1d3e57, 0.7)
</style>
```

▲ 圖 7-26　room-id-chip 更換底色和字體

利用 base-polygon 元件加上點裝飾。

➜ src\components\room-id-chip.vue

```
<template>
  <div ... >
    <div class="chip rounded-full flex items-center gap-6">
      <base-polygon
        size="1rem"
        shape="pentagon"
        fill="solid"
        opacity="0.9"
        class="polygon"
      />
      {{ id }}
      <base-polygon
        size="1rem"
        shape="pentagon"
        fill="solid"
        opacity="0.9"
        class="polygon"
      />
```

```
    </div>
  </div>
</template>

<script setup lang="ts">
...
import BasePolygon from './base-polygon.vue';
...
</script>

<style scoped lang="sass">
...
.polygon
  animation: polygon 3s infinite linear

  margin-top: -4px

@keyframes polygon
  0%
    transform: rotate(0deg)
    transform-origin: 50% 60%
  100%
    transform: rotate(360deg)
</style>
```

▲ 圖 7-27 room-id-chip 加入裝飾

再加一點背景吧。(◦'ω`◦)

➜ src\components\room-id-chip.vue

```
<template>
  <div class=" relative overflow-hidden">
    ...
    <div class="absolute w-full h-full top-0 left-0 rounded-full overflow-hidden">
      <base-polygon
        class=" absolute -left-[4.5rem]"
        size="6.8rem"
        shape="round"
        fill="fence"
        opacity="0.1"
        rotate="45deg"
      />

      <base-polygon
        class=" absolute -right-[4.5rem]"
        size="6.8rem"
        shape="round"
        fill="fence"
        opacity="0.1"
        rotate="45deg"
      />
    </div>
  </div>
</template>
...
```

▲ 圖 7-28 room-id-chip 加入背景

完成！是不是看起來好多了！✧*｡٩(´ ロ ` *)ﻭ✧*｡

接下來讓我們新增 player-list 元件，用來呈現已加入房間的玩家。

➜ src\components\player-list.vue

```
<template>
  <div class="">
    player-list
  </div>
</template>

<script setup lang="ts">
import { ref } from 'vue';

const emit = defineEmits<{
  (e: 'update:modelValue', value: string): void;
}>();
</script>

<style scoped lang="sass">
</style>
```

在 the-game-console-lobby 引入，邊開發邊觀察樣式。

➜ src\views\the-game-console-lobby.vue

```
<template>
  <game-menu ... />
  <player-list class=" absolute w-full left-0 bottom-0 z-0" />
</template>

<script setup lang="ts">
...
import PlayerList from '../components/player-list.vue';
...
</script>
```

預計已加入房間的玩家會出現在畫面下方並一字排開，並有各自的顏色。

第一步讓我們重構一下型別定義，將 src\stores\game-console.store.ts 內的型別定義資料集中至 src\types\game.type.ts 中，並追加 Player 參數欄位。

→ src\types\game.type.ts

```ts
import { RouteName } from "../router/router";

/** 遊戲名稱 */
export enum GameName {
  THE_FIRST_PENGUIN = 'the-first-penguin',
}

export interface GameInfo {
  name: `${GameName}`;
  title: string;
  description: string;
  routeName: `${RouteName}`;
}

/** 遊戲狀態 */
export enum GameConsoleStatus {
  /** 首頁 */
  HOME = 'home',
  /** 大廳 */
  LOBBY = 'lobby',
  /** 遊戲中 */
  PLAYING = 'playing',
}

/** 玩家 */
export interface Player {
  /** 唯一 ID */
  readonly clientId: string;
  /** 表示玩家手機端允許的 API 清單 */
  permissions?: any;
}
```

permissions 表示 Web API 清單，預期未來其他更多互動的遊戲使用，例如震動馬達、加速度計、陀螺儀等等。

接著新增 player-list-avatar 元件，用來代表每個玩家，預期有兩個參數：

- player：來自伺服器提供的玩家完整資料

- codeName：代號，像是 1P、2P 這種簡單的玩家代號。

➜ src\components\player-list-avatar.vue

```
<template>
  <div class="frame">
    <div
      class="avatar"
    >
      {{ props.codeName }}
    </div>
  </div>
</template>

<script setup lang="ts">
import { computed, ref } from 'vue';
import { Player } from '../types';

interface Props {
  player: Player;
  codeName: string;
}
const props = withDefaults(defineProps<Props>(), {});
</script>

<style scoped lang="sass">
</style>
```

接著在 player-list 引入 avatar。

➜ src\components\player-list.vue

```
<template>
  <transition-group
    tag="div"
    class=" overflow-hidden px-10 pointer-events-none"
    name="avatar"
  >
    <player-list-avatar
      v-for="player, i in players"
      :key="player.clientId"
      :player="player"
      :code-name="`${i + 1}P`"
    />
  </transition-group>
</template>

<script setup lang="ts">
import { ref } from 'vue';

import PlayerListAvatar from './player-list-avatar.vue';
import { Player } from '../types';
import { nanoid } from 'nanoid';

const emit = defineEmits<{
  (e: 'update:modelValue', value: string): void;
}>();

const players = ref<Player[]>([
  { clientId: nanoid() },
  { clientId: nanoid() },
]);
</script>

<style scoped lang="sass">
</style>
```

真的有夠醜。(>´ω`<)

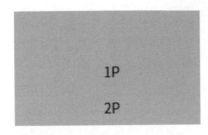

▲ 圖 7-29 引入 player-list-avatar

讓我們加樣式吧。

→ src\components\player-list-avatar.vue

```
<template>
  <div class="frame">
    <div
      class="avatar relative rounded-full flex justify-center p-2 overflow-hidden"
      :class="classes"
    >
      {{ props.codeName }}
    </div>
  </div>
</template>

<script setup lang="ts">
import { computed } from 'vue';
import { Player } from '../types';
import { getPlayerColor } from '../common/utils';
interface Props {
  player: Player;
  codeName: string;
}
const props = withDefaults(defineProps<Props>(), {});

const classes = computed(() => {
  const colorName = getPlayerColor({ codeName: props.codeName });
```

```
    return [`bg-${colorName}`];
});
</script>

<style scoped lang="sass">
.frame
  display: inline-block
  margin-right: 0.4rem
  height: 14rem
.avatar
  width: 7rem
  height: 7rem
  color: white
  text-shadow: 1px 1px 4px rgba(#000, 0.4)
  transform: translateY(50%)
  font-size: 2rem
</style>
```

 Tips

因為預期未來頭像上方還可以顯示別的訊息，所以我們把 frame 撐高一點。

▲ 圖 7-30　player-list-avatar 基礎樣式

看起來正常多了，讓我們加點質感。(•ω•)✧

引入個 base-polygon 吧。

➔ src\components\player-list-avatar.vue

```
<template>
  <div class="frame">
```

```
  <div ... >
    {{ props.codeName }}

    <base-polygon
      v-bind="polygonParams"
      class=" absolute -top-4 -left-4"
    />
  </div>
</div>
</template>

<script setup lang="ts">
...
import BasePolygon, { FillType, ShapeType } from './base-polygon.vue';
import { random, sample } from 'lodash-es';
...
const polygonParams = computed<InstanceType<typeof BasePolygon>['$props']>(() => {
  const shape = sample(Object.values(ShapeType)) ?? 'round';
  const fill = sample(Object.values(FillType)) ?? 'solid';

  return {
    shape,
    fill,
    size: '4rem',
    opacity: 0.3,
    rotate: `${random(0, 180)}deg`,
  }
});
</script>
...
```

▲ 圖 7-31 player-list-avatar 加入裝飾

現在讓我們到 player-list 中完成 transition-group 的群組動畫吧。

在 animate.sass 中加入動畫用的 class。

➜ src\style\animate.sass

```
...
.avatar
  &-move
    transition-duration: 0.8s
  &-enter-active, &-leave-active
    transition-duration: 0.4s
  &-enter-active
    transition-timing-function: cubic-bezier(0.000, 1.650, 0.415, 0.935)
  &-leave-active
    transition-timing-function: cubic-bezier(0.220, -0.385, 0.815, -0.385)
    z-index: 1
  &-enter-from, &-leave-to
    transform: translateY(50%) scale(0.1, 1.9) !important
  &-leave-active
    position: absolute !important
```

接著在 player-list 增減 player 的功能，實際觀察一下列表動畫效果。

➜ src\components\player-list.vue

```
...
<script setup lang="ts">
...

setInterval(() => {
  players.value.push({
    clientId: nanoid(),
    permissions: '',
  });

  setTimeout(() => {
    const index = random(0, players.value.length) - 1;
    players.value.splice(index, 1);
  }, 1000);
```

```
}, 2000);
</script>
...
```

▲ 圖 7-32 player-list-avatar 切換動畫

會看到目前顏色會瞬間切換，看起來不太好，加個顏色過度吧。(´„•ω•„)

➜ src\components\player-list-avatar.vue

```
...
<style scoped lang="sass">
...
.avatar
  ...
  transition-duration: 0.6s
  transition-delay: 0.4s
</style>
```

▲ 圖 7-33 player-list-avatar 顏色加入過度

看起來好多了！✧*。٩(´ロ`*)ﻭ✧*。

最後讓我們把增減 player 的程式刪除，players 的內容改由 store 引入。

➜ src\components\player-list.vue

```
...
<script setup lang="ts">
import { } from 'vue';
import { storeToRefs } from 'pinia';

import PlayerListAvatar from './player-list-avatar.vue';

import { useGameConsoleStore } from '../stores/game-console.store';

const gameConsoleStore = useGameConsoleStore();
const { players } = storeToRefs(gameConsoleStore);

</script>
...
```

最後我們可以發現一件事情，一開始鏡頭在遙遠的高空中，讓選單按鈕看起來相當突兀，所以我們讓選單按鈕一開始隱藏，運鏡到特定時機後再顯示。

在動工前先讓我們慢下來，觀察一下 game-menu 的程式內容，會發現有一大部分的程式可以再獨立封裝，各位讀者們有沒有注意到是哪一個部份呢？

答案就是 canvas，也就是 3D 背景的部分！ヾ(◍'◡'◍)ﾉﾞ

讓我們將背景的部份抽離，封裝為 game-menu-background 元件，並進行以下調整。

- template 只需保留 canvas

- selectedGame 變數改由 props 輸入。

➜ src\components\game-menu-background.vue

```
<template>
  <canvas ref="canvas" />
</template>
```

```
<script setup lang="ts">
...

interface Props {
  selectedGame: `${GameName}`;
}
const props = withDefaults(defineProps<Props>(), {});
...
</script>
```

game-menu 則引入 game-menu-background 並移除相關程式碼。

➔ src\components\game-menu.vue

```
<template>
  <div class="flex">
    <game-menu-background
      :selected-game="selectedGame"
      class="outline-none absolute w-full h-full z-0"
    />
    ...
  </div>
</template>

<script setup lang="ts">
...
import GameMenuBackground from './game-menu-background.vue';

interface Props {
  label?: string;
}
const props = withDefaults(defineProps<Props>(), {
  label: '',
});

const emit = defineEmits<{
  (e: 'completed'): void;
}>();
```

```
const selectedGame = ref<`${GameName}`>('the-first-penguin');

</script>
...
```

會發現 game-menu 的內容變得十分清爽，這就是元件模組化的一大好處。
、(●`∀´●)/

現在讓我回到剛剛的主題上，由於 3D 背景的內容被封裝成獨立元件，所以
背景元件需要 emit 事件，讓父元件知道目前發生了甚麼事，就剛剛的需求來看，
我們還需要增加運鏡開始與運鏡結束這兩個事件才行。

所以讓我們進到 game-menu-background 中增加事件吧。

➔　src\components\game-menu-background.vue

```
...
<script setup lang="ts">
...
const emit = defineEmits<{
  (e: 'completed'): void;
  (e: 'camera-movement-start'): void;
  (e: 'camera-movement-end'): void;
}>();
...
async function moveCamera(first = false) {
  ...
  const speed = first ? 0.15 : 0.5;

  emit('camera-movement-start');
  const ref = scene.beginDirectAnimation(camera, animations, 0, results[0].frameRate,
 false, speed);
  await ref.waitAsync();
  emit('camera-movement-end');

  ...
}
...
```

```
async function init() {
  ...
  /** 發出完成事件，表示畫面初始化完成 */
  emit('completed');
}
...
</script>
```

現在 game-menu 可以透過事件得知運鏡是否結束了，讓我們實作選單出現動畫吧。

這裡希望元素出現有時間差，讓動態豐富一點，這裡我們使用 gsap 實現動畫。

首先讓我們安裝 gsap。

```
npm i -D gsap
```

接著用於 class 作為選擇器目標。

➔ src\components\game-menu.vue

```
<template>
  <div class="flex">
    ...
    <div class=" ...">
      <room-id-chip class="menu-item" />

      <base-btn
        v-slot="{ hover }"
        label=" 開始遊戲 "
        class=" w-[28rem] menu-item"
        ...
      >
        ...
      </base-btn>
```

```
      <base-btn
        v-slot="{ hover }"
        label=" 結束派對 "
        class=" w-[28rem] menu-item"
        ...
      >
        ...
      </base-btn>
    </div>

    ...
  </div>
</template>
...
```

初始化 gsap 動畫物件。

➜ src\components\game-menu.vue

```
...
<script setup lang="ts">
...
let menuTween: ReturnType<typeof gsap.fromTo>;
onMounted(() => {
  menuTween = gsap.fromTo('.menu-item', {
    opacity: 0,
    translateY: '50%',
  }, {
    opacity: 1,
    translateY: '0%',
    duration: 0.6,
    ease: 'back.out(1.7)',
    /** 多個元素間隔時間 */
    stagger: 0.2,
    /** 動畫完成後刪除 inline-style，以免影響其他 CSS */
    clearProps: 'transform',
  });
  menuTween.pause();
});
```

```
</script>
...
```

最後在運鏡結束時播放動畫就完成了！

➜ src\components\game-menu.vue

```
<template>
  <div class="flex">
    <game-menu-background
      ...
      @camera-movement-end="showMenu()"
    />
    ...
  </div>
</template>

<script setup lang="ts">
...
function showMenu() {
  menuTween.play();
}
</script>
...
```

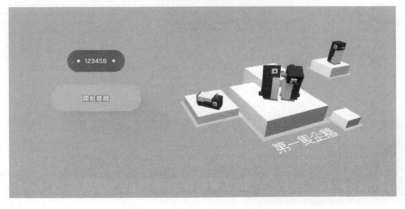

▲ 圖 7-34 menu 進入動畫

現在可以看到一開始進入畫面時，左側的選單不存在，當運鏡完成後才會出現選單。

最後讓我們實際把 game-menu 元件放到 the-game-console-lobby 大廳頁面中吧。

首先刪除 App.vue 中的 game-menu。

➜ src\App.vue

```
<template>
  <router-view />
  <loading-overlay />
</template>
...
```

在 the-game-console-lobby 引入 game-menu。

➜ src\views\the-game-console-lobby.vue

```
<template>
  <game-menu class=" w-full h-full absolute" />
</template>

<script setup lang="ts">
...
import GameMenu from '../components/game-menu.vue';
...
</script>
...
```

在 game-menu 追加處理初始化完成事件邏輯，這樣才能知道甚麼時候要結束 loading。

➜ src\components\game-menu.vue

```
<template>
  <div class="flex">
    <game-menu-background
```

```
    ...
    @camera-movement-end="showMenu()"
    @completed="handleCompleted()"
  />
    ...
  </div>
</template>
<script setup lang="ts">
...
function handleCompleted() {
  emit('completed');
}
</script>
...
```

　　最後讓我們把 the-game-console-lobby 的 loading.hide() 改成在 game-menu
載入完成後再觸發。

➜ src\views\the-game-console-lobby.vue

```
<template>
  <game-menu
    ...
    @completed="handleCompleted()"
  />
</template>

<script setup lang="ts">
...

function handleCompleted() {
  loading.hide();
}

</script>
...
```

現在在首頁按下「建立派對」後，會先出現讀取畫面，接著出現企鵝島的 3D 背景，最後出現遊戲選單了！✧*｡٩(´ロ｀*)و✧*｡

鱈魚：「截至目前為止，我們成功完成選單的外觀了！ヽ(•ω•)ﾉ」

助教：「才完成外觀而已喔！ ╭(ﾟАﾟ°)╮」

鱈魚：「當然啦，選取遊戲、開始遊戲的邏輯都還沒寫捏。讓我們繼續加油吧！∠(ᐛ 」∠)_」

7.5 加入遊戲

在開始遊戲前我們必須先讓玩家加入才行，讓我們完成首頁的「加入遊戲」按鈕。

（為了避免混淆，首頁的「加入派對」按鈕改為「建立派對」）

7.5.1 輸入房間 ID 視窗

此章節程式碼可以在以下連結取得。

- https://gitlab.com/deepmind1/animal-party/web/-/tree/feat/join-game

預期功能流程如下：

1. 玩家使用手機開啟首頁網頁

2. 按下「加入遊戲」，跳出輸入視窗

3. 輸入房間編號，進入指定房間並跳轉至搖桿畫面。

現在讓我們用手機尺寸開啟首頁看看，這裡我們使用瀏覽器的 DevTool 模擬功能。

開啟 DevTool 點擊下圖按鈕。

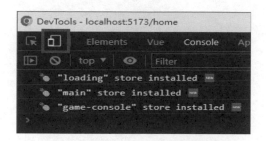

▲ 圖 7-35 模擬手機版畫面

畫面會變成如下圖，這時候可以在畫面左上角的選單，選擇想要模擬的裝置。

▲ 圖 7-36 手機版畫面

▲ 圖 7-37 手機版畫面

可以看到畫面完美的噴出去啦！(ﾟ∀ﾟ)

別緊張，由於我們的單位都是 rem，所以有一個簡單的處理方法，讓我們到 App.vue 加入以下 CSS。

➜ src\App.vue

```
...

<style lang="sass">
...

html
  font-size: 1.71vmin
```

```
@media screen and (orientation: portrait) and (max-width: 360px)
  font-size: 6px
@media screen and (orientation: portrait) and (min-width: 1200px)
  font-size: 30px
</style>
```

原理很簡單，就是根據裝置的畫面最小長度尺寸調整 html 的 font-size，由於所有的單位都是 rem，所以會一起跟著 html 的 font-size 調整。

▲ 圖 7-38 手機版畫面

完成！瞬間回復世界和平！✧*。٩(´ ꒳ `*)و✧*。

不過標題和選單好像離太遠了，讓我們調整一下 the-home 的樣式。

➜ src\views\the-home.vue

```
<template>
  ...

  <div class="absolute inset-0 flex justify-center items-center content-center gap-32">
    ...
  </div>
</template>
```

▲ 圖 7-39 首頁調整完成

助教：「目前只有看到一種尺寸而已，會不會高興太早？(´·_·`)」

鱈魚：「也是捏，讓我們使用 Responsive Viewer 一次確認吧！φ(°▽°*)♪」

Tips

在以下連結安裝瀏覽器外掛：

◆ https://chrome.google.com/webstore/detail/responsive-viewer/inmopeiepgfljkpk idclfgbgbmfcennb

Responsive Viewer 是 Chrome 的外掛，可以一次檢視指定尺寸視窗，開發 RWD 網頁不能沒有它。

安裝完成後啟動 Responsive Viewer。

▲ 圖 7-40　啟動 Responsive Viewer

啟動後畫面會變成如下圖。

▲ 圖 7-41　Responsive Viewer 檢視畫面

仔細看會發現標題文字描邊在小尺寸時感覺太粗了，讓我們調整一下。

➜ src\components\title-logo.vue

```
<template>
  <div class=" relative leading-none">

    ...

    <svg ... >
      <defs>
        <filter :id="svgFilterId">
          <feMorphology
            operator="dilate"
            :radius="radius"
          />
          ...
        </filter>
      </defs>
    </svg>
```

```
  </div>
</template>

<script setup lang="ts">
...
const radius = computed(() => {
  if ($q.screen.width <= 600) {
    return 6;
  }

  return 10;
});
</script>
...
```

　　這裡使用 Quasar 提供的 API，可以直接取得螢幕寬度，我們讓寬度小於等於 600 時，讓描邊尺寸從 10 變成 6。

▲ 圖 7-42　調整 title-logo 描邊尺寸

　　現在看起來好多了，讓我們新增 dialog-join-game 元件，用於讓手機端玩家輸入房間 ID。

首先新增邏輯部分。

- 新增 targetRoomId 變數，儲存正在輸入的 ID。

- 依照 Quasar Dialog 的規範引入 Dialog 功能。

- 新增 submit function，處理送出邏輯。

➜ src\components\dialog-join-game.vue

```
...

<script setup lang="ts">
import { useDialogPluginComponent, useQuasar } from 'quasar';
import { ref } from 'vue';

const {
  dialogRef, onDialogHide, onDialogOK, onDialogCancel
} = useDialogPluginComponent<string>()

const emit = defineEmits({
  ...useDialogPluginComponent.emitsObject
});

const $q = useQuasar();

const targetRoomId = ref('');

async function submit() {
  if (!/^[0-9]{6}$/.test(targetRoomId.value)) {
    $q.notify({
      type: 'negative',
      message: '請輸入 6 位數字'
    });
    return;
  }

  /** 產生 loading 效果 */
  const notifyRef = $q.notify({
    type: 'ongoing',
```

```
    message: ' 加入房間中 '
  });

  onDialogOK();
}
</script>
...
```

Tips

Quasar 提供了可以簡單複用 Dialog 的方法，想要深入了解的讀者可以參考以下連結：

- https://quasar.dev/quasar-plugins/dialog#sfc-with-script-setup-and-composition-api-variant

接著加入模板與樣式。

→ src\components\dialog-join-game.vue

```
<template>
  <q-dialog
    ref="dialogRef"
    class="rounded-5xl"
    @hide="onDialogHide"
  >
    <div class="card flex flex-col p-20 gap-20 overflow-hidden">
      <div class="text-5xl font-bold text-center text-[#2a3832]">
        輸入派對房間 ID
      </div>

      <q-input
        v-model="targetRoomId"
        type="number"
        color="secondary"
        outlined
        rounded
```

```
          placeholder=" 請輸入 6 位數字 "
          input-class="text-center"
          @keyup.enter="submit"
        />

        <q-btn
          unelevated
          rounded
          color="secondary"
          class="p-7 overflow-hidden"
          label=" 加入 "
          @click="submit"
        >
        </q-btn>
      </div>
    </q-dialog>
</template>

<script setup lang="ts">
...
</script>

<style scoped lang="sass">
.card
  border-radius: 2.5rem !important
  background: rgba(white, 0.7)
  backdrop-filter: blur(8px)
</style>
```

最後依照 Quasar 的規定，安裝 Dialog，這樣才能呼叫 Dialog。

➔ src\main.ts

```
..
import { Quasar, Notify, Dialog } from 'quasar';
...
createApp(App)
  .use(Quasar, {
    plugins: {
      Notify, Dialog
```

```
    },
    lang: quasarLang,
  })
...
```

Tips

想要深入了解的讀者可以參考以下連結：

- https://quasar.dev/quasar-plugins/dialog#installation

　　現在讓我們在 the-home 新增 joinGame function，讓我們呼叫 dialog-join-game 吧。

➜ src\views\the-home.vue

```
<template>
  ...
  <div ...>
    ...
    <div class="flex flex-col flex-center gap-20">
      ...
      <base-btn
        v-slot="{ hover }"
        label=" 加入遊戲 "
        ...
        @click="joinGame"
      >
        ...
      </base-btn>
    </div>
  </div>
</template>

<script setup lang="ts">
...
import DialogJoinGame from '../components/dialog-join-game.vue';
...
```

```
async function joinGame() {
  $q.dialog({
    component: DialogJoinGame,
  });
}
</script>
...
```

現在點擊「加入遊戲」，就會跳出如下圖視窗了！

▲ 圖 7-43　dialog-join-game 視窗

看起來有點單調，讓我們加點裝飾吧。

➔ src\components\dialog-join-game.vue

```
<template>
  <q-dialog ... >
    <div class="card flex flex-col p-20 gap-20 overflow-hidden">
      <base-polygon
        class=" absolute -left-32 -top-40 -z-10"
        size="20rem"
        rotate="30deg"
        opacity="0.6"
      />
      <base-polygon
        class=" absolute -right-[14rem] -bottom-[20rem] -z-10"
```

```
        size="30rem"
        shape="pentagon"
        fill="solid"
        rotate="-30deg"
        opacity="0.5"
      />

      ...

      <q-btn
        ...
        label=" 加入 "
        ...
      >
        <base-polygon
          class=" absolute -left-10 -top-16"
          size="8rem"
          opacity="0.7"
          rotate="45deg"
        />
        <q-icon
          class=" absolute -right-[1.4rem] -bottom-[2.6rem] -rotate-[90deg]
opacity-80"
          size="7.5rem"
          name="celebration"
        />
      </q-btn>
    </div>
  </q-dialog>
</template>

<script setup lang="ts">
...
import BasePolygon from './base-polygon.vue';
...
</script>

<style scoped lang="sass">
.card
```

```
  border-radius: 2.5rem !important
  background: rgba(white, 0.8)
  backdrop-filter: blur(8px)
</style>
```

▲ 圖 7-44　dialog-join-game 加入裝飾

看起來好多了！、(≧∀≦)、

目前這個視窗只有外觀，並不能實際連線，現在讓我們跳到 server 專案，提供實際連線的功能吧。

7.5.2　加入遊戲 API

此章節程式碼可以在以下連結取得。

- https://gitlab.com/deepmind1/animal-party/server/-/tree/feat/join-room

現在讓我們前往伺服器專案，實作加入房間功能吧。

在 room.service 新增 joinRoom method 用來處理玩家加入遊戲功能。

→ src\room\room.service.ts

```
...
@Injectable()
export class RoomService {
  ...

  async joinRoom(roomId: string, clientId: string) {
    const room = this.roomsMap.get(roomId);

    if (!room) {
      return Promise.reject(`不存在 ID 為 ${roomId} 的房間`);
    }

    const isJoined = room.playerIds.includes(clientId);
    if (isJoined) {
      return room;
    }

    room.playerIds.push(clientId);
    return room;
  }
}
```

接著在 OnEevets 事件中加入 player:join-room 事件，並加入 export 匯出。

→ types\socket.type.ts

```
...
export interface OnEvents {
  'player:join-room': (data: Room) => void;
}

export interface EmitEvents {
  'game-console:room-created': (data: Room) => void;
}
...
```

　　Socket.IO 除了透過 emit 發射事件以外，也可以在 on 事件中透過 callback 直接回應，如同 HTTP 請求一般，可以簡單方便的處理明確一來一回的通訊，讓我們定義一下回應標準格式。

Tips

> 想要深入了解的讀者可以參考以下連結：
>
> ◆ https://socket.io/docs/v4/emitting-events/#acknowledgements

➜ types\socket.type.ts

```
...
export type SocketResponse<T = undefined> =
  | {
      status: 'err';
      message: string;
      error?: any;
    }
  | {
      status: 'suc';
      message: string;
      data?: T;
    };

export type ClientSocket = Socket<OnEvents, EmitEvents>;
```

　　最後回到 room.gateway 中，處理玩家加入事件，並使用 SocketResponse 回應結果。

➜ src\room\room.gateway.ts

```
import { SubscribeMessage, WebSocketGateway } from '@nestjs/websockets';
import { Room, RoomService } from './room.service';
...
import { ClientSocket, OnEvents, SocketResponse } from 'types';
import { Logger } from '@nestjs/common';
```

```
import to from 'await-to-js';

@WebSocketGateway()
export class RoomGateway {
  private logger: Logger = new Logger(RoomGateway.name);
  ...
  @SubscribeMessage<keyof OnEvents>('player:join-room')
  async handlePlayerJoinRoom(
    socket: ClientSocket,
    roomId: string,
  ): Promise<SocketResponse<Room>> {
    this.logger.log(`socketId : ${socket.id}`);
    this.logger.log(`roomId : `, roomId);

    if (!this.roomService.hasRoom(roomId)) {
      const result: SocketResponse = {
        status: 'err',
        message: ' 指定房間不存在 ',
      };
      return result;
    }

    const client = this.wsClientService.getClient({
      socketId: socket.id,
    });
    if (!client) {
      const result: SocketResponse = {
        status: 'err',
        message: 'Socket Client 不存在，請重新連線 ',
      };
      return result;
    }

    const [err, room] = await to(this.roomService.joinRoom(roomId, client.id));
    if (err) {
      const result: SocketResponse = {
        status: 'err',
        message: ' 加入房間發生異常 ',
        error: err,
```

```
    };
    return result;
  }

  // 加入 socket room
  socket.join(roomId);

  const result: SocketResponse<Room> = {
    status: 'suc',
    message: ' 成功加入房間 ',
    data: room,
  };

  return result;
  }
}
```

其中：

- logger 是 NestJS 紀錄伺服器資訊的標準實踐。

- Promise 使用 await-to-js 這個套件進行包裝，變為 early return 形式。

Tips

想要深入了解的讀者可以參考以下連結：

- https://gomakethings.com/the-early-return-pattern-in-javascript/

以上 server 準備好讓玩家加入指定房間（遊戲）了！(´▽`)/

現在讓我們回到網頁專案，在 socket.type 中追加事件與型別定義。

→ src\types\socket.type.ts

```
...

interface EmitEvents {
  'player:join-room': (roomId: string, callback?: (err: any, res: SocketResponse<Room>)
```

```
=> void) => void;
}
export interface SocketResponse<T = undefined> {
  status: 'err' | 'suc';
  message: string;
  data: T;
  error: any;
}
...
```

可以注意到伺服器的 OnEvents、EmitEvents 與網頁的定義剛好會反過來。

接著同 game-console 概念一般，新增 use-client-player 負責處理各類與玩家相關功能。

➜ src\composables\use-client-player.ts

```
import { useMainStore } from "../stores/main.store";
import { Room } from "../types";

export function useClientPlayer() {
  const mainStore = useMainStore();

  function joinRoom(roomId: string): Promise<Room> {
    // client 尚未連線，先進行連線
    if (!mainStore.client?.connected) {
      const client = mainStore.connect('player');

      return new Promise((resolve, reject) => {
        client.once('connect', () => {
          client.removeAllListeners();

          emitJoinRoom(client, roomId)
            .then(resolve)
            .catch(reject)
        });

        // 發生連線異常
        client.once('connect_error', (error) => {
```

```
        client.removeAllListeners();
        reject(error);
      });
    });
  }

  // client 已經連線，直接發出事件
  return emitJoinRoom(mainStore.client, roomId);
}

/** 發送 player:join-room 事件至 server 並設定 3 秒超時 */
function emitJoinRoom(client: NonNullable<typeof mainStore['client']>, roomId:
string): Promise<Room> {
  return new Promise((resolve, reject) => {
    client.timeout(3000).emit('player:join-room', roomId, (err, res) => {
      if (err) {
        return reject(err);
      }

      if (res.status === 'err') {

        return reject(res);
      }

      resolve(res.data);
    });
  });
}

return {
  joinRoom,
}
}
```

最後在 dialog-join-party 使用 use-client-player，實際發出加入請求吧。

➜ src\components\dialog-join-game.vue

```
...
<script setup lang="ts">
```

```
...
import { useClientPlayer } from '../composables/use-client-player';
...
async function submit() {
  ...
  /** 產生 loading notify */
  const notifyRef = $q.notify({
    type: 'ongoing',
    message: ' 加入房間中 '
  });

  const [err, room] = await to(player.joinRoom(targetRoomId.value));

  /** 關閉 notify */
  notifyRef();

  if (err) {
    $q.notify({
      type: 'negative',
      message: ` 加入房間失敗 : ${err?.message}`
    });
    console.error(` 加入房間失敗 : `, err);
    return;
  }

  console.log(`[ joinRoom ] room : `, room);
  $q.notify({
    type: 'positive',
    message: ` 加入 ${room.id} 房間成功 `
  });
  onDialogOK();
}
</script>
...
```

　　現在讓我們打開兩個不同的瀏覽器，一個先建立派對，另一個則嘗試加入房間。

若打錯房號會如下圖出現錯誤通知。

▲ 圖 7-45　房間不存在

房號正確則會出現成功通知。

▲ 圖 7-46　成功加入遊戲房間

並在 console 中看到印出房間訊息。

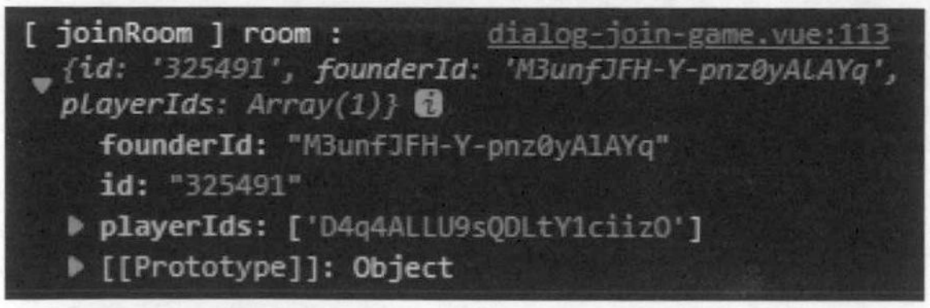

```
[ joinRoom ] room :            dialog-join-game.vue:113
▼ {id: '325491', founderId: 'M3unfJFH-Y-pnz0yAlAYq',
  playerIds: Array(1)} ⓘ
    founderId: "M3unfJFH-Y-pnz0yAlAYq"
    id: "325491"
  ▶ playerIds: ['D4q4ALLU9sQDLtY1ciizO']
  ▶ [[Prototype]]: Object
```

▲ 圖 7-47　console 資訊

　　恭喜我們成功讓玩家加入房間了！接下來讓我們把手機變成搖桿吧！
ヽ(≧∀≦)ノ

第 8 章

手機變搖桿

　　玩家加入房間成功後，可不能只有顯示加入房間成功，應該要讓玩家畫面跳轉至搖桿畫面才行。

　　現在來讓我們打造玩家的搖桿吧！(´▽`)ﾉ.

　　預計不同的狀態會有不同的專用搖桿，例如：遊戲室中有遊戲室用搖桿，企鵝遊戲有企鵝遊戲專用搖桿等等。

8.1　搖桿頁面

　　此章節程式碼可以在以下連結取得。

- https://gitlab.com/deepmind1/animal-party/web/-/tree/feat/player-gamepad-lobby

讓我們從網頁專案開始，首先來建立最外層的搖桿容器頁面。

➜ src\views\the-player-gamepad.vue

```
<template>
  <div class="w-full h-full bg-black">
    <router-view />
  </div>
</template>
```

router-view 留給不同的搖桿使用，讓我們接著新增大廳搖桿的元件。

➜ src\views\the-player-gamepad-lobby.vue

```
<template>
  <div class="w-full h-full flex text-white select-none">
    大廳搖桿頁面
  </div>
</template>

<script setup lang="ts">
import { useLoading } from '../composables/use-loading';

const loading = useLoading();

function init() {
  loading.hide();
}
init();
</script>
```

助教：「程式碼這麼少，是不是偷摸魚？ \(・ω’・\)」

鱈魚：「程式碼少才方便理解啊。ᕙ(ᗜ ᕗ)」

助教：（拿出棍子。(╬⊙д⊙)）

鱈魚：「冷靜冷靜，一步一步來，後來程式碼就會多起來啦。(˃´ω`˂)」

現在讓我們將頁面加入 router 中。

➔ src\router\router.ts

```
...
export enum RouteName {
  ...

  PLAYER_GAMEPAD = 'player-gamepad',
  PLAYER_GAMEPAD_LOBBY = 'player-gamepad-lobby',
}

const routes: RouteRecordRaw[] = [
  ...
  {
    path: `/player-gamepad`,
    name: RouteName.PLAYER_GAMEPAD,
    component: () => import('../views/the-player-gamepad.vue'),
    children: [
      {
        path: `lobby`,
        name: RouteName.PLAYER_GAMEPAD_LOBBY,
        component: () => import('../views/the-player-gamepad-lobby.vue')
      },
    ]
  },

  {
    path: '/:pathMatch(.*)*',
    redirect: '/'
  },
]
...
```

現在從程式碼來看，可以很明確的知道就算進入 the-player-gamepad 元件，也會因為元件內沒有任何程式邏輯，導致畫面不會變動。

現在讓我們加入跳轉至對應搖桿的邏輯吧。

➔ src\views\the-player-gamepad.vue

```
...
<script setup lang="ts">
import { RouteName } from '../router/router';

import { useRouter } from 'vue-router';
import { useGameConsoleStore } from '../stores/game-console.store';

const gameConsoleStore = useGameConsoleStore();
const router = useRouter();

function init() {
  if (!gameConsoleStore.roomId) {
    router.push({
      name: RouteName.HOME
    });
    return;
  }

  if (gameConsoleStore.status === 'lobby') {
    router.push({
      name: RouteName.PLAYER_GAMEPAD_LOBBY
    });
  }
}
init();
</script>
```

但是實際嘗試加入遊戲後，會發現畫面卡在 loading，沒辦法跳到搖桿畫面。

原因很簡單，因為搖桿畫面依照遊戲機狀態進行跳轉，但我們還沒實作同步遊戲機狀態功能，所以現在來讓我們完成最重要的部分：「同步遊戲機狀態」。

首先在 socket.type 新增事件定義。

➜ src\types\socket.type.ts

```
...
interface OnEvents {
  ...
  'game-console:state-update': (data: Required<UpdateParams>) => void;
}

interface EmitEvents {
  ...
  'game-console:state-update': (data: UpdateParams) => void;
}
...
```

接著在 use-client-game-console 新增設定遊戲狀態的 function。

➜ src\composables\use-client-game-console.ts

```
...
export function useClientGameConsole() {
  const mainStore = useMainStore();
  const { connect, disconnect } = mainStore;
  ...

  function setStatus(status: `${GameConsoleStatus}`) {
    gameConsoleStore.updateState({
      status
    });

    if (!mainStore.client?.connected) {
      return Promise.reject('client 尚未連線');
    }

    mainStore.client.emit('game-console:state-update', {
      status
    });
  }
  function setGameName(gameName: `${GameName}`) {
```

```
    gameConsoleStore.updateState({
      gameName
    });

    if (!mainStore.client?.connected) {
      return Promise.reject('client 尚未連線 ');
    }

    mainStore.client.emit('game-console:state-update', {
      gameName
    });
  }

  return {
    ...
    /** 設定遊戲狀態，會自動同步至房間內所有玩家 */
    setStatus,
    /** 設定遊戲名稱，會自動同步至房間內所有玩家 */
    setGameName,
  }
}
```

現在我們可以在 game-console-lobby 中使用 useClientGameConsole，進行
狀態同步了。ㄟ(´∀`ㄟ)

➜ src\views\the-game-console-lobby.vue

```
...

<script setup lang="ts">
...
import { useClientGameConsole } from '../composables/use-client-game-console';

const loading = useLoading();
const gameConsole = useClientGameConsole();

function handleCompleted() {
  init();
}
```

```
function init() {
  gameConsole.setStatus('lobby');
  loading.hide();
}

</script>
```

以上 game-console 設定狀態的部分基本上 OK 了，接下來我們還需要：

- 伺服器對同一個房間內的所有玩家廣播 game-console 狀態更新事件

- 玩家搖桿要監聽狀態變更事件，並要有主動發出請求狀態的能力。

現在讓我們移駕到伺服器專案！(≧∇≦)/

8.2　伺服器同步 game-console 狀態

此章節程式碼可以在以下連結取得。

- https://gitlab.com/deepmind1/animal-party/server/-/tree/feat/player-gamepad-lobby

第一步是讓我們新增 game-console 模組，專門處理 game-console 相關邏輯，讓我們調整 CLI 自動生成內容並引入重要模組。

➡️ src\game-console\game-console.module.ts

```ts
import { Module } from '@nestjs/common';
import { GameConsoleService } from './game-console.service';
import { GameConsoleGateway } from './game-console.gateway';
import { WsClientModule } from 'src/ws-client/ws-client.module';
import { RoomModule } from 'src/room/room.module';

@Module({
  imports: [WsClientModule, RoomModule],
  providers: [GameConsoleGateway, GameConsoleService],
  exports: [GameConsoleService],
})
export class GameConsoleModule {
  //
}
```

➡️ src\game-console\game-console.gateway.ts

```ts
import { SubscribeMessage, WebSocketGateway } from '@nestjs/websockets';
import { GameConsoleService } from './game-console.service';
import { UtilsService } from 'src/utils/utils.service';
import { WsClientService } from 'src/ws-client/ws-client.service';
import { Logger } from '@nestjs/common';

@WebSocketGateway()
export class GameConsoleGateway {
  private logger: Logger = new Logger(GameConsoleGateway.name);

  constructor(
    private readonly gameConsoleService: GameConsoleService,
    private readonly utilsService: UtilsService,
    private readonly wsClientService: WsClientService,
  ) {
    //
  }
}
```

➜ src\game-console\game-console.service.ts

```
import { Injectable, Logger } from '@nestjs/common';
import { RoomService } from 'src/room/room.service';
import { WsClientService } from 'src/ws-client/ws-client.service';

@Injectable()
export class GameConsoleService {
  private logger: Logger = new Logger(GameConsoleService.name);

  constructor(
    private readonly roomService: RoomService,
    private readonly wsClientService: WsClientService,
  ) {
    //
  }
}
```

新增 game-console.type 定義資料型別。

➜ src\game-console\game-console.type.ts

```
export enum GameConsoleStatus {
  /** 首頁 */
  HOME = 'home',
  /** 大廳等待中 */
  LOBBY = 'lobby',
  /** 遊戲中 */
  PLAYING = 'playing',
}

export interface Player {
  clientId: string;
}

export interface GameConsoleState {
  status: `${GameConsoleStatus}`;
  gameName?: string;
  players: Player[],
}
```

```
export type UpdateGameConsoleState = Partial<GameConsoleState>;
```

接著新增 socket 事件定義！ㄟ(•ω•)ㄏ

➡ types\socket.type.ts

```
...
export interface OnEvents {
  ...
  'game-console:state-update': (data: UpdateGameConsoleState) => void;
}

export interface EmitEvents {
  ...
  'game-console:state-update': (data: GameConsoleState) => void;
}
...
```

　　檔案與定義都有了，現在讓我們來實作 game-console 模組邏輯吧。

　　首先 game-console.service 負責儲存或提供 game-console 狀態，我們使用 lodash 協助我們精簡程式，輸入以下指令安裝 lodash。

```
npm i lodash
```

　　還需要安裝型別，方便開發。

```
npm i -D @types/lodash
```

　　來實作功能吧。

➡ src\game-console\game-console.service.ts

```
import { Injectable, Logger } from '@nestjs/common';
import { defaultsDeep, cloneDeep } from 'lodash';
import { RoomService } from 'src/room/room.service';
import { WsClientService } from 'src/ws-client/ws-client.service';
```

```typescript
import { GameConsoleState, Player } from './game-console.type';

/** GameConsoleService 儲存之狀態不需包含 players
 * 因為 players 數值由 roomService 提供，不須重複儲存，所以這裡忽略
 */
type GameConsoleData = Omit<GameConsoleState, 'players'>;

const defaultState: GameConsoleData = {
  status: 'home',
  gameName: undefined,
};

@Injectable()
export class GameConsoleService {
  private logger: Logger = new Logger(GameConsoleService.name);
  /** key 為 founder 之 clientId */
  private readonly gameConsolesMap = new Map<string, GameConsoleData>();

  constructor(
    private readonly roomService: RoomService,
    private readonly wsClientService: WsClientService,
  ) {
    //
  }

  setState(founderId: string, state: Partial<GameConsoleData>) {
    const oriState = this.gameConsolesMap.get(founderId);

    const newState: GameConsoleData = defaultsDeep(
      state,
      oriState,
      defaultState,
    );

    this.gameConsolesMap.set(founderId, newState);
  }

  getState(founderId: string) {
    /** 使用 cloneDeep 複製物件，以免原始物件被意外修改 */
```

```
    const data = cloneDeep(this.gameConsolesMap.get(founderId));

    // 取得房間
    const room = this.roomService.getRoom({
      founderId,
    });
    if (!room) {
      return undefined;
    }

    // 加入玩家
    const players: Player[] = room.playerIds.map((playerId) => ({
      clientId: playerId,
    }));

    const state: GameConsoleState = defaultsDeep(data, {
      status: 'home',
      players,
    });

    return state;
  }
}
```

Tips

defaultsDeep 可以簡單地將物件資料合併，主要用於提供物件內每個欄位的預設值，想要深入了解的讀者可以參考以下連結：

◆ https://www.lodashjs.com/docs/lodash.defaultsDeep

game-console.gateway 接收狀態更新事件。

➔ src\game-console\game-console.gateway.ts

```
...
export class GameConsoleGateway {
  ...
  constructor( ... ) {}
```

```
  @SubscribeMessage<keyof OnEvents>('game-console:state-update')
  async handleGameConsoleStateUpdate(
    socket: ClientSocket,
    state: UpdateGameConsoleState,
  ) {
    const client = this.wsClientService.getClient({
      socketId: socket.id,
    });
    if (!client) return;

    const { status, gameName } = state;

    this.gameConsoleService.setState(client.id, { status, gameName });
  }
}
```

　　現在我們已經可以從遊戲機網頁發送狀態更新至伺服器，並在伺服器儲存遊戲機網頁狀態。

　　最後還差狀態更新時，伺服器廣播至房間內所有玩家功能。

　　新增廣播用 method。

➜ src\game-console\game-console.service.ts

```
...
import { EmitEvents, OnEvents } from 'types';
import { Server } from 'socket.io';
...
@Injectable()
export class GameConsoleService {
  ...
  async broadcastState(
    founderId: string,
    server: Server<OnEvents, EmitEvents>,
  ) {
    const room = this.roomService.getRoom({
      founderId,
```

```
  });

  if (!room) {
    this.logger.warn(`此 founderId 未建立任何房間 : ${founderId}`);
    return;
  }

  const state = this.getState(founderId);
  if (!state) {
    this.logger.warn(`此 founderId 不存在 state : ${founderId}`);
    return;
  }

  const sockets = await server.in(room.id).fetchSockets();
  sockets.forEach((socketItem) => {
    socketItem.emit('game-console:state-update', state);
  });
}
}
```

最後在 game-console.gateway 呼叫廣播功能，由於需要直接呼叫 socket server，所以同時新增 server 成員。

➡ src\game-console\game-console.gateway.ts

```
import {
  SubscribeMessage,
  WebSocketGateway,
  WebSocketServer,
} from '@nestjs/websockets';
import { ClientSocket, EmitEvents, OnEvents } from 'types';
import { Server } from 'socket.io';
...

@WebSocketGateway()
export class GameConsoleGateway {
  ...
  /** 使用 WebSocketServer 裝飾器取得 WebSocket Server */
  @WebSocketServer()
```

```
private server!: Server<OnEvents, EmitEvents>;
...
@SubscribeMessage<keyof OnEvents>('game-console:state-update')
async handleGameConsoleStateUpdate( ... ) {
  ...
  this.gameConsoleService.setState(client.id, { status, gameName });

  // 廣播狀態
  this.gameConsoleService.broadcastState(client.id, this.server);
}
}
```

以上我們完成「伺服器對同一個房間內的所有玩家廣播 game-console 狀態更新事件」，現在還差「玩家搖桿要監聽狀態變更事件，並要有主動發出請求狀態的能力」。

讓我們繼續努力！、(●´∀´●)/，讓我們回到網頁專案。

 ## 8.3 完成跳轉大廳搖桿功能

我們先來重構一下 GameConsoleState 型別定義。

新增 GameConsoleState 型別。

➜ src\types\game.type.ts

```
...
/** 玩家 */
export interface Player { ... }

export interface GameConsoleState {
  status: `${GameConsoleStatus}`;
  gameName: `${GameName}`;
  players: Player[];
}
```

變更 game-console.store 之 UpdateParams 命名與定義。

→ src\stores\game-console.store.ts

```
...
import {
  GameConsoleState, GameConsoleStatus,
  GameName, Player
} from '../types';

export type UpdateStateParams = Partial<GameConsoleState>;
...
```

現在讓我們在 use-client-player 中提供 game-console 狀態更新觸發的 hook 吧！

→ src\composables\use-client-player.ts

```
import { GameConsoleState, Room } from "../types";
...
import { createEventHook } from "@vueuse/core";
import { onBeforeUnmount } from "vue";

export function useClientPlayer() {
  ...
  const stateUpdateHook = createEventHook<GameConsoleState>();
  /** 元件解除安裝前，移除 Listener 以免記憶體洩漏 */
  onBeforeUnmount(() => {
    mainStore.client?.removeListener('game-console:state-update', stateUpdateHook.
trigger);
  });

  return {
    joinRoom,

    onGameConsoleStateUpdate: () => {
      /** 監聽 game-console:state-update 事件 */
      mainStore.client?.on('game-console:state-update', stateUpdateHook.trigger);
      return stateUpdateHook.on;
    },
```

```
  }
}
```

　　在 player-gamepad 中使用此 hook，持續接收 game-console 狀態更新，並依照狀態跳轉至指定頁面。

➜ src\views\the-player-gamepad.vue

```
...
<script setup lang="ts">
...
import { useClientPlayer } from '../composables/use-client-player';
...
const player = useClientPlayer();

function init() {
  if (!gameConsoleStore.roomId) {
    router.push({
      name: RouteName.HOME
    });
    return;
  }

  player.onGameConsoleStateUpdate((state) => {
    const { status } = state;

    console.log(`[ onGameConsoleStateUpdate ] state : `, state);
    gameConsoleStore.updateState(state);

    if (status === 'home') {
      router.push({
        name: RouteName.HOME
      });
    }
    if (status === 'lobby') {
      router.push({
        name: RouteName.PLAYER_GAMEPAD_LOBBY
      });
    }
```

```
  });
}
init();
</script>
```

如此便可以在遊戲機網頁狀態更新時，玩家的網頁也會自動跳轉了！

助教：「終於好了嗎？快睡著惹…(˃́ω˂̀)」

鱈魚：「但是現在玩家加入遊戲後還是一樣會卡在 loading 畫面呦。ᕕ(°∀°)ᕗ」

助教：「所以是怎樣啦！ \(・ω´・ \)」

別氣別氣，這是因為玩家加入遊戲，根本就不會觸發遊戲機的狀態更新，所以還差最後一步，「玩家網頁可以請求取得遊戲機狀態」的功能。

一樣讓我們在 socket.type 新增事件。

➜ types\socket.type.ts

```
...
export interface OnEvents {
  ...
  'player:request-game-console-state': () => void;
  ...
}
...
```

接著在 use-client-player 中提供請求遊戲機狀態的 function。

➜ src\composables\use-client-player.ts

```
...
export function useClientPlayer() {
  ...
  async function requestGameConsoleState() {
    if (!mainStore.client?.connected) {
      return Promise.reject('client 尚未連線 ');
```

```
    }
    mainStore.client.emit('player:request-game-console-state');
  }

  return {
    ...
    requestGameConsoleState,
  }
}
```

最後在 player-gamepad 呼叫，也就是註冊 onGameConsoleStateUpdate 之後呼叫。

→ src\views\the-player-gamepad.vue

```
...
<script setup lang="ts">
...
function init() {
  ...
  player.requestGameConsoleState();
}
init();
</script>
```

網頁的部分完成了，現在我們只要在伺服器回應玩家網頁的請求就完成了！(´▽`)/

讓我們回到伺服器專案，第一步就是先新增事件定義。

→ types\socket.type.ts

```
...
export interface OnEvents {
  ...
  'player:request-game-console-state': () => void;
  ...
}
...
```

並在 game-console.gateway 處理事件。

➜ src\game-console\game-console.gateway.ts

```
...
export class GameConsoleGateway {
  ...
  @SubscribeMessage<keyof OnEvents>('player:request-game-console-state')
  async handleRequestState(socket: ClientSocket) {
    const client = this.wsClientService.getClient({
      socketId: socket.id,
    });
    if (!client) {
      const result: SocketResponse = {
        status: 'err',
        message: '此 socket 不存在 client',
      };
      return result;
    }

    const room = this.roomService.getRoom({
      playerId: client.id,
    });
    if (!room) {
      const result: SocketResponse = {
        status: 'err',
        message: 'client 未加入任何房間 ',
      };
      return result;
    }

    const state = this.gameConsoleService.getState(room.founderId);
    if (!state) {
      const result: SocketResponse = {
        status: 'err',
        message: ' 此房間之 game-console 不存在 state',
      };
      return result;
    }

    socket.emit('game-console:state-update', state);
```

```
const result: SocketResponse<GameConsoleState> = {
  status: 'suc',
  message: '取得 state 成功',
  data: state,
};
return result;
}
}
```

現在是真的完成了！✧*。٩(ˊᗜˋ*)و✧*。

現在我們加入遊戲後，手機的畫面會如下圖呈現。

大廳搖桿頁面

▲ 圖 8-1　進入大廳搖桿頁面

在 devtool 中也會看到狀態更新資訊。

```
[ onGameConsoleStateUpdate ] state :
▼ {status: 'lobby', players: Array(1)} ⓘ
  ▼ players: Array(1)
    ▶ 0: {clientId: 'RfoTfHRtmVByXg_y-B_wa'}
      length: 1
    ▶ [[Prototype]]: Array(0)
    status: "lobby"
  ▶ [[Prototype]]: Object
```

▲ 圖 8-2　進入大廳搖桿頁面

助教：「所以⋯我說那個搖桿呢？(´・ω・`)」

鱈魚：「啊⋯(⊙ ᯅ ⊙')」

所以我說那個搖桿呢？

此章節程式碼可以在以下連結取得。

- https://gitlab.com/deepmind1/animal-party/web/-/tree/feat/create-gamepad

跳轉至玩家大廳搖桿畫面後，現在可以讓我們開始完成搖桿的內容了。

預期外觀如下圖。

▲ 圖 9-1　大廳搖桿外觀

基本組成為：

- 左邊為方向鍵

- 右邊為確認按鈕

- 上方為玩家代號

- 背景是玩家對應顏色

可以發現方向鍵實際上也由四個按鈕組成，所以第一步讓我們新增按鈕元件吧。

定義參數與事件。

➜ src\components\gamepad-btn.vue

```
...
<script setup lang="ts">
import { ref } from 'vue';

interface Props {
  /** 尺寸 */
  size?: string;
  /** 按鈕內 icon 名稱 */
  icon?: string;
  /** 按鈕底色 */
  color?: string;
  /** 按鈕觸發底色 */
  activeColor?: string,
}
const props = withDefaults(defineProps<Props>(), {
  size: '2rem',
  icon: undefined,
  color: 'grey-10',
  activeColor: 'grey-3',
});

const emit = defineEmits<{
  (e: 'up'): void;
```

```
  (e: 'down'): void;
  (e: 'trigger', status: boolean): void;
  (e: 'click'): void;
}>();
</script>
...
```

新增狀態變數、狀態數值與事件。

➜ src\components\gamepad-btn.vue

```
...
<script setup lang="ts">
import { computed, ref, watch } from 'vue';
...
const status = ref(false);
watch(status, (value) => emit('trigger', value));

const color = computed(() =>
  status.value ? props.activeColor : props.color
);

function handleUp(e: TouchEvent | MouseEvent) {
  e.preventDefault();

  status.value = false;
  emit('up');
  emit('click');
}
function handleDown(e: TouchEvent | MouseEvent) {
  e.preventDefault();

  status.value = true;
  emit('down');
}
</script>
...
```

最後把資料與事件綁定於 template 吧。

➜ src\components\gamepad-btn.vue

```
<template>
  <q-btn
    round
    unelevated
    :size="props.size"
    :icon="props.icon"
    :color="color"
    @mouseup="handleUp"
    @mousedown="handleDown"
    @touchend="handleUp"
    @touchstart="handleDown"
    @contextmenu="(e: any) => e?.preventDefault()"
  >
    <slot />
  </q-btn>
</template>
...
```

這裡使用 slot 保留彈性，並綁定 @contextmenu 以防觸控長按時，意外開啟右鍵選單。

這樣按鈕就完成了！讓我們實際擺到畫面中看看吧。(´▽`)╯

src\views\the-player-gamepad-lobby.vue

```
<template>
  <div class="w-full h-full flex text-white select-none">
    <gamepad-btn
      class="absolute bottom-10 right-20"
      size="6rem"
      icon="done"
    />
  </div>
</template>
```

```
<script setup lang="ts">
import GamepadBtn from '../components/gamepad-btn.vue';
...
</script>
```

按鈕出現了。(′„•ω•„)

▲ 圖 9-2 第一顆按鈕

讓我們趁勝追擊,繼續完成方向鍵吧。

新增元件、預期參數與事件。

➜ src\components\gamepad-d-pad.vue

```
...
<script setup lang="ts">
import { ref } from 'vue';

type KeyName = 'up' | 'left' | 'right' | 'down';

interface Props {
  /** 尺寸.直徑 */
  size?: string;
  btnSize?: string;
}
const props = withDefaults(defineProps<Props>(), {
  size: '34rem',
  btnSize: '3rem'
});
```

```
const emit = defineEmits<{
  (e: 'click', keyName: KeyName): void;
  (e: 'trigger', data: { keyName: KeyName, status: boolean }): void;
}>();
</script>
```

　　預期引用剛剛建立的 gamepad-btn 建立按紐，設計 function 接收按鈕觸發事件。

　　加入 gamepad-btn 並完成 template 與 CSS。

➜ src\components\gamepad-d-pad.vue

```
<template>
  <div class="d-pad rounded-full bg-grey-10">
    <gamepad-btn
      class="btn up"
      color="grey-9"
      icon="arrow_drop_up"
      size="3rem"
      @trigger="(status) => handleBtnTrigger('up', status)"
    />
    <gamepad-btn
      class="btn left"
      color="grey-9"
      icon="arrow_left"
      size="3rem"
      @trigger="(status) => handleBtnTrigger('left', status)"
    />
    <gamepad-btn
      class="btn right"
      color="grey-9"
      icon="arrow_right"
      size="3rem"
      @trigger="(status) => handleBtnTrigger('right', status)"
    />
    <gamepad-btn
      class="btn down"
      color="grey-9"
```

```
      icon="arrow_drop_down"
      size="3rem"
      @trigger="(status) => handleBtnTrigger('down', status)"
    />
  </div>
</template>

<script setup lang="ts">
...

import GamepadBtn from './gamepad-btn.vue';
...
</script>

<style scoped lang="sass">
.d-pad
  width: v-bind('props.size')
  height: v-bind('props.size')

.btn
  position: absolute
  &.up
    left: 50%
    top: 0%
    transform: translate(-50%, 20%)
  &.left
    left: 0%
    top: 50%
    transform: translate(20%, -50%)
  &.right
    right: 0%
    top: 50%
    transform: translate(-20%, -50%)
  &.down
    left: 50%
    bottom: 0%
    transform: translate(-50%, -20%)
</style>
```

最後把方向鍵放到畫面中吧。、(●`∀´●)/

➡ src\views\the-player-gamepad-lobby.vue

```
<template>
  <div
    ...
    @touchmove="(e) => e.preventDefault()"
  >
    <gamepad-d-pad class="absolute bottom-5 left-8" />
    <gamepad-btn ... />
  </div>
</template>

<script setup lang="ts">
import GamepadBtn from '../components/gamepad-btn.vue';
import GamepadDPad from '../components/gamepad-d-pad.vue';
...
</script>
```

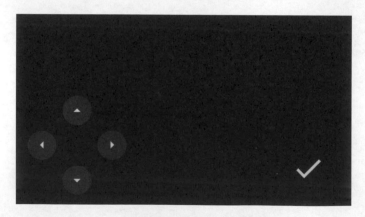

▲ 圖 9-3　加入方向鍵

追加一個取消 touchmove 事件，避免觸控時不小心拖動畫面。

再來讓我們加上「提示玩家將手機橫握的提示」。

➜ src\views\the-player-gamepad-lobby.vue

```
<template>
  <div ... >
    ...

    <q-dialog
      v-model="isPortrait"
      persistent
    >
      <q-card class="p-8">
        <q-card-section class="flex flex-col items-center gap-6">
          <q-spinner-box
            color="primary"
            size="10rem"
          />
          <div class="text-4xl">
            請將手機轉為橫向
          </div>
          <div class="text-base">
            轉為橫向後，此視窗會自動關閉
          </div>
        </q-card-section>
      </q-card>
    </q-dialog>
  </div>
</template>

<script setup lang="ts">
...
import { useScreenOrientation } from '@vueuse/core';

const { orientation } = useScreenOrientation();
...

const isPortrait = computed(() => orientation.value?.includes('portrait') ?? false);
</script>
```

現在只要直握手機，都會出現以下畫面 ㄟ(•ω•)ㄏ

▲ 圖 9-4 直握警告訊息

最後讓我們加上顯示玩家代號的部分吧！

玩家相關功能基本上都集中在 use-client-player 中，讓我們在此新增以下功能：

- codeName 表示玩家代號。

- colorName 表示玩家顏色。

➜ src\composables\use-client-player.ts

```
...
import { getPlayerColor } from "../common/utils";

export function useClientPlayer() {
  ...
  const codeName = computed(() => {
    const index = gameConsoleStore.players.findIndex((player) =>
      player.clientId === mainStore.clientId
    );

    if (index < 0) {
      return 'unknown';
    }
```

```
    return `${index + 1}P`;
  });

  const colorName = computed(() => getPlayerColor({ codeName: codeName.value }));

  return {
    ...
    codeName,
    colorName,
  }
}
```

回到搖桿元件引入 codeName 與 colorName。

→ src\views\the-player-gamepad-lobby.vue

```
...
<script setup lang="ts">
...
import { useClientPlayer } from '../composables/use-client-player';
...
const { codeName, colorName } = useClientPlayer();
...
const bgClass = computed(() => `bg-${colorName.value}`);
</script>
...
```

最後完成 template 與 CSS。

→ src\views\the-player-gamepad-lobby.vue

```
<template>
  <div
    class="w-full h-full flex text-white select-none"
    :class="bgClass"
    @touchmove="(e) => e.preventDefault()"
  >
    ...
```

```
    <div class="code-name">
      {{ codeName }}
    </div>

    <q-dialog ... >
      ...
    </q-dialog>
  </div>
</template>
...
<style scoped lang="sass">
.code-name
  position: absolute
  top: 0
  left: 50%
  transform: translateX(-50%)
  display: flex
  justify-content: center
  padding: 0.1rem
  font-size: 10rem
  text-shadow: 0px 0px 2rem rgba(#000, 0.5)
</style>
```

我們完成大廳搖桿畫面了！✧*｡٩(ˊ ω ˋ*)و✧*｡

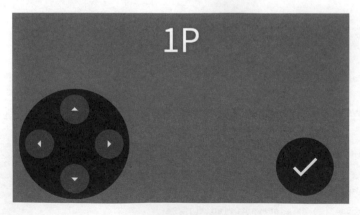

▲ 圖 9-5　1P 大廳搖桿畫面

如果這個時候再開另一個網頁加入成為 2P，就會如下圖。

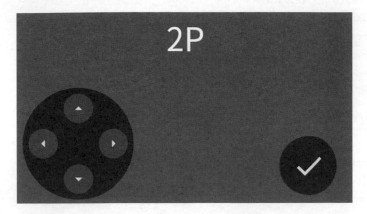

▲　圖 9-6　2P 大廳搖桿畫面

現在讓我們把搖桿資料傳輸給伺服器吧！(•ω•)✧

9.1　傳輸搖桿資料

現在讓我們設計一下 GamepadData 的資料定義，新增 player.type。

➜ src\types\player.type.ts

```
/** 按鍵類型 */
export enum KeyName {
  UP = 'up',
  LEFT = 'left',
  RIGHT = 'right',
  DOWN = 'down',

  CONFIRM = 'confirm',
}

/** 數位訊號
 *
 * 只有開和關兩種狀態
```

```
 */
export interface DigitalData {
  name: `${KeyName}`;
  value: boolean;
}

/** 類比訊號
 *
 * 連續數字組成的訊號，例如：類比搖桿、姿態感測器訊號等等
 */
export interface AnalogData {
  name: `${KeyName}`;
  value: number;
}

export type SignalData = DigitalData | AnalogData;

/** 搖桿資料 */
export interface GamepadData {
  playerId: string;
  keys: SingleData[];
}
```

記得交給 index.ts 匯出。

➜ src\types\index.ts

```
export * from './main.type';
export * from './socket.type';
export * from './game.type';
export * from './player.type';
```

接著新增 socket 事件定義。

➜ src\types\socket.type.ts

```
...

interface OnEvents {
```

```
  ...
  'player:gamepad-data': (data: GamepadData) => void;
}

interface EmitEvents {
  ...
  'player:gamepad-data': (data: GamepadData) => void;
  ...
}
...
```

現在我們需要完成以下項目：

- 玩家搖桿發出控制訊號

- 遊戲機監聽並接收搖桿控制訊號

分別在 use-client-player 與 use-client-game-console 新增對應功能。

首先是 use-client-player 新增 emitGamepadData，用於發射控制訊號。

➔ src\composables\use-client-player.ts

```
...
import { GameConsoleState, Room, SignalData } from "../types";
...

export function useClientPlayer() {
  ...
  async function emitGamepadData(data: SignalData[]) {
    if (!mainStore.client?.connected) {
      return Promise.reject('client 尚未連線 ');
    }

    mainStore.client.emit('player:gamepad-data', {
      playerId: mainStore.clientId,
      keys: data,
    })
  }
```

```
return {
  ...
  emitGamepadData,
}
}
```

最後是 use-client-game-console 新增 onGamepadData，監聽控制訊號。

➜ src\composables\use-client-game-console.ts

```
...
export function useClientGameConsole() {
  ...
  const gamepadDataHook = createEventHook<GamepadData>();
  onBeforeUnmount(() => {
    mainStore.client?.removeListener('player:gamepad-data', gamepadDataHook.trigger);
  });

  return {
    ...
    /** 搖桿控制訊號事件 */
    onGamepadData: (fn: Parameters<typeof gamepadDataHook['on']>[0]) => {
      mainStore.client?.on('player:gamepad-data', gamepadDataHook.trigger);
      return gamepadDataHook.on(fn);
    },
  }
}
```

以上我們準備好「發送搖桿訊號」與「遊戲機接收控制訊號」的功能了！
✧*｡٩(ˊᗜˋ*)و✧*｡

9.2 玩家一起粗乃玩！

此章節程式碼可以在以下連結取得。

- https://gitlab.com/deepmind1/animal-party/server/-/tree/feat/sync-player-data

現在我們可以讓玩家加入房間，也準備好傳輸搖桿資料了，現在我們把遊戲大廳內的玩家頭像對應實際上的玩家吧！(`•ω´•)✧

必須在玩家數量發生變更時，發送玩家數量更新事件才行，讓我們前往伺服器專案，追加「玩家加入房間」與「玩家斷線」對應功能。

首先新增事件定義。

➔ types\socket.type.ts

```
...
import {
  UpdateGameConsoleState,
  GameConsoleState,
  Player,
} from 'src/game-console/game-console.type';
...
export interface EmitEvents {
  ...
  'game-console:player-update': (data: Player[]) => void;
}
...
```

　　並在 room.service 新增 emitPlayers 功能，用於發送發送指定房間之玩家數量。

➜ src\room\room.service.ts

```
import { ClientId, WsClientService } from 'src/ws-client/ws-client.service';
import { Server } from 'socket.io';
import { EmitEvents, OnEvents } from 'types';
...
export class RoomService {
  ...

  constructor(private readonly wsClientService: WsClientService) { }
  ...
  /** 發送指定房間之玩家資料 */
  async emitPlayers(founderId: string, server: Server<OnEvents, EmitEvents>) {
    const room = this.getRoom({ founderId });
    if (!room) return;

    const players = room.playerIds.map((playerId) => ({
      clientId: playerId,
    }));

    // 對房間內所有人發送玩家更新事件
    server.to(room.id).emit('game-console:player-update', players);
  }
}
```

　　並在 room.gateway 處理事件。

➜ src\room\room.gateway.ts

```
...
export class RoomGateway {
  ...
  @WebSocketServer()
  private server!: Server<OnEvents, EmitEvents>;
  ...
  handleDisconnect(socket: ClientSocket) {
```

```
    ...

    if (client.type === 'player') {
      // 取得此玩家所處房間
      const room = this.roomService.getRoom({ playerId: client.id });

      this.roomService.deletePlayer(client.id);

      // 若房間存在則發送玩家資料更新
      if (room) {
        this.roomService.emitPlayers(room.founderId, this.server);
      }
      return;
    }
  }

  @SubscribeMessage<keyof OnEvents>('player:join-room')
  async handlePlayerJoinRoom(
    socket: ClientSocket,
    roomId: string,
  ): Promise<SocketResponse<Room>> {
    ...

    // 發送玩家資料更新
    this.roomService.emitPlayers(room.founderId, this.server);

    const result: SocketResponse<Room> = {...};
    return result;
  }
}
```

現在伺服器會在房間人數發生變更時，自動推送訊息至遊戲機網頁。

最後差點忘記還需要加入「同步搖桿資料功能」惹。(´„•ω•„)

將網頁專案的 player.type 檔案複製過來伺服器專案。

➜ types\player.type.ts

```ts
/** 按鍵類型 */
export enum KeyName {
  UP = 'up',
  LEFT = 'left',
  RIGHT = 'right',
  DOWN = 'down',

  CONFIRM = 'confirm',
}

/** 數位訊號
 *
 * 只有開和關兩種狀態
 */
export interface DigitalData {
  name: `${KeyName}`;
  value: boolean;
}

/** 類比訊號
 *
 * 連續數字組成的訊號，例如：類比搖桿、姿態感測器訊號等等
 */
export interface AnalogData {
  name: `${KeyName}`;
  value: number;
}

export type SignalData = DigitalData | AnalogData;

/** 搖桿資料 */
export interface GamepadData {
  playerId: string;
  keys: SignalData[];
}
```

並從 index.ts 導出。

➜ types\index.ts

```
export * from './socket.type';
export * from './player.type';
```

接著在 socket.type 新增搖桿資料事件。

➜ types\socket.type.ts

```
...
import { GamepadData } from 'types';

export interface OnEvents {
  ...
  'player:gamepad-data': (data: GamepadData) => void;
  ...
}

export interface EmitEvents {
  'player:gamepad-data': (data: GamepadData) => void;
  ...
}
...
```

最後在 game-console.gateway 處理事件。

➜ src\game-console\game-console.gateway.ts

```
...
export class GameConsoleGateway {
  ...
  @SubscribeMessage<keyof OnEvents>('player:gamepad-data')
  async handlePlayerGamepadData(socket: ClientSocket, data: GamepadData) {
    const client = this.wsClientService.getClient({ socketId: socket.id });
    if (!client) {
      const result: SocketResponse = {
        status: 'err',
        message: '此 socket 不存在 client',
```

```
  };
  return result;
}


const room = this.roomService.getRoom({
  playerId: client.id,
});
if (!room) {
  const result: SocketResponse = {
    status: 'err',
    message: 'client 未加入任何房間 ',
  };
  return result;
}


const founderClient = this.wsClientService.getClient({
  clientId: room.founderId,
});
if (!founderClient) {
  const result: SocketResponse = {
    status: 'err',
    message: ' 此 socket 不存在 client',
  };
  return result;
}


const targetSocket = this.server.sockets.sockets.get(
  founderClient.socketId,
);
if (!targetSocket) {
  const result: SocketResponse = {
    status: 'err',
    message: ' 不存在 room founder 對應之 Client',
  };
  return result;
}


targetSocket.emit('player:gamepad-data', data);
```

```
    const result: SocketResponse = {
      status: 'suc',
      message: '傳輸搖桿資料成功 ',
      data: undefined,
    };
    return result;
  }
}
```

現在伺服器具備了「主動通知玩家人數變化」與「轉傳搖桿控制訊號」的
功能了！

9.3　網頁偵測玩家更新事件

讓我們回到網頁專案，並在 use-client-game-console 加入玩家數量變化事件
的 hook 吧！

首先新增事件定義。

➔ src\types\socket.type.ts

```
...

interface OnEvents {
  ...
  'game-console:player-update': (data: Player[]) => void;
}
...
```

接著新增 onPlayerUpdate

➔ src\composables\use-client-game-console.ts

```
...
export function useClientGameConsole() {
  ...
```

```
const gamepadDataHook = createEventHook<GamepadData>();
const playerUpdateHook = createEventHook<Player[]>();

onBeforeUnmount(() => {
  mainStore.client?.removeListener('player:gamepad-data', gamepadDataHook.trigger);
  mainStore.client?.removeListener('game-console:player-update', playerUpdateHook.
trigger);
});

return {
  ...
  /** 玩家變更事件，例如玩家加入或斷線等等 */
  onPlayerUpdate: (fn: Parameters<typeof playerUpdateHook['on']>[0]) => {
    mainStore.client?.on('game-console:player-update', playerUpdateHook.trigger);
    return playerUpdateHook.on(fn);
  },
}
}
```

接著在 the-game-console 頁面中使用此 hook，這樣遊戲機網頁就會在每次玩家資料更新時，同時取得並更新資料了。

→ src\views\the-game-console.vue

```
...
<script setup lang="ts">
...
import { useClientGameConsole } from '../composables/use-client-game-console';
...
const gameConsole = useClientGameConsole();

function init() {
  ...

  gameConsole.onPlayerUpdate((players) => {
    gameConsoleStore.updateState({ players });
  });

  // 跳轉至遊戲大廳
```

```
  router.push({...});
}
init();
</script>
```

這樣遊戲機網頁就可以在玩家加入或斷線時，變更玩家數量了。

現在讓我們建立派對，並開啟另一個瀏覽器加入遊戲，就會像下圖一般，左下角會跑出玩家頭像！✧*｡٩(´ ▽ ｀*)و✧*｡

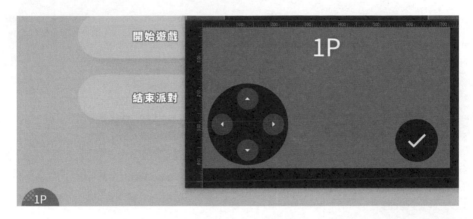

▲ 圖 9-7 同步玩家資料

現在讓玩家搖桿實際發出訊號，並於遊戲大廳取得搖桿資料看看吧！ヽ(≧∀≦)ﾉ

首先是搖桿發出訊號，概念很簡單，就是接收按鈕元件的 trigger 事件即可。

➜ src\views\the-player-gamepad-lobby.vue

```
<template>
  <div ... >
    <gamepad-d-pad
      ...
      @trigger="({ keyName, status }) => handleBtnTrigger(keyName, status)"
    />
    <gamepad-btn
      ...
```

```
      @trigger="(status) => handleBtnTrigger('confirm', status)"
    />

    ...
  </div>
</template>

<script setup lang="ts">
...
import { KeyName } from '../types';

...
const { codeName, colorName, emitGamepadData } = useClientPlayer();

...

function handleBtnTrigger(keyName: `${KeyName}`, status: boolean) {
  console.log(`[ handleBtnTrigger ] : `, { keyName, status });

  emitGamepadData([{
    name: keyName,
    value: status,
  }]);
}
</script>
...
```

最後在 the-game-console-lobby 頁面，也就是遊戲機頁面中接收訊號吧。

➔ src\views\the-game-console-lobby.vue

```
...
<script setup lang="ts">
...
function init() {
  gameConsole.setStatus('lobby');
  loading.hide();

  gameConsole.onGamepadData((data) => {
    console.log(`[ onGamepadData ] data : `, data);
```

```
    });
  }
</script>
```

現在來按按看搖桿上的按鍵，會發現除了搖桿網頁的 console 中會出現訊息外，遊戲機網頁中的 console 也會及時出現搖桿訊號。

▲ 圖 9-8　傳輸搖桿訊號

以上我們完成遊戲機網頁與玩家搖桿網頁的即時資料通訊了！♪(´ω`)و

9.4　是誰偷按確定？

現在玩家有了，訊號也進來了，讓搖桿與畫面產生實際互動效果吧！

9.4.1　對話泡泡

玩派對遊戲的時候，常常發生不知道是誰在選單偷按的問題，現在讓我們在玩家頭像中，新增「說出觸發按鍵」的功能，以後誰偷按都無所遁形啦！ヽ(´∀`)ノ前往 player-list-avatar 加入對話泡泡的功能，加入以下內容：

- 定義訊息資料

- 顯示與隱藏泡泡用的 function。

- 映射按鈕名稱與顯示 Icon 的 function。

➜ src\components\player-list-avatar.vue

```
...
<script setup lang="ts">
...
import { KeyName, Player } from '../types';
import { debounce, random, sample } from 'lodash-es';
import { nanoid } from 'nanoid';
...
const messageInfo = reactive({
  id: '',
  text: '',
});

/** debounce 會在被呼叫後指定時間內觸發，輕鬆達成時間到自動消失效果 */
const hideBalloon = debounce(() => {
  messageInfo.text = '';
}, 2000);

function showBalloon(text: string) {
  const id = nanoid();

  messageInfo.id = id;
  messageInfo.text = text;

  hideBalloon();
}

/** 定義按鈕名稱與 icon */
const keyToIcon = [
  {
    keyName: KeyName.UP,
```

```
    icon: 'arrow_drop_up'
  },
  {
    keyName: KeyName.LEFT,
    icon: 'arrow_left'
  },
  {
    keyName: KeyName.RIGHT,
    icon: 'arrow_right'
  },
  {
    keyName: KeyName.DOWN,
    icon: 'arrow_drop_down'
  },
  {
    keyName: KeyName.CONFIRM,
    icon: 'done'
  },
]
function getIconName(name: string) {
  const target = keyToIcon.find(({ keyName }) => keyName === name);
  return target?.icon ?? 'question_mark';
}

</script>
...
```

最後透過 defineExpose 提供外部使用。

➜ src\components\player-list-avatar.vue

```
...
<script setup lang="ts">
...
defineExpose({
  playerId: props.player.clientId,
  showBalloon
});
</script>
...
```

邏輯完成了，現在讓我們完成樣式吧。

messageInfo.text 換成 KeyName.UP，不然泡泡會出不來。

➡ src\components\player-list-avatar.vue

```
<template>
  <div class="frame relative">
    ...
    <div class="balloon-box">
      <transition name="balloon">
        <div
          v-if="messageInfo.text"
          :key="messageInfo.id"
          class="balloon"
        >
          <q-icon
            color="grey-9"
            size="4rem"
            :name="getIconName(messageInfo.text)"
          />
        </div>
      </transition>
    </div>
  </div>
</template>

<script setup lang="ts">
...
const messageInfo = reactive({
  id: '',
  text: `${KeyName.UP}`,
});
...
</script>

<style scoped lang="sass">
...
.balloon-box
  position: absolute
```

```
    top: 30%
    left: 0
    width: 100%
    height: 100%
.balloon
  position: absolute
  background: white
  box-shadow: 5px 5px 10px rgba(#000, 0.1)
  border-radius: 9999px
  padding: 0.4rem 1.2rem
  &::before
    content: ''
    width: 2rem
    height: 2rem
    position: absolute
    left: 30%
    bottom: 0
    transform: translateX(-30%) rotate(30deg)
    background: white
    box-shadow: 5px 5px 10px rgba(#000, 0.01)
</style>
```

▲ 圖 9-9 完成對話泡泡樣式

現在讓我們實際觸發泡泡看看，首先透過 ref 取得所有的玩家頭像。

➜ src\components\player-list.vue

```
<template>
  <transition-group
    tag="div"
    class=" overflow-hidden px-10 pointer-events-none"
    name="avatar"
  >
    <player-list-avatar
      v-for="player, i in players"
      ref="playerRefs"
      ...
    />
  </transition-group>
</template>

<script setup lang="ts">
import { ref } from 'vue';
...
const playerRefs = ref<InstanceType<typeof PlayerListAvatar>[]>([]);

</script>
```

接著在 onGamepadData 事件中觸發 player-avatar 的 showBalloon()。

➜ src\components\player-list.vue

```
...
<script setup lang="ts">
...
import { useClientGameConsole } from '../composables/use-client-game-console';

const gameConsole = useClientGameConsole();
...

gameConsole.onGamepadData((data) => {
  const lastDatum = data.keys.at(-1);
  if (!lastDatum || lastDatum.value) return;
```

```
  const targetPlayer = playerRefs.value.find(({ playerId }) => playerId === data.
playerId);
  if (!targetPlayer) return;

  targetPlayer.showBalloon(lastDatum.name);
});
</script>
...
```

　　現在點擊搖桿按鈕，會發現泡泡內容與觸發按鈕相同，但是沒有進入和
離開動畫效果，看起來實在有點死板，讓我們加入 transition 動畫 class 吧。
(´▽`)/

→ src\components\player-list-avatar.vue

```
...
<style scoped lang="sass">
...
.balloon-enter-active, .balloon-leave-active
  transition-duration: 0.4s
  transition-timing-function: cubic-bezier(0.150, 1.535, 0.625, 1.015)
.balloon-leave-active
  transition-timing-function: cubic-bezier(1.000, 0.005, 0.150, 1.005)
.balloon-enter-from, .balloon-leave-to
  transform: translateY(100%) rotate(-30deg) !important
  opacity: 0 !important
.balloon-leave-to
  transform: translateY(100%) scale(0.4) !important
</style>
```

Tips

Transition 是 Vue 的內建元件，透過定義指定 class，就可以輕鬆寫意的完成
DOM 的進入、移除動畫，想要深入了解的讀者可以參考以下連結：

- https://cn.vuejs.org/guide/built-ins/transition.html

以上完成話泡泡功能了，我們成功讓玩家說話了！(´ᗜ`)ノヽ(´ᗜ`)

9.4.2 搖桿控制選單

最後讓我們透過搖桿選擇或點選主選單按鈕，新增 use-gamepad-navigator 用於綁定按鈕並觸發相關控制。

首先制定控制元件基本介面，用來說明、限制此功能能夠支援使用的控制元件。

➔ src\composables\use-gamepad-navigator.ts

```
import { Ref, ref } from 'vue';

export interface ControllableElement {
  click(): void;
  hover(): void;
  leave(): void;
  isHover(): boolean;
}

export function useGamepadNavigator<T extends ControllableElement>(elements: Ref<T[]>
= ref([])) {
  return {}
}
```

也就是說被綁定的元件必須提供 click、hover、leave、isHover 方法才行。

Tips

想要深入了解泛型限制的讀者可以參考以下連結：

- https://ithelp.ithome.com.tw/m/articles/10266542

接著加入各類 function。

➜ src\composables\use-gamepad-navigator.ts

```
...
export function useGamepadNavigator<T extends ControllableElement>(elements: Ref<T[]>
 = ref([])) {

  /** 綁定控制元件 */
  function mountElement(el: T | null) {
    if (!el) return;

    elements.value.push(el);
  }

  /** hover 指定元件 */
  function hoverElement(index: number) {
    elements.value.forEach((el) => el.leave());
    elements.value?.[index]?.hover();
  }

  /** 目前 hover 元件的 index */
  const currentIndex = computed(() =>
    elements.value.findIndex(({ isHover }) => isHover())
  );

  /** 上一個元件 */
  function prev() {
    if (currentIndex.value < 0) {
      return hoverElement(0);
    }

    let targetIndex = currentIndex.value - 1;
    if (targetIndex < 0) {
      targetIndex += elements.value.length;
    }

    return hoverElement(targetIndex);
  }
```

```javascript
/** 下一個元件 */
function next() {
  if (currentIndex.value < 0) {
    return hoverElement(0);
  }

  const targetIndex = (currentIndex.value + 1) % elements.value.length;
  return hoverElement(targetIndex);
}

/** 點擊目前 hover 元件 */
function click() {
  if (currentIndex.value < 0) {
    hoverElement(0);
    return elements.value?.[0]?.click();
  }

  const targetIndex = currentIndex.value;
  hoverElement(targetIndex);
  return elements.value[targetIndex].click();
}

/** 自動 hover 第一個元件 */
onMounted(() => {
  elements.value?.[0]?.hover();
});

return {
  mountElement,
  next,
  prev,
  click,
}
}
```

　　回到 game-menu 元件中引用 use-gamepad-navigator，讓選單按鈕變成可以透過程式控制，而非必須使用者實際點選。

將要可用的按鈕加入型別後，綁定至 use-gamepad-navigator 中。

➜ src\components\game-menu.vue

```
<template>
  <div class="flex">
    ...
    <div ... >
      ...
      <base-btn
        :ref="mountButton"
        ...
      >
        ...
      </base-btn>

      <base-btn
        :ref="mountButton"
        ...
      >
        ...
      </base-btn>
    </div>
  </div>
</template>

<script setup lang="ts">
...
import { useGamepadNavigator } from '../composables/use-gamepad-navigator';
...
const { mountElement, click, next, prev } = useGamepadNavigator();

/** 將按鈕綁定 */
function mountButton(el: unknown) {
  const controlElement = el as InstanceType<typeof BaseBtn>;
  mountElement(controlElement)
}

/** 讓其他元件可以控制選單 */
defineExpose({
```

```
  click, next, prev,
});
</script>
...
```

會忽然發現 TypeScript 警告引數指派類型錯誤。Σ(´Д｀;)

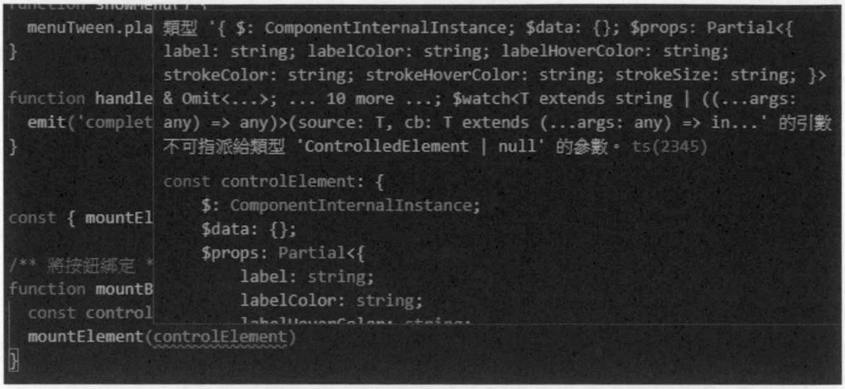

▲ 圖 9-10　綁定元件型別錯誤

　　這是因為 base-btn 元件對外提供的介面不符合 use-gamepad-navigator 規定的 ControllableElement 介面，現在讓我們新增 base-btn 的 defineExpose，讓介面相符吧。(´▽`)/

➡ src\components\base-btn.vue

```
...

<script setup lang="ts">
...
defineExpose<ControllableElement>({
  click: handleClick,
  isHover: () => state.value.hover,
  hover: handleMouseenter,
  leave: handleMouseleave,
});
</script>
...
```

現在會發現剛剛的錯誤消失了，最後讓我們在 the-game-console-lobby 頁面的 onGamepadData 中呼叫 game-menu 提供的功能吧。。

➜ src\views\the-game-console-lobby.vue

```
<template>
  <game-menu
    ref="menu"
    ...
  />
  ...
</template>

<script setup lang="ts">
...
const menu = ref<InstanceType<typeof GameMenu>>();

function init() {
  gameConsole.setStatus('lobby');
  loading.hide();

  gameConsole.onGamepadData((data) => {
    console.log(`[ onGamepadData ] data : `, data);

    const datum = data.keys.at(-1);
    if (!datum) return;

    const { name, value } = datum;
    if (value) return;

    if (name === 'up') {
      menu.value?.prev();
      return;
    }

    if (name === 'down') {
      menu.value?.next();
      return;
    }
```

```
    if (name === 'confirm') {
      menu.value?.click();
      return;
    }
  });
}
</script>
```

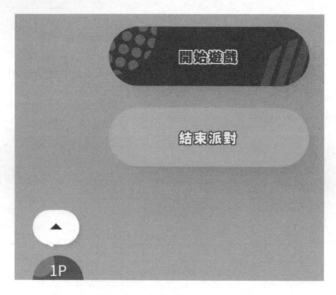

▲ 圖 9-11 標題 Logo 完成

現在會發現我們可以透過搖桿控制選單了！ヾ(◕'ヮ'◕)ﾉﾞ

但是有一個小問題，就是按下確認後，沒有觸發 base-btn 點擊動畫，讓我們改進一下 handleClick。

➜ src\components\base-btn.vue

```
...
<script setup lang="ts">
...
import { promiseTimeout } from '@vueuse/core';
...
```

```
function handleClick(showEffect = false) {
  emit('click');

  if (showEffect) {
    processClick();
  }
}
async function processClick() {
  state.value.hover = true;
  state.value.active = true;

  await promiseTimeout(200);

  state.value.active = false;
}
...
defineExpose<ControllableElement>({
  click(effect = true) {
    handleClick(effect)
  },
  ...
});
</script>
...
```

現在可以透過搖桿隔空觸發按鈕點擊動畫了！✧*。٩(´ﾛ`*)۶✧*。

再來準備讓我們進入重頭戲，開發遊戲的部分了！(ㅍ ﹏ ㅍ)✧

第10章

第一隻企鵝

現在讓我們進入開發遊戲的環節吧！(´▽`)ノ

阿德利企鵝在下水前，會將最前頭的企鵝踢下水，確認水中沒有天敵後才會下水，努力不要被踢下水吧！

10.1 完成遊戲選單

此章節程式碼可以在以下連結取得。

- https://gitlab.com/deepmind1/animal-party/web/-/tree/feat/game-the-first-penguin

在上一個章節中 game-menu 可以控制選單位置，但是還差了選擇遊戲和實際開始遊戲的功能。

第一步讓我們新增遊戲頁面，新增 games 資料夾，用來集中未來所有遊戲的頁面。

接著讓我們新增遊戲畫面元件與頁面元件。

➜ src\games\the-first-penguin\game-scene.vue

```
<template>
  <canvas
    ref="canvas"
    class="outline-none"
  />
</template>

<script setup lang="ts">
import { ref } from 'vue';

const canvas = ref<HTMLCanvasElement>();
</script>
```

➜ src\views\the-game-console-the-first-penguin.vue

```
<template>
  <game-scene class=" w-full h-full" />
</template>

<script setup lang="ts">
import { ref } from 'vue';

import GameScene from '../games/the-first-penguin/game-scene.vue';

</script>
```

並將頁面元件註冊至 router。

➜ src\router\router.ts

```
...
export enum RouteName {
  ...
  GAME_CONSOLE_THE_FIRST_PENGUIN = 'game-console-the-first-penguin',
  ...
}
```

```
const routes: RouteRecordRaw[] = [
  ...
  {
    path: `/game-console`,
    name: RouteName.GAME_CONSOLE,
    component: () => import('../views/the-game-console.vue'),
    children: [
      ...
      {
        path: `the-first-penguin`,
        name: RouteName.GAME_CONSOLE_THE_FIRST_PENGUIN,
        component: () => import('../views/the-game-console-the-first-penguin.vue')
      },
    ]
  },
  ...
]
...
```

現在讓我們完成 game-menu 完整功能，使其可以跳轉至遊戲頁面吧。

- 新增遊戲清單資料定義。

- 新增 currentIndex 變數，表示目前選擇遊戲

- 新增選擇遊戲用的 function 並使用 defineExpose 匯出。

➜ src\components\game-menu.vue

```
<template>
  <div class="flex">
    <game-menu-background
      :selected-game="selectedGame.name"
      ...
    />
    ...
  </div>
</template>
```

```ts
<script setup lang="ts">
...
interface GameInfo {
  name: `${GameName}`;
  routeName: `${RouteName}`;
}
...

const games: GameInfo[] = [
  {
    name: 'the-first-penguin',
    routeName: 'game-console-the-first-penguin',
  },
];

const currentIndex = ref(0);
const selectedGame = computed(() => games[currentIndex.value]);
function prevGame() {
  currentIndex.value--;
  if (currentIndex.value < 0) {
    currentIndex.value += games.length;
  }
}
function nextGame() {
  currentIndex.value++;
  currentIndex.value %= games.length;
}
...
/** 讓其他元件可以控制選單 */
defineExpose({
  click, next, prev, prevGame, nextGame,
});
</script>
...
```

助教：「也才一個遊戲，真是選了個寂寞…(๑•᎑•๑`)」

鱈魚：「還會有其他遊戲啦 … (´;ω;`)」

　　最後讓我們新增 startGame function，用於開始遊戲並轉跳頁面，同時綁定在選單按鈕上。

➜ src\components\game-menu.vue

```
<template>
  <div class="flex">
    ...
    <div ... >
      ...
      <base-btn
        ...
        label=" 開始遊戲 "
        ...
        @click="startGame()"
      >
        ...
      </base-btn>
      ...
    </div>
  </div>
</template>

<script setup lang="ts">
...
import { debounce } from 'lodash-es';
...
import { useClientGameConsole } from '../composables/use-client-game-console';
import { useRouter } from 'vue-router';
import { useLoading } from '../composables/use-loading';
...

const gameConsole = useClientGameConsole();
const router = useRouter();
const loading = useLoading();
...
const startGame = debounce(async () => {
  gameConsole.setStatus('playing');
  gameConsole.setGameName(selectedGame.value.name);
```

```
  await loading.show();

  router.push({
    name: selectedGame.value.routeName
  });
}, 3000, {
  leading: true,
  trailing: false,
});

...
</script>
...
```

Tips

startGame() 加上 debounce，是為了防止玩家按太多下，造成異常。

現在在大廳中按下「開始遊戲」後，就會跳轉至企鵝的遊戲畫面了！

接下來讓我們正式開始打造遊戲吧！ヽ(●ˋ∀ˊ●)/

10.2 大海、浮冰、企鵝勒？

讓我們利用 babylon.js 打造 3D 遊戲吧！ ٩(•ω•)۶

預期會有一群企鵝在浮冰上互撞，玩家們可以使用類比搖桿控制企鵝方向，也可以使出旋轉攻擊，被撞出浮冰外則算出界，存活到最後的玩家獲勝！

讓我們複習一下 babylon.js 產生遊戲畫面需要哪些要素。

1. 取得 canvas DOM

2. 建立 Engine

3. 建立 Scene

4. 建立至少一個光源

5. 建立至少一個 camera

讓我們新增建立場景所需的 function 吧。

- 建立引擎基本上就是傳入 canvas DOM 與開啟反鋸齒，沒有其他特別參數。

- 場景這裡力求簡單，我們直接使用預設光源

- 相機部分，沒有任何需求，所以使用最簡單的 ArcRotateCamera

➜ src\games\the-first-penguin\game-scene.vue

```
...
<script setup lang="ts">
...
import { ArcRotateCamera, Engine, Scene, Vector3 } from '@babylonjs/core';

const canvas = ref<HTMLCanvasElement>();

let engine: Engine;
let scene: Scene;

function createEngine(canvas: HTMLCanvasElement) {
  const engine = new Engine(canvas, true);
  return engine;
}
function createScene(engine: Engine) {
  const scene = new Scene(engine);
  /** 使用預設光源 */
  scene.createDefaultLight();
  return scene;
}
function createCamera(scene: Scene) {
  const camera = new ArcRotateCamera(
    'camera',
```

```
    Math.PI / 2,
    Math.PI / 4,
    34,
    new Vector3(0, 0, 2),
    scene
  );

  return camera;
}
</script>
```

 Tips

想要深入了解的讀者可以參考以下連結：

◆ Arc Rotate Camera：https://doc.babylonjs.com/features/featuresDeepDive/
  cameras/camera_introduction#arc-rotate-camera

◆ Create Default Light：https://doc.babylonjs.com/features/featuresDeepDive/
  scene/fastBuildWorld#create-default-light

接著增加 init() 內容，開始建立場景。

➜ src\games\the-first-penguin\game-scene.vue

```
<script setup lang="ts">
...
function init() {
  if (!canvas.value) {
    console.error(' 無法取得 canvas DOM');
    return;
  }

  engine = createEngine(canvas.value);
  scene = createScene(engine);
  createCamera(scene);

  /** 反覆渲染場景，這樣畫面才會持續變化 */
```

```
engine.runRenderLoop(() => {
  scene.render();
});
}
</script>
```

現在讓我們在 onMounted 中呼叫 init() 吧。

➜ src\games\the-first-penguin\game-scene.vue

```
<script setup lang="ts">
...
function init() {...}

onMounted(() => {
  init();
});
</script>
```

熟悉 Vue 的讀者們一定都知道會在 onMounted 中才初始化，是因為這樣才取得到 canvas DOM。

接著是當元件銷毀前，需要關閉引擎。

➜ src\games\the-first-penguin\game-scene.vue

```
<script setup lang="ts">
...
function init() {...}

onMounted(() => {
  init();
});
</script>
```

再加入考慮畫面隨視窗自動縮放的功能。

➔ src\games\the-first-penguin\game-scene.vue

```ts
<script setup lang="ts">
...
function init() {...}

onMounted(() => {
  init();
  window.addEventListener('resize', handleResize);
});

onBeforeUnmount(() => {
  engine.dispose();
  window.removeEventListener('resize', handleResize);
});

function handleResize() {
  engine.resize();
}
</script>
```

最後讓我們定義初始化事件，用來通知場景初始化完成。

➔ src\games\the-first-penguin\game-scene.vue

```ts
...
<script setup lang="ts">
...

const emit = defineEmits<{
  (e: 'init'): void;
}>();
...

onMounted(() => {
  init();
  window.addEventListener('resize', handleResize);

  emit('init');
});
```

```
...
</script>
```

在 the-game-console-the-first-penguin 頁面中接收事件，並隱藏 loading 效果。

➡ src\views\the-game-console-the-first-penguin.vue

```
<template>
  <game-scene
    class=" w-full h-full"
    @init="handleInit"
  />
</template>

<script setup lang="ts">
...
import { useLoading } from '../composables/use-loading';

const loading = useLoading();

function handleInit() {
  loading.hide();
}
</script>
```

現在按下開始遊戲後就會進入遊戲場景了！

▲ 圖 10-1　初始化企鵝遊戲場景

鱈魚：「出現了！ ᕙ(˚ ∀ ˚)ᕗ」

助教：「出個毛線！這坨深藍色我用 CSS 都能產生！ \(°д° \)」

鱈魚：「那是背景預設的顏色啦，讓我們慢慢來嘛。(´„•ω•„)」

組成一個完整 3D 物件最基本構成是 Mesh 與 Material，前者負責組成「骨架」，後者負責「皮膚」。

開始之前，先讓我們把 game-scene 放到 App.vue 中邊做邊看，方便開發。

➜　src\App.vue

```
<template>
  <router-view />
  <loading-overlay />

  <game-scene class=" absolute w-full h-full" />
</template>

<script setup lang="ts">
import LoadingOverlay from './components/loading-overlay.vue';
import GameScene from './games/the-first-penguin/game-scene.vue';
</script>
...
```

現在讓我們建立一片海洋（其實就是一片地板），過程如下：

1. 建立海洋要用的網格

   這裡使用 MeshBuilder.CreateGround，可以產生一個不具厚度的網格，因為地板不需要厚度。

2. 建立材質

   將顏色設定為藍色後附加於網格上。

Tips

想要深入了解的讀者可以參考以下連結：

◆ Creating A Ground：https://doc.babylonjs.com/features/featuresDeepDive/ mesh/creation/set/ground

◆ Background Material：https://doc.babylonjs.com/features/featuresDeepDive/ environment/backgroundMaterial

以上步驟便可以完成海洋地板了，現在建立新增海洋的 function。

➜ src\games\the-first-penguin\game-scene.vue

```
...
<script setup lang="ts">
...
import {
  ArcRotateCamera, BackgroundMaterial, Color3,
  Engine, MeshBuilder, Scene, Vector3
} from '@babylonjs/core';
...
function createCamera(scene: Scene) { ... }

function createSea(scene: Scene) {
  const sea = MeshBuilder.CreateGround('sea', { height: 1000, width: 1000 });

  const material = new BackgroundMaterial("seaMaterial", scene);
  material.useRGBColor = false;
  material.primaryColor = new Color3(0.57, 0.70, 0.83);

  sea.material = material;

  return sea;
}

function init() {
  ...
```

```
  createCamera(scene);
  createSea(scene);
  ...
}
...
</script>
```

▲ 圖 10-2　建立海洋

鱈魚：「現在是一片淺藍色了！ヽ(•ω•)ᕗ」

助教：「所以我說那個 3D 呢？(○´･д･)ﾉ」

鱈魚：「那就讓我們建立浮冰吧！(•ω•)✧」

➔ src\games\the-first-penguin\game-scene.vue

```
...

<script setup lang="ts">
...
import {
  ArcRotateCamera, BackgroundMaterial, Color3,
  Engine, MeshBuilder, Scene, StandardMaterial, Vector3
} from '@babylonjs/core';
...
```

```
function createSea(scene: Scene) { ... }

function createIce(scene: Scene) {
  const ice = MeshBuilder.CreateBox('ice', {
    width: 30,
    depth: 30,
    height: 4,
  });
  ice.material = new StandardMaterial('iceMaterial', scene);

  return ice;
}

function init() {
  ...
  createSea(scene);
  createIce(scene);
...
}
...
</script>
```

冰塊出現了！

▲ 圖 10-3 建立浮冰

鱈魚：「是不是和吃冰一樣簡單啊！♪(´ω`)و)」

助教：「所以我說企鵝哩？(｡_｡)」

鱈魚：「…(⊙ω⊙)و)」

10.3 企鵝登場

現在有海也有冰，就只差企鵝了。(´▽`)ノ

經過掐指一算，可以通靈出企鵝的功能一定很複雜，所以讓我們把企鵝用 Class 裝起來。

- 企鵝有待命（idle）、走路（walk）、攻擊（attack），三種狀態。

- ownerId 為玩家 ID，用於辨識企鵝所屬的玩家。

➜ src\games\the-first-penguin\penguin.ts

```
import {
  Scene, Color3, Vector3,
  SceneLoader, AbstractMesh,
} from '@babylonjs/core';
/** 引入 loaders，這樣才能載入 glb 檔案 */
import '@babylonjs/loaders';
import { defaultsDeep } from 'lodash-es';

export interface PenguinParams {
  /** 起始位置 */
  position?: Vector3;
  ownerId: string;
}

type State = 'idle' | 'walk' | 'attack';

export class Penguin {
  mesh?: AbstractMesh;
```

```
  name: string;
  scene: Scene;
  params: Required<PenguinParams> = {
    position: new Vector3(0, 0, 0),
    ownerId: '',
  };

  state: State = 'walk';

  constructor(name: string, scene: Scene, params?: PenguinParams) {
    this.name = name;
    this.scene = scene;
    this.params = defaultsDeep(params, this.params);
  }

  async init() {
    const result = await SceneLoader.ImportMeshAsync('', '/games/the-first-penguin/',
'penguin.glb', this.scene);

    const penguin = result.meshes[0];
    penguin.position = this.params.position;

    return this;
  }
}
```

回到 game-scene 中，建立企鵝吧。

➔ src\games\the-first-penguin\game-scene.vue

```
<script setup lang="ts">
...
import { Penguin } from './penguin';
...
async function createPenguin(id: string, index: number) {
  const penguin = await new Penguin(`penguin-${index}`, scene, {
    position: new Vector3(0, 2, 0),
    ownerId: id,
  }).init();
```

```
  return penguin;
}

async function init() {
  ...
  createIce(scene);
  await createPenguin('', 1);
  ...
}
...
</script>
```

現在會看到畫面中央出現一個發瘋的企鵝！ᐠ(˚∀˚)ᐟ

▲　圖 10-4　建立企鵝

會一直轉是因為預設攻擊動畫持續循環播放的關係，現在讓我們初始化所有動畫，使企鵝冷靜冷靜。(˚∀˚)

➜　src\games\the-first-penguin\penguin.ts

```
import {
  Scene, Color3, Vector3,
  SceneLoader, AbstractMesh, AnimationGroup,
} from '@babylonjs/core';
...
```

```
interface AnimationMap {
  idle?: AnimationGroup,
  walk?: AnimationGroup,
  attack?: AnimationGroup,
}
...
export class Penguin {
  ...
  state: State = 'walk';
  private animation: AnimationMap = {
    idle: undefined,
    walk: undefined,
    attack: undefined,
  };

  constructor(name: string, scene: Scene, params?: PenguinParams) { ... }

  private initAnimation(animationGroups: AnimationGroup[]) {
animationGroups.forEach((group) => group.stop());

    const attackAni = animationGroups.find(({ name }) => name === 'attack');
    const walkAni = animationGroups.find(({ name }) => name === 'walk');
    const idleAni = animationGroups.find(({ name }) => name === 'idle');

    this.animation.attack = attackAni;
    this.animation.walk = walkAni;
    this.animation.idle = idleAni;
  }

  async init() {
    ...

    this.initAnimation(result.animationGroups);

    return this;
  }
}
```

現在我們得到一隻冷靜的企鵝了。(´▽`)/

接下來讓企鵝動起來吧！關鍵是要加入物理系統，首先先幫企鵝設定 Hit Box，並將企鵝的模型綁在 Hit Box 上。

Tips

Hit Box 顧名思義就是用來碰撞的箱子，也就是用來定義企鵝實際上在物理系統中會發生碰撞的部分。

➜ src\games\the-first-penguin\penguin.ts

```
import {
  Scene, Color3, Vector3,
  SceneLoader, AbstractMesh, AnimationGroup, MeshBuilder, PhysicsImpostor,
} from '@babylonjs/core';
...
export class Penguin {
  ...
  constructor(name: string, scene: Scene, params?: PenguinParams) {... }

  ...
  private createHitBox() {
    const hitBox = MeshBuilder.CreateBox(`${this.name}-hit-box`, {
      width: 2, depth: 2, height: 4
    });
    hitBox.position = this.params.position;
    // 設為半透明方便觀察
    hitBox.visibility = 0.5;

    /** 使用物理效果 */
    const hitBoxImpostor = new PhysicsImpostor(
      hitBox,
      PhysicsImpostor.BoxImpostor,
      { mass: 1, friction: 0.7, restitution: 0.7 },
      this.scene
    );

    hitBox.physicsImpostor = hitBoxImpostor;
```

```
    return hitBox;
  }

  async init() {
    const result = await SceneLoader.ImportMeshAsync('', '/the-first-penguin/',
'penguin.glb', this.scene);
    this.initAnimation(result.animationGroups);

    // 產生 hitBox
    const hitBox = this.createHitBox();
    this.mesh = hitBox;

    // 將企鵝綁定至 hitBox
    const penguin = result.meshes[0];
    penguin.setParent(hitBox);
    penguin.position = new Vector3(0, -2, 0);


    return this;
  }
}
```

然後回到場景中，把企鵝產生的位置移高一點，讓我們完整的觀察一下 Hit Box 與企鵝目前的位置。

讓我們將 Y 從 2 改成 10。

➜ src\games\the-first-penguin\game-scene.vue

```
...
<script setup lang="ts">
...
async function createPenguin(id: string, index: number) {
  const penguin = await new Penguin(`penguin-${index}`, scene, {
    position: new Vector3(0, 10, 0),
    ownerId: id,
  }).init();

  return penguin;
}
```

```
...
</script>
```

現在可以看到企鵝周圍多了一個半透明長方體，那個就是企鵝的 hit box 了。

▲ 圖 10-5　企鵝與 Hit Box

助教：「為甚麼不用企鵝模型當作 Hit Box 就好了？」

鱈魚：「其實也可以，只是適當的簡化可以有效提升效能。(´▽`)/‿」

助教：「不是因為想偷懶？('๑ ๑`)」

鱈魚：「並沒有好嗎。\(・ω´・ \)」

現在引入物理引擎，讓企鵝降落吧！ヽ(≧∀≦)ﾉ

第一步先來安裝一下物理引擎套件，這裡我們使用 cannon。

```
npm i cannon-es
```

回到場景中，引入並啟用物理引擎。

➜ src\games\the-first-penguin\game-scene.vue

```
...
<script setup lang="ts">
import * as CANNON from 'cannon-es';
import { onBeforeUnmount, onMounted, ref } from 'vue';
import {
  ArcRotateCamera, BackgroundMaterial, CannonJSPlugin, Color3,
  Engine, MeshBuilder, Scene, StandardMaterial, Vector3
} from '@babylonjs/core';
...
function createScene(engine: Engine) {
  const scene = new Scene(engine);
  /** 使用預設光源 */
  scene.createDefaultLight();

  const physicsPlugin = new CannonJSPlugin(true, 8, CANNON);
  scene.enablePhysics(new Vector3(0, -9.81, 0), physicsPlugin);

  return scene;
}
...
</script>
```

現在讓我們恭請企鵝駕到！(ˆ•ᵕˆ•)

500 英尺、100 英尺 ...

▲ 圖 10-6 企鵝穿過浮冰

啊啊啊啊啊！降落過頭啦！Σ(っ °Д °;)っ

10-23

第 **10** 章　第一隻企鵝

這是會穿過浮冰一路往下掉落，是因為浮冰沒有加入碰撞箱，所以無法乘載企鵝，趕緊追加一下。

➜ src\games\the-first-penguin\game-scene.vue

```
...
<script setup lang="ts">
...
import {
  ArcRotateCamera, BackgroundMaterial, CannonJSPlugin, Color3,
  Engine, MeshBuilder, PhysicsImpostor, Scene, StandardMaterial, Vector3
} from '@babylonjs/core';
...

function createIce(scene: Scene) {
  const ice = MeshBuilder.CreateBox('ice', {
    width: 30,
    depth: 30,
    height: 4,
  });
  ice.material = new StandardMaterial('iceMaterial', scene);
  // mass 設為 0，就可以固定在原地不動
  ice.physicsImpostor = new PhysicsImpostor(ice, PhysicsImpostor.BoxImpostor,
    { mass: 0, friction: 0, restitution: 0 }, scene
  );

  return ice;
}
...
</script>
```

現在企鵝佇立於冰層之上了！

▲ 圖 10-7 企鵝佇立於冰層之上

讓我們加上浮在企鵝頭上的小徽章，可以自由設定顏色，讓玩家辨識自己的企鵝，同時隱藏 hit box 吧。

➜ src\games\the-first-penguin\penguin.ts

```ts
import {
  Scene, Color3, Vector3,
  SceneLoader, AbstractMesh, AnimationGroup, MeshBuilder, PhysicsImpostor,
StandardMaterial,
} from '@babylonjs/core';
...

export interface PenguinParams {
  ...
  /** 玩家顏色 */
  color?: Color3;
}
...

export class Penguin {
 ...
  params: Required<PenguinParams> = {
    ...
    color: new Color3(0.9, 0.9, 0.9),
  };

  ...
```

```
constructor(name: string, scene: Scene, params?: PenguinParams) {... }

...
private createBadge() {
  const badge = MeshBuilder.CreateBox(`${this.name}-badge`, {
    width: 0.5, depth: 0.5, height: 0.5
  });
  const material = new StandardMaterial('badgeMaterial', this.scene);
  material.diffuseColor = this.params.color;
  badge.material = material;

  const deg = Math.PI / 4;
  badge.rotation = new Vector3(deg, 0, deg);
  badge.visibility = 0.9;

  return badge;
}

async init() {
  ...

  // 建立 badge
  const badge = this.createBadge();
  badge.setParent(hitBox);
  badge.position = new Vector3(0, 3, 0);

  return this;
}
}
```

▲ 圖 10-8 企鵝徽章

最後在徽章加個旋轉動畫吧。

➜ src\games\the-first-penguin\penguin.ts

```
import {
  Scene, Color3, Vector3,
  SceneLoader, AbstractMesh, AnimationGroup,
  MeshBuilder, PhysicsImpostor, StandardMaterial,
  Animation,
} from '@babylonjs/core';
...
export class Penguin {
  ...
  constructor(name: string, scene: Scene, params?: PenguinParams) {... }

  ...
  private createBadge() {
    ...

    // 建立動畫
    const frameRate = 10;
    const badgeRotate = new Animation(
      'badgeRotate',
      'rotation.y',
      frameRate / 5,
      Animation.ANIMATIONTYPE_FLOAT,
      Animation.ANIMATIONLOOPMODE_CYCLE
    );

    const keyFrames = [
      {
        frame: 0,
        value: 0
      },
      {
        frame: frameRate,
        value: 2 * Math.PI
      }
    ],
```

```
    badgeRotate.setKeys(keyFrames);
    badge.animations.push(badgeRotate);

    this.scene.beginAnimation(badge, 0, frameRate, true);

    return badge;
  }
  ...
}
```

▲ 圖 10-9　徽章加入旋轉動畫

企鵝外觀完成！再來就差讓企鵝動起來了！ヽ(•ω•)ノ

10.4 爆走企鵝

此章節程式碼可以在以下連結取得。

- https://gitlab.com/deepmind1/animal-party/web/-/tree/feat/game-penguin-dynamics

讓企鵝動起來吧！ㄟ(•ω•)ㄏ

由於在建立冰塊時，已經將摩擦力（friction）設定為 0，所以企鵝只要動起來就不會停下來。

要模擬出在光滑表面上操作的感覺很簡單，只要讓玩家的方向操作，是在企鵝上加「加速度」，而不是「速度」就可以了。

讓我們新增一個移動用的 method。

➔ src\games\the-first-penguin\penguin.ts

```ts
import {
  Scene, Color3, Vector3,
  SceneLoader, AbstractMesh, AnimationGroup,
  MeshBuilder, PhysicsImpostor, StandardMaterial,
  Animation,
} from '@babylonjs/core';
...

export class Penguin {
  ...

  /** 指定移動方向與力量 */
  walk(force: Vector3) {
    if (!this.mesh) {
      throw new Error('未建立 Mesh');
    }

    /** 施加力量 */
    this.mesh.physicsImpostor?.applyForce(force, Vector3.Zero());
  }
}
```

Tips

applyForce 表示施加力量，第一個參數為力量向量，第二個參數表示作用位置。

這裡指定 0 點（Vector3.Zero() 等同於 new Vector3(0, 0, 0)）表示作用於企鵝中心點，這樣才不會產生力矩，導致企鵝旋轉，讀者們可以改成其他數值實際體驗看看效果。

更多資訊可參考以下連結：

◆ Forces：https://doc.babylonjs.com/features/featuresDeepDive/physics/forces

讓我們用鍵盤控制看看效果如何，首先回到遊戲場景加入以下內容：

- 偵測鍵盤事件。

- 將企鵝物件用變數存起來。

➔ src\games\the-first-penguin\game-scene.vue

```
<script>
...
import {
  ArcRotateCamera, BackgroundMaterial, CannonJSPlugin, Color3,
  Engine, KeyboardEventTypes, MeshBuilder, PhysicsImpostor, Scene, StandardMaterial,
  Vector3
} from '@babylonjs/core';
...
async function init() {
  ...
  const penguin = await createPenguin('', 1);

  scene.onKeyboardObservable.add((keyboardInfo) => {
    /** 忽略按下以外的事件 */
    if (keyboardInfo.type !== KeyboardEventTypes.KEYDOWN) return;

  });

  ...
```

```
}

</script>
```

所以現在要怎麼判斷鍵盤方向鍵與企鵝移動的方向呢？

已知 babylon.js 坐標系是左手坐標系，而：

- 正 X 方向是畫面左方

- 正 Z 方向是畫面後方

- 正 Y 方向是往上

所以程式如下。

→ src\games\the-first-penguin\game-scene.vue

```
<script>
...
import {
  ArcRotateCamera, BackgroundMaterial, CannonJSPlugin, Color3,
  Engine, KeyboardEventTypes, MeshBuilder, PhysicsImpostor, Scene, StandardMaterial,
Vector3
} from '@babylonjs/core';
...
async function init() {
  ...
  const penguin = await createPenguin('', 1);

  scene.onKeyboardObservable.add((keyboardInfo) => {
    /** 忽略按下以外的事件 */
    if (keyboardInfo.type !== KeyboardEventTypes.KEYDOWN) return;

    switch (keyboardInfo.event.key) {
      case 'ArrowLeft': {
        penguin.walk(new Vector3(100, 0, 0));
        break;
      }
      case 'ArrowUp': {
```

```
      penguin.walk(new Vector3(0, 0, -100));
      break;
    }
    case 'ArrowRight': {
      penguin.walk(new Vector3(-100, 0, 0));
      break;
    }
    case 'ArrowDown': {
      penguin.walk(new Vector3(0, 0, 100));
      break;
    }
  }
});

...
}

</script>
```

現在可以用鍵盤方向鍵移動企鵝了。

▲ 圖 10-10　方向鍵控制企鵝移動

我們得到一隻急速漂移的企鵝！ ୧(˚ ∀˚)୨

為了避免企鵝跑太快，讓我們加個速度限制吧。

➜ src\games\the-first-penguin\penguin.ts

```typescript
...
export class Penguin {
  ...
  private readonly maxSpeed = 15;

  constructor(name: string, scene: Scene, params?: PenguinParams) {...}
...
  private limitMaxVelocity() {
    if (!this.mesh || !this.mesh.physicsImpostor) return;

    /** 取得目前速度向量 */
    const velocity = this.mesh.physicsImpostor.getLinearVelocity();
    if (!velocity) return;

    /** 若目前速度向量大於最大限制，則調整目前速度 */
    const currentSpeed = velocity.length();
    if (currentSpeed > this.maxSpeed) {
      const newVelocity = velocity.normalize().scaleInPlace(this.maxSpeed);
      this.mesh.physicsImpostor?.setLinearVelocity(newVelocity);
    }
  }

  async init() {
    ...

    // 持續在每個 frame render 之前呼叫
    this.scene.registerBeforeRender(() => {
      this.limitMaxVelocity();
    });

    return this;
  }

  ...
}
```

助教：「只有這樣不給力捏（ ’ ·ω·` ）」

鱈魚：「那就讓企鵝轉個方向吧！ ᕙ(╹∀╹)ᕗ」

只要求出力向量與人物向量夾角，再讓企鵝轉到指定角度即可。

兩向量夾角可以利用向量夾角公式求得，公式為：

$$\cos\theta = \frac{a \cdot b}{|\vec{a}||\vec{b}|}$$

依公式設計一個計算受力角度的 method，並在施加力量後讓企鵝轉向指定角度。

➔ src\games\the-first-penguin\penguin.ts

```
...
export class Penguin {
  ...
  constructor(name: string, scene: Scene, params?: PenguinParams) {...}

  /** 取得力與人物的夾角 */
  private getForceAngle(force: Vector3) {
    if (!this.mesh) {
      throw new Error('未建立 Mesh');
    }

    const forceVector = force.normalize();
    /** 企鵝面相正 Z 軸方向 */
    const characterVector = new Vector3(0, 0, 1);
    const deltaAngle = Math.acos(Vector3.Dot(forceVector, characterVector));

    /** 反餘弦求得角度範圍為 0~180 度，需要自行判斷負角度部分。
     *  力向量 X 軸分量為負時，表示夾角為負。
     */
    if (forceVector.x < 0) {
      return deltaAngle * -1;
    }

    return deltaAngle;
  }
```

```
  ...

  /** 指定移動方向與力量 */
  walk(force: Vector3) {
    ...

    // 依據力方向轉向
    const targetAngle = this.getForceAngle(force);
    this.mesh.rotation = new Vector3(0, targetAngle, 0);
  }
}
```

助教：「能不能再給力一點？(˙ω˙)」

鱈魚：「那就讓轉向加上動畫吧！(´∇`)/」

使用先前建立的 createAnimation 產生動畫吧，新增 rotate method 用於企鵝旋轉。

→ src\games\the-first-penguin\penguin.ts

```
import {
  Scene, Color3, Vector3,
  SceneLoader, AbstractMesh, AnimationGroup,
  MeshBuilder, PhysicsImpostor, StandardMaterial,
  Animation,
} from '@babylonjs/core';
...
import { createAnimation } from '../../common/utils';

interface AnimationMap {
  idle?: AnimationGroup,
  walk?: AnimationGroup,
  attack?: AnimationGroup,
}
...

export class Penguin {
  ...
```

```
constructor(name: string, scene: Scene, params?: PenguinParams) {...}
...
private rotate(angle: number) {
  const { animation, frameRate } = createAnimation(this.mesh, 'rotation', new
Vector3(0, angle, 0), {
    speedRatio: 3,
  });

  this.scene.beginDirectAnimation(this.mesh, [animation], 0, frameRate);
}
...
/** 指定移動方向與力量 */
walk(force: Vector3) {
  if (!this.mesh) {
    throw new Error(' 未建立 Mesh');
  }

  /** 施加力量 */
  this.mesh.physicsImpostor?.applyForce(force, Vector3.Zero());

  // 依據力方向轉向
  const targetAngle = this.getForceAngle(force);
  this.rotate(targetAngle);
}
}
```

旋轉吧！企鵝！

▲ 圖 10-11　旋轉企鵝

助教：「轉是會轉了，不過為甚麼最後從左邊轉到下面的時候會轉那麼大圈啊！ᕕ(°Ａ,°)ﾉ」

鱈魚：「我也想和你解釋，但我只是隻魚(.•ᴗ•.)」

助教：（拿出菜刀。(╬⊙д⊙)）

鱈魚：「冷靜冷靜 Σ(´Д`;)，讓我們來探討一下。」

探討以下例子，以企鵝向上看為 0 度位置：

- 若起始角度為 180 度（向下），往左轉時應該要轉到 270 度才會自然（轉 90 度）。

- 若起始角度為 -90 度（向左），往下轉時應該要轉到 -180 度才會自然（轉 -90 度）。

但是 getForceAngle 求得的角度 a 之範圍為 $-180 \leq a < 180$，所以結果會變成：

- 若起始角度為 180 度（向下），往左轉時轉到 -90 度，轉了 -270 度。

- 若起始角度為 -90 度（向左），往下轉時轉到 180 度，轉了 270 度。

看起來就是轉了很大一圈。

這裡有個很簡單的解決辦法，就是判斷如果旋轉角度大於 180 度，就先讓企鵝瞬間轉到目標角度 a 的兩倍補角位置，這樣考慮以上案例就會是：

- 若起始角度為 -90 度（向左），目標轉到 180 度（向下），先瞬間轉到 270 度（看起來也是向左）再轉到目標，結果就是從 270 度轉到 180 度，轉了 -90 度。

看圖可能會比較好理解。

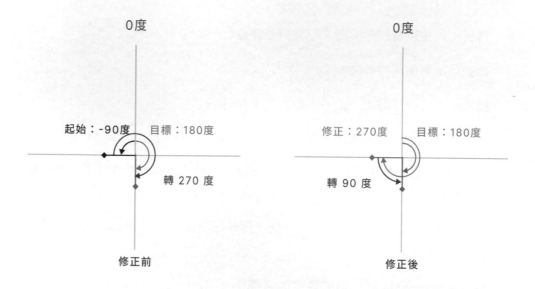

▲ 圖 10-12　修正旋轉問題

讓我們調整 rotate 內容。

➔ src\games\the-first-penguin\penguin.ts

```
...
export class Penguin {
  ...
  private rotate(angle: number) {
    if (!this.mesh) return;

    /** 若角度超過 180 度，則先直接切換至兩倍補角處，讓轉向更自然 */
    const currentAngle = this.mesh.rotation.y;
    if (Math.abs(angle - currentAngle) > Math.PI) {
      const supplementaryAngle = Math.PI * 2 - Math.abs(currentAngle);
      if (currentAngle < 0) {
        this.mesh.rotation = new Vector3(0, supplementaryAngle, 0);
      } else {
        this.mesh.rotation = new Vector3(0, -supplementaryAngle, 0);
```

```
      }
    }

    const { animation, frameRate } = createAnimation(this.mesh, 'rotation', new
Vector3(0, angle, 0), {
        speedRatio: 3,
    });

    this.scene.beginDirectAnimation(this.mesh, [animation], 0, frameRate);
  }
  ...
}
```

現在轉向變自然了！✧*｡٩(´ꇴ`*)ﻭ✧*｡

10.5　完全體企鵝

助教：「都沒動作的移動很像幽靈欸，能不能再給力一點？」

鱈魚：「那就來點動畫吧。(´„•ω•„)」

實作概念為：

1. 玩家觸發企鵝動作

2. 動作會改變狀態

3. 程式根據狀態自動播放指定動畫

首先我們先新增 setState method 用來改變企鵝的狀態。

➜ src\games\the-first-penguin\penguin.ts

```
...
export class Penguin {
  ..
  /** 設定人物狀態 */
```

```
  private setState(value: State) {
    this.state = value;
  }
  ...
}
```

　　讓我們呼叫模型中的動畫，利用動畫權重混合功能，讓兩種動畫之間切換滑順。

　　原理很簡單，把播放狀態改成以 0 到 1 的權重呈現，原有動畫為 1，目標動畫為 0。

　　動畫切換時，原有動畫逐步減少至 0，目標動畫逐步增加至 1，如此便可以在兩種動畫之間平滑切換。

 Tips

想要深入了解的讀者可以參考以下連結：

◆ babylon.js - Animation blending：https://doc.babylonjs.com/features/
featuresDeepDive/animation/advanced_animations#animation-blending

　　新增處理動畫與混合動畫的 method，並在 walk 中呼叫 setState('walk')、init 中呼叫 setState('idle')。

➜ src\games\the-first-penguin\penguin.ts

```
...
export class Penguin {
  ..
  /** 設定人物狀態 */
  private setState(value: State) {...}

  /** 處理狀態動畫
   *
   * 利用 [runCoroutineAsync API](https://doc.babylonjs.com/features/featuresDeepDive/
events/coroutines) 實現
   */
```

```
  private processStateAnimation(newState: State) {
    if (newState === this.state) return;

    const playingAnimation = this.animation[this.state];
    const targetAnimation = this.animation[newState];

    this.state = newState;
    if (!targetAnimation || !playingAnimation) return;

    /** 攻擊動畫不循環播放 */
    const loop = this.state !== 'attack';
    /** 切換至攻擊動畫速度要快一點 */
    const offset = this.state === 'attack' ? 0.3 : undefined;

    /** 清除先前的動畫過渡 */
    this.scene.onBeforeRenderObservable.cancelAllCoroutines();
    this.scene.onBeforeRenderObservable.runCoroutineAsync(this.animationBlending(playi
ngAnimation, targetAnimation, loop, offset));
  }
  /** 動畫混合
   * 讓目前播放動畫的權重從 1 到 0，而目標動畫從 0 到 1。
   * 利用 Generator Function 配合 babylon 的 [runCoroutineAsync API](https://doc.
babylonjs.com/features/featuresDeepDive/events/coroutines)
   * 迭代，達成融合效果。
   */
  private * animationBlending(fromAnimation: AnimationGroup, toAnimation:
 AnimationGroup, loop = true, step = 0.1) {
    let currentWeight = 1;
    let targetWeight = 0;

    toAnimation.play(loop);

    while (targetWeight < 1) {
      targetWeight += step;
      currentWeight -= step;

      toAnimation.setWeightForAllAnimatables(targetWeight);
      fromAnimation.setWeightForAllAnimatables(currentWeight);
      yield;
    }
```

```
    toAnimation.play(loop);
    fromAnimation.stop();
  }

  ...

  async init() {
    ...

    this.setState('idle');

    return this;
  }


  /** 指定移動方向與力量 */
  walk(force: Vector3) {
    ...
    this.setState('walk');
  }

  ...
}
```

現在企鵝會在移動時播放走路動畫了。

▲ 圖 10-13 企鵝走路動畫

Tips

想要深入了解 Generator Function 的讀者可以參考以下連結：

◆ https://pjchender.dev/javascript/js-generator/

　　但是企鵝一旦走起來，走路動畫就停不下來了。現在希望 walk 停止觸發後狀態會回歸 idle，由於 walk method 會被持續觸發，所以這裡我們可以使用 debounce 簡單完成。

➜ src\games\the-first-penguin\penguin.ts

```
...
export class Penguin {
  ...
  /** 指定移動方向與力量 */
  walk(force: Vector3) {
    ...

    this.setState('walk');
    this.walkToIdleDebounce();
  }
  /** 停止呼叫後 500ms 設為 idle 狀態 */
  private walkToIdleDebounce = debounce(async () => {
    this.setState('idle');
  }, 500)
}
```

　　鱈魚：「現在企鵝不會走個不停了！」

　　助教：「看起來不錯，但好像少了點甚麼？(´･ω･`)」

　　鱈魚：「那就讓企鵝增加攻擊技能吧！ヽ(●´∀´●)/」

增加攻擊功能，並加入以下限制：

- 每 2 秒才能攻擊一次

- 攻擊時無法變換方向和移動

- 攻擊需要至少耗費 1 秒

➜ src\games\the-first-penguin\penguin.ts

```
...
export class Penguin {
  ...
  /** 指定移動方向與力量 */
  walk(force: Vector3) {
    // 攻擊時無法移動
    if (this.state === 'attack') return;
    ...
  }
  /** 停止呼叫後 500ms 設為 idle 狀態 */
  private walkToIdleDebounce = debounce(async () => {
    this.setState('idle');
  }, 500)

  /** 攻擊，限制攻擊頻率，2 秒一次 */
  attack = throttle(() => {
    this.setState('attack');
    this.attackToIdleDebounce();
    this.walkToIdleDebounce.cancel();
  }, 2000, {
    leading: true,
    trailing: false,
  })
  /** 攻擊結束後 1 秒時，回到 idle 狀態 */
  private attackToIdleDebounce = debounce(() => {
    this.setState('idle');
  }, 1000, {
    leading: false,
    trailing: true,
  })
}
```

接著透過空白鍵發動攻擊吧。

➡ src\games\the-first-penguin\game-scene.vue

```
...
<script>
...
async function init() {
  ...

  scene.onKeyboardObservable.add((keyboardInfo) => {
    /** 忽略按下以外的事件 */
    if (keyboardInfo.type !== KeyboardEventTypes.KEYDOWN) return;

    switch (keyboardInfo.event.key) {
      ...
      case ' ': {
        penguin.attack();
        break;
      }
    }
  });

  ...
}
...
</script>
```

現在我們得到可以幹架的完全體企鵝了！ヽ(≧∀≦)ﾉ

▲ 圖 10-14 企鵝攻擊動畫

接著讓我們實作被攻擊效果吧，新增一個被攻擊的 method 並微調部分程式。

概念很簡單，就是被攻擊時會施加一個指定的力量即可。

➜ src\games\the-first-penguin\penguin.ts

```
...
export class Penguin {
  ...

  private readonly maxSpeed = 15;
  private readonly assaultedForce = 20;

  constructor(name: string, scene: Scene, params?: PenguinParams) {...}
  ...
  /** 被攻擊
   * @param direction 移動方向
   */
  assaulted = throttle((direction: Vector3) => {
    if (!this.mesh) {
      throw new Error(' 未建立 Mesh');
    }

    // 計算力量
    const force = direction.normalize().scaleInPlace(this.assaultedForce);
    this.mesh.physicsImpostor?.applyImpulse(force, Vector3.Zero());
  }, 500, {
    leading: true,
    trailing: false,
  })
}
```

現在來實作攻擊判定，回到場景，調整一下程式：

- 新增 penguins 變數，儲存已建立企鵝

- 調整 createPenguin，新增更多設定參數，調整每個企鵝位置

- 建立兩隻企鵝，一隻當沙包

- 鍵盤控制第一隻企鵝

➔ src\games\the-first-penguin\game-scene.vue

```
...
<script>
...
const penguins: Penguin[] = [];

interface CreatePenguinParams {
  position: Vector3;
}
async function createPenguin(id: string, index: number, params: CreatePenguinParams) {
  const penguin = await new Penguin(`penguin-${index}`, scene, {
    position: params.position,
    ownerId: id,
  }).init();

  return penguin;
}

async function init() {
  ...
  createIce(scene);

  const result = await Promise.allSettled([
    createPenguin('', 0, {
      position: new Vector3(0, 10, 0),
    }),
    createPenguin('', 1, {
      position: new Vector3(5, 10, 0),
    }),
  ])
  result.forEach((data) => {
    if (data.status !== 'fulfilled') return;
    penguins.push(data.value);
  });
```

```
  const penguin = penguins[0];
  scene.onKeyboardObservable.add((keyboardInfo) => {...});


  ...
}

...
</script>
```

現在畫面上有兩隻企鵝了。

▲ 圖 10-15　建立兩隻企鵝

接著加入攻擊碰撞偵測。

→ src\games\the-first-penguin\game-scene.vue

```
...
<script>
...
// 偵測企鵝碰撞事件
function detectCollideEvents(penguins: Penguin[]) {
  const length = penguins.length;
  for (let i = 0; i < length; i++) {
    for (let j = i; j < length; j++) {
      if (i === j) continue;
```

```
      const aMesh = penguins[i].mesh;
      const bMesh = penguins[j].mesh;
      if (!aMesh || !bMesh) continue;

      if (aMesh.intersectsMesh(bMesh)) {
        handleCollideEvent(penguins[i], penguins[j]);
      }
    }
  }
}

function handleCollideEvent(aPenguin: Penguin, bPenguin: Penguin) {
  if (!aPenguin.mesh || !bPenguin.mesh) return;

  const aState = aPenguin.state;
  const bState = bPenguin.state;
  // 沒有企鵝在 attack 狀態，不須動作
  if (![aState, bState].includes('attack')) return;

  const direction = bPenguin.mesh.position.subtract(aPenguin.mesh.position);
  if (aState === 'attack') {
    bPenguin.assaulted(direction);
  } else {
    aPenguin.assaulted(direction.multiply(new Vector3(-1, -1, -1)));
  }
}

async function init() {
  ...
  /** 持續運行指定事件 */
  scene.registerAfterRender(() => {
    detectCollideEvents(penguins);
  });

  /** 反覆渲染場景，這樣畫面才會持續變化 */
  ...
}
...
</script>
```

▲ 圖 10-16　企鵝攻擊擊退效果

助教：「能不能再給力一 ...」

鱈魚：「不行，窩沒力了 ... (ﾉ´ω`)」

10.6　建立控制搖桿

此章節程式碼可以在以下連結取得。

- https://gitlab.com/deepmind1/animal-party/web/-/tree/feat/game-the-first-penguin-gamepad

企鵝準備好了，現在讓我們建立企鵝遊戲用的玩家搖桿頁面吧。

企鵝遊戲搖桿基本上和大廳搖桿相同，只差在不是方向鍵，而是類比搖桿。

預期外觀如下圖。

▲ 圖 10-17 企鵝搖桿草稿

外圈最大範圍為 pad

- 內圈隨著手指移動的部分為 thumb

- 讓我們複製大廳搖桿並刪除方向鍵部分。

➜ src\views\the-player-gamepad-the-first-penguin.vue

```
<template>
  <div
    class="w-full h-full flex text-white select-none overflow-hidden"
    :class="bgClass"
    @touchmove="(e) => e.preventDefault()"
  >
    <gamepad-btn
      ...
    />

    <div class="code-name">
      {{ codeName }}
    </div>

    <q-dialog
      v-model="isPortrait"
```

```
    persistent
  >
    ...
  </q-dialog>
 </div>
</template>
...
```

接著將頁面新增至 Router 中。

➡ src\router\router.ts

```
...
export enum RouteName {
  ...
  PLAYER_GAMEPAD_THE_FIRST_PENGUIN = 'player-gamepad-the-first-penguin',
}

const routes: RouteRecordRaw[] = [
  ...
  {
    path: `/player-gamepad`,
    name: RouteName.PLAYER_GAMEPAD,
    component: () => import('../views/the-player-gamepad.vue'),
    children: [
      ...
      {
        path: `the-first-penguin`,
        name: RouteName.PLAYER_GAMEPAD_THE_FIRST_PENGUIN,
        component: () => import('../views/the-player-gamepad-the-first-penguin.vue')
      },
    ]
  },
  ...
]
...
```

現在如建立遊戲場景時一般，先將頁面放在 App.vue 裡面，方便開發。

讓我們刪除 game-scene，加入 the-player-gamepad-the-first-penguin。

➜ src\App.vue

```
<template>
  <router-view />
  <loading-overlay />

  <player-gamepad class=" absolute w-full h-full" />
</template>

<script setup lang="ts">
import LoadingOverlay from './components/loading-overlay.vue';
import PlayerGamepad from './views/the-player-gamepad-the-first-penguin.vue';
</script>
...
```

目前應該會看到如下圖畫面。

▲ 圖 10-18　準備建立企鵝搖桿

可以發現只差類比搖桿並調整一下按鈕類型就好了！(´▽`)/.

如同按鈕元件一般，我們來建立類比搖桿元件吧，首先是 Props 與事件部分。

➜ src\components\gamepad-analog-stick.vue

```ts
<script setup lang="ts">
interface Props {
  /** 尺寸，直徑 */
  size?: string
}
const props = withDefaults(defineProps<Props>(), {
  size: '34rem'
});

const emit = defineEmits<{
  (e: 'trigger', data: { x: number, y: number }): void;
}>();
</script>
```

trigger 發出的數值即為拇指控制類比搖桿的移動量。

接著加入 template 與樣式部分。

➜ src\components\gamepad-analog-stick.vue

```
<template>
  <div
    class="pad rounded-full bg-grey-10"
    @contextmenu="(e) => e.preventDefault()"
  >
    <div class="thumb" />
  </div>
</template>
...
<style scoped lang="sass">
.pad
  width: v-bind('props.size')
  height: v-bind('props.size')
  display: flex
  justify-content: center
  align-items: center
.thumb
  width: 40%
```

```
  height: 40%
  background: white
  border-radius: 9999px
  opacity: 0.4
</style>
```

引入搖桿元件，看看外觀吧。

➡ src\views\the-player-gamepad-the-first-penguin.vue

```
<template>
  <div
    ...
  >
    <gamepad-analog-stick class="absolute bottom-5 left-8" />
    ...
  </div>
</template>

<script setup lang="ts">
import { computed } from 'vue';

import GamepadBtn from '../components/gamepad-btn.vue';
import GamepadAnalogStick from '../components/gamepad-analog-stick.vue';
...
</script>
```

▲ 圖 10-19 類比搖桿外觀

　　讓我們依序加入功能吧，最重要的部分是偵測拉動事件，這裡使用 Quasar 提供的 Touch Pen 指令輕鬆實現！﹨(≧∀≦)﹨

Tips

想要深入了解的讀者可以參考以下連結：

◆ Quasar Touch Pan Directive：https://quasar.dev/vue-directives/touch-pan

　　由於 Quasar 不知道為甚麼沒有提供 Touch Pen 的資料定義，所以先讓我們根據文檔定義一下。

➜ src\components\gamepad-analog-stick.vue

```
...
<script setup lang="ts">
interface TouchPenDetails {
  touch: boolean;
  mouse: boolean;
  position: {
    top: number;
    left: number;
  };
  direction: 'up' | 'right' | 'down' | 'left';
  isFirst: boolean;
  isFinal: boolean;
  duration: number;
  distance: {
    x: number;
    y: number;
  };
  offset: {
    x: number;
    y: number;
  };
  delta: {
    x: number;
    y: number;
  };
```

```
}
...
</script>
...
```

接著加入指令與事件。

➜ src\components\gamepad-analog-stick.vue

```
<template>
  <div
    v-touch-pan.prevent.mouse="handleTouch"
    class="pad rounded-full bg-grey-10"
    @contextmenu="(e) => e.preventDefault()"
  >
    <div class="thumb" />
  </div>
</template>

<script setup lang="ts">
...
function handleTouch(details: TouchPenDetails) {
  const { position } = details;
  console.log(`position : `, position);
}
</script>
...
```

現在在類比搖桿上拉動，應該會在 console 中出現如下圖訊息。

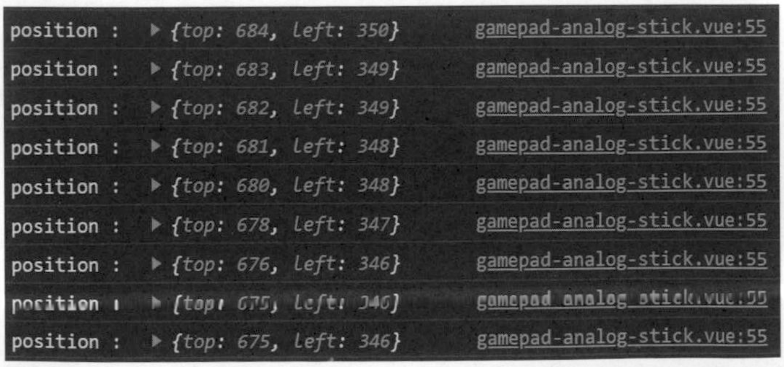

```
position :  ▶ {top: 684, Left: 350}          gamepad-analog-stick.vue:55
position :  ▶ {top: 683, Left: 349}          gamepad-analog-stick.vue:55
position :  ▶ {top: 682, Left: 349}          gamepad-analog-stick.vue:55
position :  ▶ {top: 681, Left: 348}          gamepad-analog-stick.vue:55
position :  ▶ {top: 680, Left: 348}          gamepad-analog-stick.vue:55
position :  ▶ {top: 678, Left: 347}          gamepad-analog-stick.vue:55
position :  ▶ {top: 676, Left: 346}          gamepad-analog-stick.vue:55
position :  ▶ {top: 675, Left: 346}          gamepad-analog-stick.vue:55
position :  ▶ {top: 675, Left: 346}          gamepad-analog-stick.vue:55
```

▲ 圖 10-20 Touch Pen 資料

Quasar 真方便。ヽ(≧∀≦)ゝ

現在來實作 thumb 隨著手指移動的功能。

由於 Quasar 取得之 Touch Position 之基於畫面最左上角為原點，所以我們必須先取得 pad 的位置與尺寸，才有辦法換算出 thumb 需要偏移的距離。

首先取得 pad 位置與尺寸，並計算出 pad 中心點的 top 與 left。

➜ src\components\gamepad-analog-stick.vue

```
<template>
  <div
    ref="pad"
    ...
  >
    ...
  </div>
</template>

<script setup lang="ts">
import { useElementSize } from '@vueuse/core';
...
const pad = ref<HTMLElement>();
const { width, height } = useElementSize(pad);

const padCenterPosition = computed(() => {
  const top = pad.value?.offsetTop ?? 0;
  const left = pad.value?.offsetLeft ?? 0;

  return {
    top: top + height.value / 2,
    left: left + width.value / 2,
  }
});
...
</script>
...
```

接著新增 thumb 相關資料變數：

- 儲存偏移量與是否活動

- 計算 CSS 樣式：利用 transform 產生偏移效果

➜ src\components\gamepad-analog-stick.vue

```
...
<script setup lang="ts">
...
const thumb = ref({
  offset: {
    x: 0,
    y: 0
  },
  active: false,
});
const thumbStyle = computed(() => ({
  transform: `translate(${thumb.value.offset.x}px, ${thumb.value.offset.y}px)`,
  opacity: thumb.value.active ? 0.8 : undefined,
}));
...
</script>
...
```

調整 handleTouch() 內容。

➜ src\components\gamepad-analog-stick.vue

```
...
<script setup lang="ts">
...
function handleTouch(details: TouchPenDetails) {
  const { position, isFirst, isFinal } = details;

  const x = position.left - padCenterPosition.value.left;
  const y = position.top - padCenterPosition.value.top;

  thumb.value.offset = {
```

```
    x, y
  };

  if (isFirst) {
    thumb.value.active = true;
  }

  if (isFinal) {
    thumb.value = {
      offset: { x: 0, y: 0 },
      active: false
    }
  }
}
</script>
...
```

最後把 thumbStyle 綁定至 template 中的 thumb 吧。

➜ src\components\gamepad-analog-stick.vue

```
<template>
  <div
    ...
  >
    <div
      class="thumb"
      :style="thumbStyle"
    />
  </div>
</template>
...
```

現在 thumb 會和手指位置一起跑了！(´▽`)╱

▲ 圖 10-21　thumb 隨鼠標拖動移動

但是有一個問題，thumb 跑出 pad 範圍啦。(´●ω●`)

讓我們限制一下 thumb 移動範圍，調整一下 handleTouch 內容。

➜ src\components\gamepad-analog-stick.vue

```
<script setup lang="ts">
import { clamp } from 'lodash-es';
...
function handleTouch(details: TouchPenDetails) {
  const { position, isFirst, isFinal } = details;

  const x = position.left - padCenterPosition.value.left;
  const y = position.top - padCenterPosition.value.top;

  /** 計算位移的長度 */
  const vectorLength = Math.sqrt(Math.pow(x, 2) + Math.pow(y, 2));

  /** 計算目前偏移最大值 */
  const xMax = (Math.abs(x) / vectorLength) * (width.value / 2);
  const yMax = (Math.abs(y) / vectorLength) * (height.value / 2);

  /** 利用 clamp 限制偏移數值在 -max 與 max 之間 */
  thumb.value.offset = {
    x: clamp(x, -xMax, xMax),
    y: clamp(y, -yMax, yMax),
```

```
  };

  if (isFirst) {
    thumb.value.active = true;
  }

  if (isFinal) {
    thumb.value = {
      offset: { x: 0, y: 0 },
      active: false
    }
  }
}
</script>
...
```

現在 thumb 會乖乖待在 pad 範圍內了。(´ ▽ `)╱

最後讓我們加點回彈動畫，增加細節。

➜ src\components\gamepad-analog-stick.vue

```
<template>
  <div
    ...
  >
    <div
      class="thumb"
      :class="{ 'active': thumb.active }"
      :style="thumbStyle"
    />
  </div>
</template>
...
<style scoped lang="sass">
...
.thumb
  ...
  transition-duration: 0.3s
```

```
  transition-timing-function: cubic-bezier(0.000, 1.650, 0.190, 1.005)
  &.active
    transition-duration: 0s
</style>
```

最後讓我們 emit 資料出去吧。

只要 offset 不為 0，則定時發送資料。

➜ src\components\gamepad-analog-stick.vue

```
...
<script setup lang="ts">
import { Vector2 } from '@babylonjs/core';
import { useElementSize, useIntervalFn } from '@vueuse/core';
...
function handleTouch(details: TouchPenDetails) {
  ...

  if (isFinal) {
    thumb.value = {
      offset: { x: 0, y: 0 },
      active: false
    }

    emit('trigger', { x: 0, y: 0 });
  }
}


useIntervalFn(() => {
  const { x, y } = thumb.value.offset;
  if (x === 0 && y === 0) return;

  /** 轉為單位向量，讓 x、y 的範圍介於 -1 至 1 之間 */
  const vector = new Vector2(x, y).normalize();
  emit('trigger', { x: vector.x, y: vector.y });
}, 50);
</script>
...
```

現在讓我們回到玩家搖桿畫面，來接收一下類比搖桿 emit 出來的資料。

新增 handleAnalogStickTrigger() 接收資料。

➜ src\views\the-player-gamepad-the-first-penguin.vue

```
<template>
  <div ... >
    <gamepad-analog-stick
      class="absolute bottom-5 left-8"
      @trigger="handleAnalogStickTrigger"
    />
    ...
  </div>
</template>

<script setup lang="ts">
...
function handleAnalogStickTrigger(data: { x: number, y: number }) {
  console.log(`[ handleAnalogStickTrigger ] : `, data);
}
</script>
...
```

現在拉動看看類比搖桿，會發現資料出現了！(·ω·)ノ ㄏ(·ω·)

```
[ handleAnalogStickTrigger ] :
▶ {x: 0.7369873557211645, y: -0.6759065301557057}
[ handleAnalogStickTrigger ] :
▶ {x: 0.7369873557211645, y: -0.6759065301557057}
[ handleAnalogStickTrigger ] :
▶ {x: 0.7369873557211645, y: -0.6759065301557057}
[ handleAnalogStickTrigger ] :    ▶ {x: 0, y: 0}
```

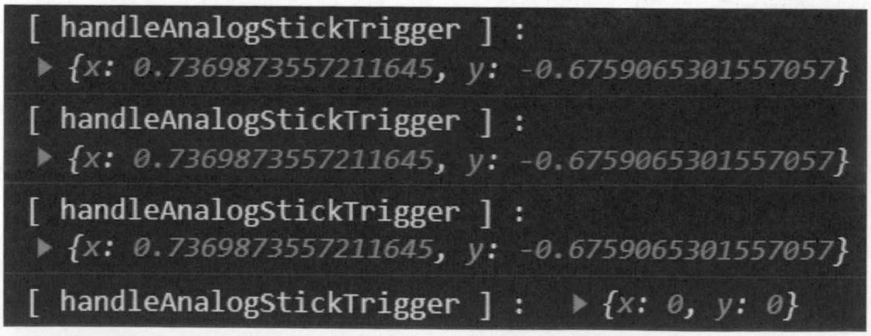

▲ 圖 10-22 類比搖桿資料

現在讓我們追加一下搖桿訊號的 KeyName 定義。

- x-axis：X 軸訊號

- y-axis：Y 軸訊號

- a：通用按鍵，在企鵝遊戲中是攻擊按鍵

➔ src\types\player.type.ts

```ts
/** 按鍵類型 */
export enum KeyName {
  ...
  A = 'a',
  X_AXIS = 'x-axis',
  Y_AXIS = 'y-axis',
}
...
```

讓我們加入將搖桿訊號發送至遊戲機的部分。

➔ src\views\the-player-gamepad-the-first-penguin.vue

```vue
...
<script setup lang="ts">
...
function handleAnalogStickTrigger(data: { x: number, y: number }) {
  console.log(`[ handleAnalogStickTrigger ] : `, data);

  emitGamepadData([
    {
      name: 'x-axis',
      value: data.x,
    },
    {
      name: 'y-axis',
      value: data.y,
    }
  ]);
}
</script>
```

...

可以發現只要加上 emitGamepadData 就完成了！ヽ(≧∀≦)ノ

最後將原本的確定按鈕改成 A 按鈕。

→ src\views\the-player-gamepad-the-first-penguin.vue

```
<template>
  <div
    ...
  >
    ...

    <gamepad-btn
      class="absolute bottom-10 right-20"
      size="6rem"
      @trigger="(status) => handleBtnTrigger('a', status)"
    >
      <div class="text-9xl mb-8">
        A
      </div>
    </gamepad-btn>

    ...
  </div>
</template>
...
```

目前我們成功完成企鵝遊戲的搖桿了，現在讓我們加入最重要的部分！ヽ(*´︶`*)ノ

回到最外層的玩家搖桿容器中，加入跳轉至企鵝遊戲搖桿的邏輯吧。

→ src\views\the-player-gamepad.vue

```
<script setup lang="ts">
...
function init() {
  ...
```

```
player.onGameConsoleStateUpdate((state) => {
  const { status, gameName } = state;
  ...
  if (status === 'lobby') {...}

  if (status !== 'playing') return;

  console.log(`[ onStateUpdate ] gameName : `, gameName);
  if (gameName === 'the-first-penguin') {
    router.push({
      name: RouteName.PLAYER_GAMEPAD_THE_FIRST_PENGUIN
    });
  }
});

player.requestGameConsoleState();
}
init();
</script>
```

現在切換搖桿功能也準備好了，讓我們把 App.vue 中的搖桿元件刪除，讓首頁變回原本的樣子，實際跑一遍流程看看吧！

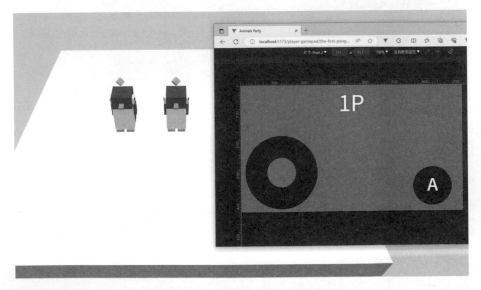

▲ 圖 10-23　跳轉至企鵝遊戲搖桿

可以發現進入遊戲後玩家搖桿畫面自動切換成類比搖桿了！◇*｡٩(´ㅇ`*)۶◇*｡

現在把企鵝遊戲場景中原本鍵盤控制事件刪除，改成直接取得玩家搖桿訊號吧。

➡ src\games\the-first-penguin\game-scene.vue

```ts
<script setup lang="ts">
...
import { useClientGameConsole } from '../../composables/use-client-game-console';
...
const gameConsole = useClientGameConsole();
...
function initGamepadEvent() {
  gameConsole.onGamepadData((data) => {
    console.log('[ gameConsole.onGamepadData ] data : ', data);
  });
}

async function init() {
  ...

  initGamepadEvent();

  /** 持續運行指定事件 */
  scene.registerAfterRender(() => {...});

  /** 反覆渲染場景，這樣畫面才會持續變化 */
  engine.runRenderLoop(() => {...});
}
...
</script>
```

現在遊戲機網頁可以收到來自玩家搖桿網頁的控制訊號了。

```
[ gameConsole.onGamepadData ] data :   ▶{pLayerId: 'pXBRgopMH00-sj345PGd5', keys: Array(2)}
[ gameConsole.onGamepadData ] data :   ▶{pLayerId: 'pXBRgopMH00-sj345PGd5', keys: Array(2)}
[ gameConsole.onGamepadData ] data :   ▼{pLayerId: 'pXBRgopMH00-sj345PGd5', keys: Array(2)} ⓘ
                                        ▼ keys: Array(2)
                                          ▶ 0: {name: 'x-axis', value: -0.3731607733308334}
                                          ▶ 1: {name: 'y-axis', value: -0.9277666933271179}
                                            length: 2
                                          ▶ [[Prototype]]: Array(0)
                                            playerId: "pXBRgopMH00-sj345PGd5"
                                          ▶ [[Prototype]]: Object
[ gameConsole.onGamepadData ] data :   ▶{pLayerId: 'pXBRgopMH00-sj345PGd5', keys: Array(2)}
```

▲ 圖 10-24 標題 Logo 完成

助教：「伺服器沒有新增搖桿資料定義沒問題嗎？(*´･д･)?」

鱈魚：「因為伺服器會轉送所有資料，就結果來說沒關係，不過當然是新增一下最好啦。(´„•ω•„) 」

到目前為止我們完成遊戲場景與搖桿了！✧*。٩(´ロ｀*)و✧*。

10.7　一人一隻才公平

此章節程式碼可以在以下連結取得。

- https://gitlab.com/deepmind1/animal-party/web/-/tree/feat/game-the-first-penguin-for-each-player

　　讓我們實際使用搖桿訊號控制器動作看看吧，先控制第一隻企鵝看看。

➜ src\games\the-first-penguin\game-scene.vue

```
<template>
  <canvas
    ref="canvas"
    class="outline-none"
  />
</template>

<script setup lang="ts">
...
import { curry } from 'lodash-es';
import { GamepadData, KeyName, SignalData } from '../../types';
...

function initGamepadEvent() {
  gameConsole.onGamepadData((data) => {
    console.log('[ gameConsole.onGamepadData ] data : ', data);

    const penguin = penguins[0];
    if (!penguin) return;

    ctrlPenguin(penguin, data);
  });
}

/** 根據 key 取得資料 */
const findSingleData = curry((keys: SignalData[], name: `${KeyName}`): SignalData |
undefined =>
  keys.find((key) => key.name === name)
);

/** 控制指定企鵝 */
function ctrlPenguin(penguin: Penguin, data: GamepadData) {
  const { keys } = data;
  const findData = findSingleData(keys);

  // 攻擊按鍵
```

```
  const attackData = findData('a');
  if (attackData) {
    penguin.attack();
    return;
  }

  // 移動按鍵
  const xData = findData('x-axis');
  const yData = findData('y-axis');

  const x = xData?.value ?? 0;
  const y = yData?.value ?? 0;

  if (x === 0 && y === 0) return;
  if (typeof x === 'number' && typeof y === 'number') {
    // 企鵝往上（正 Z 方向）是螢幕的 -Y 方向，所以要反轉
    penguin.walk(new Vector3(x, 0, -y));
  }
}

async function init() {...}
...
</script>
```

其中 curry 是 Functional Programing 的重要概念之一，利用 closure（閉包）特性，可以讓 function 形成一個鏈，變成可以依序傳入參數，最終完成運算。

個人覺得很適合用來處理資料，可以看到取得控制訊號的程式可以變得更為簡潔易讀。

 Tips

想要深入了解的讀者可以參考以下連結：

◆ Curry（柯里化）：https://jigsawye.gitbooks.io/mostly-adequate-guide/content/ch4.html

可以看到推動搖桿企鵝會往相同方向移動了！ヽ(•ω•)ノ

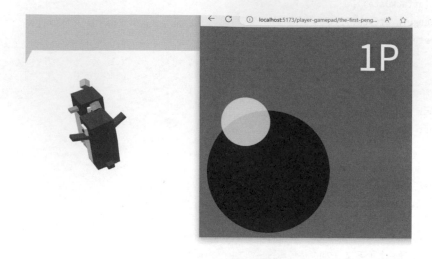

▲ 圖 10-25　搖桿順利控制企鵝

現在讓我們實際依照玩家資料建立對應的企鵝，一人一隻企鵝吧。(´▽`)/

首先我們需要一個可以依照玩家 ID 取得玩家代號的功能。

➜ src\composables\use-client-game-console.ts

```
...
export function useClientGameConsole() {
  ...
  function getPlayerCodeName(id: string) {
    const index = gameConsoleStore.players.findIndex(({ clientId }) =>
      clientId === id
    );

    if (index < 0) {
      return 'unknown ';
    }

    return `${index + 1}P`;
  }
```

```
  ...

  return {
    ...
    setGameName,
    getPlayerCodeName,
    ...
  }
}
```

接著調整 createPenguin() 內容，使用 id 建立不同顏色企鵝。

➔ src\games\the-first-penguin\game-scene.vue

```
...
<script setup lang="ts">
...
import { getPlayerColor } from '../../common/utils';
import { colors } from 'quasar';
const { getPaletteColor, textToRgb } = colors;
...
async function createPenguin(id: string, index: number, params: CreatePenguinParams) {
  /** 依照玩家 ID 取得對應顏色名稱並轉換成 rgb */
  const codeName = gameConsole.getPlayerCodeName(id);
  const color = getPlayerColor({ codeName });
  const hex = getPaletteColor(color);
  const rgb = textToRgb(hex);

  const penguin = await new Penguin(`penguin-${index}`, scene, {
    position: params.position,
    color: new Color3(rgb.r / 255, rgb.g / 255, rgb.b / 255),
    ownerId: id,
  }).init();

  return penguin;
}
...
</script>
```

然後个能讓每隻企鵝出現的位置都一樣，這樣會卡成一團企球。(°∀°)

我們讓企鵝們盡量排成矩形，設計一個計算矩型排列的 function 吧。

畫個圖規劃一下。

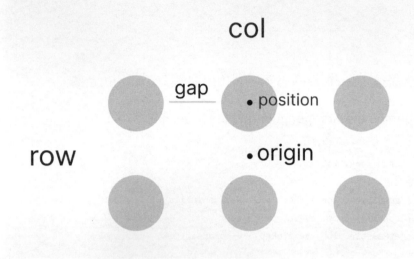

▲ 圖 10-26　矩形排列

- col 表示垂直列數量、row 表示水平欄數量。

- position 表示座標位置。

- gap 表示每個座標的間距。

- origin 表示整個矩形中心點。

別處也有可能用的到此 function，所以讓我們放在 utils 中。

➜ src\common\utils.ts

```
import { Animation, Color3, CubicEase, EasingFunction, QuarticEase, Vector3 } from
'@babylonjs/core';
import { defaults, flow, get, fill } from 'lodash-es';
...
/** 取得指定數量方形矩陣，可用於排列登場角色
 *
```

```
 * @param gap 間距
 * @param length 數量
 * @param origin 座標原點，預設 new Vector3(0, 0, 0)
 */
export function getSquareMatrixPositions(gap: number, length: number, origin = new
Vector3(0, 0, 0)) {
  /** 最大 col 為 length 開根號後無條件進位 */
  const maxCol = Math.ceil(Math.sqrt(length));
  /** 最大 row 為 length 除以 maxCol 後無條件進位 */
  const maxRow = Math.ceil(length / maxCol);

  /** flow 可以依序執行指定的 function */
  const result: Vector3[] = flow([
    /** 產生指定長度的 0 矩陣 */
    () => fill(Array(length), 0),

    /** 依序計算每個座標位置 */
    (array: number[]) => {
      return array.map((value, index) => {
        /** 根據 index 取得對應 col、row */
        const col = index % maxCol;
        const row = Math.floor(index / maxCol);

        return new Vector3(col * gap, 0, row * gap);
      });
    },

    /** 將目前中心點平移至目前原點 */
    (positions: Vector3[]) => {
      /** -1 是因為偏移量要從 0 開始算 */
      const currentCenter = new Vector3((maxCol - 1) * gap / 2, 0, (maxRow - 1) * gap
/ 2);
      return positions.map((position) => position.subtract(currentCenter));
    },

    /** 將目前中心點平移至指定 origin */
    (positions: Vector3[]) => positions.map((position) => position.add(origin)),
  ])();
```

```
    return result;
}
```

目前還沒有玩家資料，在 use-client-game-console 新增提供玩家資料。

➡ src\composables\use-client-game-console.ts

```
...
export function useClientGameConsole() {
  ...
  return {
    ...
    players: computed(() => gameConsoleStore.players),
  }
}
```

回到遊戲場景，調整建立企鵝的部分，改為依玩家資料建立企鵝。

➡ src\games\the-first-penguin\game-scene.vue

```
...
<script setup lang="ts">
import { getPlayerColor, getSquareMatrixPositions } from '../../common/utils';
...
async function init() {
  ...
  createIce(scene);

  const players = gameConsole.players.value;
  const positions = getSquareMatrixPositions(
    5, players.length, new Vector3(0, 10, 0)
  );
  const tasks = players.map(({ clientId }, index) =>
    createPenguin(clientId, index, {
      position: positions[index],
    })
  );
  const result = await Promise.allSettled(tasks)
  result.forEach((data) => {
    if (data.status !== 'fulfilled') return;
```

```
    penguins.push(data.value);
  });
  ...
}
...
</script>
```

　　現在讓我們想辦法變出兩個玩家，可以再開一個不同的瀏覽器，或是使用手機透過區網連線。

　　進入遊戲後，會發現每個玩家都有一隻企鵝了！一人一隻最公平！(ㅍ‿ㅍ)✧

▲ 圖 10-27 依照玩家資料產生企鵝

　　最後最重要的部分，就是讓玩家只能控制自己的企鵝！ˋ(•ω•)ˊ

　　調整 initGamepadEvent() 內容。

→ src\games\the-first-penguin\game-scene.vue

```
...
<script setup lang="ts">
...
function initGamepadEvent() {
  gameConsole.onGamepadData((data) => {
    const { playerId } = data;
```

```
    const penguin = penguins.find((penguin) => penguin.params.ownerId === playerId);
    if (!penguin) return;

    ctrlPenguin(penguin, data);
  });
}
...
</script>
```

現在每個玩家都能控制自己的企鵝了！✧*｡٩(´ ❂ ｀*)ﻭ✧*｡

讀者們可以用手機嘗試看看，至於手機要怎麼開啟此網頁呢？

還記得輸入「npm run dev」之後跑出來的訊息嗎？

▲ 圖 10-28　開啟 dev server

　　手機與電腦需要在同一個區域網路下（例如連接同一個 Wi-Fi），接著在手機上輸入「192.168.0.111:5173」就可以看到遊戲網頁了。

Tips

192.168.0.111 是作者的區網 IP，讀者們記得要輸入自己實際顯示的 IP 喔。

(10.8) 第一隻企鵝

此章節程式碼可以在以下連結取得。

- https://gitlab.com/deepmind1/animal-party/web/-/tree/feat/game-the-first-penguin-finish

現在讓我們依序完成所有的遊戲邏輯吧。(´▽`)ノ

首先是冰塊會逐漸縮小，讓戰況更加刺激，加入動畫即可達成效果。

在 createIce() 增加動畫內容。

➜ src\games\the-first-penguin\game-scene.vue

```
...
<script>
...
function createIce(scene: Scene) {
  const ice = MeshBuilder.CreateBox('ice', {
    width: 30,
    depth: 30,
    height: 4,
  });
  ice.material = new StandardMaterial('iceMaterial', scene);
  // mass 設為 0，就可以固定在原地不動
  ice.physicsImpostor = new PhysicsImpostor(ice, PhysicsImpostor.BoxImpostor,
    { mass: 0, friction: 0, restitution: 0 }, scene
  );
```

```
  /** 建立動畫 */
  const { animation, frameRate } = createAnimation(
    ice, 'scaling', new Vector3(0.1, 1, 0.1), {
    speedRatio: 0.05,
  });

  ice.animations.push(animation);
  scene.beginAnimation(ice, 0, frameRate);

  /** 物理碰撞也要隨著尺寸更新 */
  scene.registerBeforeRender(() => {
    ice.physicsImpostor?.setScalingUpdated();
  });

  return ice;
}
</script>
```

得到一塊逐漸縮小，看起來暖化很嚴重的浮冰！ヽ(•ω•)ﾉ

▲ 圖 10-29 浮冰逐漸縮小

接著讓我們偵測勝利企鵝的部分。

首先新增處理出界企鵝的 function。

➡ src\games\the-first-penguin\game-scene.vue

```
...
<script>
...
function handleCollideEvent(aPenguin: Penguin, bPenguin: Penguin) {...}

/** 處理出界的企鵝
 * y 軸低於 -3 判定為出界
 */
function detectOutOfBounds(penguins: Penguin[]) {
  penguins.forEach((penguin) => {
    if (!penguin.mesh) return;

    if (penguin.mesh.position.y < -3) {
      penguin.mesh.dispose();
    }
  });
}
</script>
```

接著新增偵測獲勝玩家的部分。

➡ src\games\the-first-penguin\game-scene.vue

```
...
<script>
...
/** 處理出界的企鵝
 * y 軸低於 -3 判定為出界
 */
function detectOutOfBounds(penguins: Penguin[]) {...}

const isGameOver = ref(false);
const winnerCodeName = ref('');
```

```
/** 偵測是否有贏家 */
function detectWinner(penguins: Penguin[]) {
  const alivePenguins = penguins.filter(({ mesh }) => !mesh?.isDisposed());

  if (alivePenguins.length !== 1) return;

  engine.stopRenderLoop();

  const winnerId = alivePenguins[0].params.ownerId;

  winnerCodeName.value = gameConsole.getPlayerCodeName(winnerId);
  isGameOver.value = true;
}
</script>
```

再來新增遊戲結束的 Dialog。

→　src\games\the-first-penguin\game-scene.vue

```
<template>
  <div class="overflow-hidden">
    <canvas
      ...
      class="outline-none w-full h-full"
    />

    <q-dialog
      v-model="isGameOver"
      persistent
    >
      <div class="card gap-14">
        <div class="flex items-center text-3xl text-gray-600">
          <q-icon name="emoji_events" />
          遊戲結束
        </div>
        <div class="text-3xl text-sky-700">
          玩家 {{ winnerCodeName }} 獲勝！
        </div>
```

```
        <div class="text-xl text-gray-400">
            按下 <q-icon name="done" /> 回到大廳
        </div>
      </div>
    </q-dialog>
  </div>
</template>
...
<style scoped lang="sass">
.card
  width: 30rem
  height: 24rem
  background: white
  border-radius: 2rem
  display: flex
  flex-direction: column
  justify-content: center
  align-items: center
</style>
```

最後讓每個 function 就定位吧。(´ ▽ `)∕

持續呼叫偵測用的 function 即可。

➜　src\games\the-first-penguin\game-scene.vue

```
...
<script>
...
async function init() {
  ...

  /** 持續運行指定事件 */
  scene.registerAfterRender(() => {
    detectCollideEvents(penguins);
    detectOutOfBounds(penguins);
    detectWinner(penguins);
  });
  ...
```

```
}
...
</script>
```

　　讓我們在搖桿上追加一下確認按鍵，用來結束遊戲。

➡ src\views\the-player-gamepad-the-first-penguin.vue

```
<template>
  <div ... >
    ...

    <gamepad-btn
      class="absolute top-10 right-20 opacity-90"
      size="3rem"
      icon="done"
      @trigger="(status) => handleBtnTrigger('confirm', status)"
    />

    <div class="code-name">
      {{ codeName }}
    </div>

    ...
  </div>
</template>
...
```

　　最後讓我們實作「遊戲結束後，按下確認回到大廳」的功能。

➡ src\games\the-first-penguin\game-scene.vue

```
<template>
  <div class="overflow-hidden">
    <canvas
      ref="canvas"
      class="outline-none"
    />

    <q-dialog
```

```
      v-model="isGameOver"
      persistent
    >
      <div class="card gap-14">
        <div class="flex items-center text-3xl text-gray-600">
          <q-icon name="emoji_events" />
          遊戲結束
        </div>
        <div class="text-3xl text-sky-700">
          玩家 {{ winnerCodeName }} 獲勝！
        </div>

        <div class="text-xl text-gray-400">
          按下 <q-icon name="done" /> 回到大廳
        </div>
      </div>
    </q-dialog>
  </div>
</template>

<script setup lang="ts">
...
import  { RouteName } from '../../router/router';
...
import { useRouter } from 'vue-router';
import { useLoading } from '../../composables/use-loading';
...
/** 控制指定企鵝 */
function ctrlPenguin(penguin: Penguin, data: GamepadData) {
  ...
  // 確認按鍵
  const confirmData = findData('confirm');
  if (confirmData && isGameOver.value) {
    backToLobby();
    return;
  }
  ...
}
```

```
async function backToLobby() {
  isGameOver.value = false;

  await loading.show();
  router.push({
    name: RouteName.GAME_CONSOLE_LOBBY
  });
}
...
</script>
...
```

至此我們完成企鵝遊戲的所有邏輯了，現在各位讀者可以聚集小夥伴，一起互相傷害了！✧*。٩(ˊ□ˋ*)و✧*。

10.9 完成大廳按鈕功能

遊戲狂歡後，我們現在回到大廳了。

最後讓我們完成大廳的「結束派對」按鈕吧。

在 use-client-game-console 新增結束派對的功能。

➜ src\composables\use-client-game-console.ts

```
...
export function useClientGameConsole() {
  ...
  async function startParty() {...}
  function endParty() {
    mainStore.client?.disconnect();
}
  ...

  return {
    /** 開始派對
```

```
     *
     * 建立連線，並回傳房間資料
     */
    startParty,

    endParty,

    ...
  }
}
```

現在回到選單元件，實現「結束派對」按鈕功能。

➡ src\components\game-menu.vue

```html
<template>
  <div class="flex">
    ...
    <div ...>
      ...
      <base-btn
        ...
        label=" 結束派對 "
        ...
        @click="endParty()"
      >
        ...
      </base-btn>
    </div>
  </div>
</template>

<script setup lang="ts">
...
const endParty = debounce(async () => {
  gameConsole.setStatus('home');
  await loading.show();
  gameConsole.endParty();

  router.push({
```

```
    name: RouteName.HOME
  });
}, 3000, {
  leading: true,
  trailing: false,
});
...
</script>
...
```

　　最後在首頁隱藏 loading 畫面，我們就大功告成了。ヽ(≧∀≦)ノ

➜ src\views\the-home.vue

```
...
<script setup lang="ts">
import { onMounted } from 'vue';
...
async function startParty() {
  ...
  if (err) {
    ...
    loading.hide();
    return;
  }
  ...
}
...

onMounted(() => {
  loading.hide();
});
</script>
...
```

　　現在選擇「結束派對」，不只遊戲機網頁會回到首頁，所有的玩家的搖桿頁面也會自動跳回至首頁了。

小雞快飛

現在讓我們進入下一個遊戲吧！ヾ(๑'◡'๑)ﾉ

截至目前為止，我們幾乎已經完成所有的基礎建設，例如連線至 server、搖桿跳轉等等，接下來的章節我們大部分段落都會專注在遊戲開發了！(´▽`)ﾉ

一群農場的雞利用大砲從農場逃跑了！先別管大砲怎麼來，玩家們可以利用自己的手機陀螺儀控制雞的飛行姿態，努力閃過空中的障礙物，讓小雞們逃出農場的魔掌吧！

就如同企鵝島一般，第一步先讓我們來打造遊戲選單用的小島吧。

11.1 建立小島

此章節程式碼可以在以下連結取得。

- https://gitlab.com/deepmind1/animal-party/web/-/tree/feat/game-chicken-fly-island

目前計畫雖然說是小島，但是實際上不會有島，而是小雞在天空繞圈圈。

讓我們在 game-menu-background 加入小島吧。

首先在遊戲名稱列舉新增小雞快飛。

➡ src\types\game.type.ts

```
...
/** 遊戲名稱 */
export enum GameName {
  ...
  CHICKEN_FLY = 'chicken-fly',
}
...
```

接著在 src\components\game-menu 資料夾中新增 chicken-fly 資料夾，並新增 index.ts 檔案。

➡ src\components\game-menu\chicken-fly\index.ts

```
import {
  Scene, TransformNode
} from '@babylonjs/core';
import { ModelIsland } from '../../../types';
import { ref } from 'vue';

export type ChickenFlyIsland = ModelIsland;

export async function createChickenFlyIsland(scene: Scene): Promise<ChickenFlyIsland> {
  const rootNode = new TransformNode('chicken-fly-island');

  const active = ref(false);
  function setActive(value: boolean) {
    active.value = value;
  }

  return {
    name: 'chicken-fly',
    rootNode,
    setActive,
```

```
    }
}
```

同時我們整理一下檔案目錄，把原本的 src\components\game-menu\the-first-penguin.ts 改名成 index.ts 並移動至 src\components\game-menu\the-first-penguin 資料夾中，這樣看起來更整齊了。

▲ 圖 11-1　調整目錄結構

讓我們慢慢完成小雞島的內容吧，先把 game-menu 放到 App.vue 中，方便開發。

➡ src\App.vue

```
<template>
  ...

  <game-menu class=" absolute w-full h-full" />
</template>

<script setup lang="ts">
import LoadingOverlay from './components/loading-overlay.vue';
import GameMenu from './components/game-menu.vue';
</script>
...
```

接著我們讓進場鏡頭目標改為小雞島，方便開發。

首先新增 the-game-console-chicken-fly，作為小雞遊戲頁面容器。

➜ src\views\the-game-console-chicken-fly.vue

```
<template></template>

<script setup lang="ts">
import { ref } from 'vue';

interface Props {
  label?: string;
}
const props = withDefaults(defineProps<Props>(), {
  label: '',
});

const emit = defineEmits<{
  (e: 'update:modelValue', value: string): void;
}>();
</script>

<style scoped lang="sass">
</style>
```

在 router 加入此頁面。

➜ src\router\router.ts

```
...
export enum RouteName {
  ...
  GAME_CONSOLE_CHICKEN_FLY = 'game-console-chicken-fly',
  ...
}

const routes: RouteRecordRaw[] = [
  ...
  {
    path: `/game-console`,
    ...
    children: [
      ...
      {
```

```
      path: `chicken-fly`,
      name: RouteName.GAME_CONSOLE_CHICKEN_FLY,
      component: () => import('../views/the-game-console-chicken-fly.vue')
    },
  ]
},
...
]
...
```

接著新增小雞島的鏡頭位置並將速度調快，跳過進場動畫。

➡ src\components\game-menu-background.vue

```
...
<script setup lang="ts">
...
const shots: Shot[] = [
  ...
  {
    name: 'chicken-fly',
    camera: {
      target: new Vector3(0.63, 14.81, 29.54),
      alpha: 0.247,
      beta: 1.238,
      radius: 30.8628,
    }
  },
];
...
async function moveCamera(first = false) {
  ...
  // const speed = first ? 0.15 : 0.5;
  const speed = 10;
  ...
}
...
</script>
```

最後把預設遊戲先改為 chicken-fly。

➜ src\components\game-menu.vue

```
...
<script setup lang="ts">
...
const games: GameInfo[] = [
  ...
  {
    name: 'chicken-fly',
    routeName: 'game-console-chicken-fly',
  },
];


...
const currentIndex = ref(1);
...
</script>
```

Tips

currentIndex 設為 1 是因為 index 從 0 開始算，所以這裡的 1 表示第二個遊戲。

現在畫面最終會變成這樣。

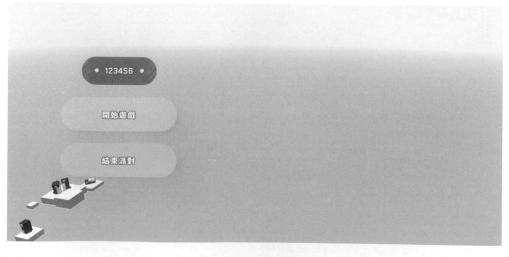

▲ 圖 11-2 調整進場鏡頭

現在鏡頭喬好了，讓我們來把小雞島做出來吧！

首先新增 public\chicken-fly 資料夾，前往 GitLab 同一個目錄下載 3D 模型並放入。

Tips

讀者可以前往 GitLab 下載模型：

- https://gitlab.com/deepmind1/animal-party/web/-/tree/feat/game-chicken-fly-island/public/chicken-fly

接著在 game-menu-background 加入小雞島吧。

➜ src\components\game-menu-background.vue

```
...
<script setup lang="ts">
...
import { createChickenFlyIsland } from './game-menu/chicken-fly';
...

/** 將所有島嶼儲存至此 */
let islands: ModelIsland[] = [];
async function createIslands(scene: Scene) {
  const results = await Promise.allSettled([
    createPenguinIsland(scene),
    createChickenFlyIsland(scene),
  ]);

  results.forEach((result) => {
    if (result.status !== 'fulfilled') return;
    islands.push(result.value);
  });
}
...
</script>
```

第 11 章　小雞快飛

不過由於小雞島目前甚麼都沒有，連個雞毛都看不到，讓我們放一個方塊看看。

同時加入 position 參數，用來指定島嶼的初始位置。

➡ src\components\game-menu\chicken-fly\index.ts

```ts
import {
  MeshBuilder,
  Scene, StandardMaterial, TransformNode, Vector3
} from '@babylonjs/core';
...
import { defaults } from 'lodash-es';
...
interface Param {
  position?: Vector3;
}
const defaultParam: Required<Param> = {
  position: new Vector3(0, 0, 0),
}

export async function createChickenFlyIsland(scene: Scene, param: Param):
 Promise<ChickenFlyIsland> {
  const { position } = defaults(param, defaultParam);

  ...

  const mesh = MeshBuilder.CreateBox('test');
  mesh.setParent(rootNode);

  rootNode.position = position;
  ...
}
```

建立小雞島追加初始位置。

➡ src\components\game-menu-background.vue

```vue
...
<script setup lang="ts">
```

11-8

```
...
async function createIslands(scene: Scene) {
  const results = await Promise.allSettled([
    createPenguinIsland(scene),
    createChickenFlyIsland(scene, {
      position: new Vector3(0.63, 14.81, 29.54),
    }),
  ]);

  results.forEach((result) => {
    if (result.status !== 'fulfilled') return;
    islands.push(result.value);
  });
}
...
</script>
```

現在會看到一個方塊出現在畫面中央！(ㅍ◞ㅍ)✧

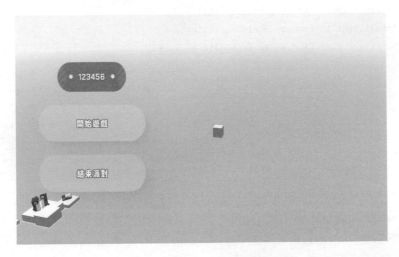

▲ 圖 11-3 小雞島方塊

助教：「方塊陰影怎麼那麼單調，功力不足？(°c_,°)」

鱈魚：「才不是勒，是因為預設光源的關係。(๑•з•๑)」

助教：「這樣牽拖不好喔。(°A,°`)」

鱈魚：「真的是因為光源啦。Σ(´д`;)」

這是因為預設光源的照射方向是從正上方垂直照下來，就像是正中午的太陽一樣。

現在讓我們調整一下光源的方向，從場景光源清單取出預設光源並調整方向。

➜ src\components\game-menu-background.vue

```
...
<script setup lang="ts">
...
function createScene(engine: Engine) {
  const scene = new Scene(engine);
  /** 使用預設光源 */
  scene.createDefaultLight();
  const defaultLight = scene.lights.at(-1);
  if (defaultLight instanceof HemisphericLight) {
    defaultLight.direction = new Vector3(0.5, 1, 0);
  }

  ...
}
...
</script>
```

預設光源型別是 HemisphericLight，而 scene.lights 矩陣內的型別是 Light。雖然 HemisphericLight 繼承於 Light，但是並非 Light 都有方向性（direction），所以我們需要用 instanceof 判斷 defaultLight 是不是 HemisphericLight，如果是 HemisphericLight 才可以設定其 direction。

Tips

想要深入了解的讀者可以參考以下連結：

◆ Introduction To Lights ：https://doc.babylonjs.com/features/featuresDeepDive/
lights/lights_introduction

現在方塊的陰影不單調了。ᕙ(*´︶`*)ᕗ

▲ 圖 11-4 調整預設光源

目前預期做一個繞著圓心繞圈飛行的雞，讓我們新增雞的檔案。

➜ src\components\game-menu\chicken-fly\chicken.ts

```
import { Vector3, Scene, SceneLoader, AnimationGroup, Color3 } from "@babylonjs/
core";
import { defaults } from "lodash-es";

interface Param {
  scaling?: number;
  position?: Vector3;
  rotation?: Vector3;
}

const defaultParam: Required<Param> = {
  scaling: 1,
  position: Vector3.Zero(),
  rotation: Vector3.Zero(),
}
```

```
export async function createChicken(name: string, scene: Scene, param?: Param) {
  const { scaling, position, rotation } = defaults(param, defaultParam);

  const result = await SceneLoader.ImportMeshAsync('', '/chicken-fly/', 'flying-chicken.
glb', scene);

  const mesh = result.meshes[0];
  mesh.name = name;
  mesh.position = position;
  mesh.rotation = rotation;
  mesh.scaling = new Vector3(scaling, scaling, scaling);

  return { mesh }
}
```

接著在小雞島中引用 createChicken，實際建立一隻雞吧。

➜ src\components\game-menu\chicken-fly\index.ts

```
...
import { createChicken } from './chicken';
...
export async function createChickenFlyIsland(scene: Scene, param: Param):
 Promise<ChickenFlyIsland> {
  ...
  async function createChickens() {
    /**
     * 從 createChicken 函數獲取第三個參數的類型（索引為 2，因為索引從 0 開始）
     */
    const list: Parameters<typeof createChicken>[2][] = [
      {
        scaling: 1,
        position: new Vector3(0, 0, 0),
        rotation: new Vector3(0, 0, 0),
      },
    ];

    /** 建立 createChicken 的 Promise 矩陣 */
    const tasks = list.map((option, i) =>
      createChicken(`chicken-${i}`, scene, option)
```

```
  );
  /** 當所有的 Promise 都完成時，回傳所有 Promise 的結果。 */
  const results = await Promise.allSettled(tasks);

  /** 將成功建立的物體綁定至 rootNode */
  results.forEach((result) => {
    if (result.status === 'rejected') return;
    result.value.mesh.setParent(rootNode);
  });
  }
  await createChickens();
  ...
}
```

飛雞出現了！ᕕ(゜∀゜)ᕗ

▲ 圖 11-5　小雞登場

助教：「翅膀這麼短還能飛喔。 ╭(°A,°)╮ 」

鱈魚：「是魔法，我用了魔法。 ᕙ(∀`ᕙ) 」

　　要讓小雞繞著某個物體旋轉很容易，使用我們已經很熟悉的 TransformNode 就可以實現了，將小雞附加在 TransformNode 上後，旋轉 TransformNode，就可以產生小雞繞的圓心旋轉的效果。

所以讓我們新增 TransformNode。

➡ src\components\game-menu\chicken-fly\chicken.ts

```
import { Vector3, Scene, SceneLoader, AnimationGroup, Color3, TransformNode } from
"@babylonjs/core";
...

export async function createChicken(name: string, scene: Scene, param?: Param) {
  const { scaling, position, rotation } = defaults(param, defaultParam);

  const rootNode = new TransformNode('chicken', scene);

  const result = await SceneLoader.ImportMeshAsync('', '/chicken-fly/', 'flying-chicken.
glb', scene);

  const mesh = result.meshes[0];
  mesh.name = name;
  mesh.scaling = new Vector3(scaling, scaling, scaling);
  mesh.setParent(rootNode);

  rootNode.position = position;
  rootNode.rotation = rotation;

  return { mesh }
}
```

這個時候會發現一個問題，小雞 mesh 的 parent 已經設為目前新增的 rootNode 上了，這樣外面島嶼 rootNode 怎麼辦呢？

很簡單，設定在小雞內的 rootNode 上即可，所以這裡讓我們改寫一下，不要 return mesh。

➡ src\components\game-menu\chicken-fly\chicken.ts

```
...
export async function createChicken(name: string, scene: Scene, param?: Param) {
  ...
  return {
```

```
        setParent: rootNode.setParent.bind(rootNode),
    }
}
```

 Tips

這裡大家可能會有個疑問，為甚麼不要寫成：

setParent: rootNode.setParent,

這樣就好了呢？

這是因為 TransformNode 的 setParent 函數內部依賴於 this 用以訪問自身物件中的屬性。如果直接寫成 setParent: rootNode.setParent，會使 setParent 內的 this 不再指向原來的 TransformNode 對象的 this，導致結果異常，而 Function. bind 可以防止這件事情發生。

想深入了解的讀者可以參考以下連結：

◆ https://ithelp.ithome.com.tw/articles/10302040

　　直接提供 setParent 這個 method 是最好的方法，意思是外層的島嶼不用管這個小雞物件究竟是誰需要 parent，反正小雞物件只要提供 setParent 這個方法，外層島嶼負責呼叫就行。

　　因為綁定方式改變了，島嶼建立小雞的程式也要跟著改寫。

➜ src\components\game-menu\chicken-fly\index.ts

```
...
export async function createChickenFlyIsland(scene: Scene, param: Param):
Promise<ChickenFlyIsland> {
    ...

    async function createChickens() {
        ...
        results.forEach((result) => {
            ...
```

```
      result.value.setParent(rootNode);
    });
  }
  ...
}
```

目前小雞是這樣。

▲ 圖 11-6 加入 TransformNode

現在讓我們把小雞與 TransformNode 加入點偏移吧，新增 radius 參數。

➜ src\components\game-menu\chicken-fly\chicken.ts

```
...
interface Param {
  ...
  radius?: number;
}

const defaultParam: Required<Param> = {
  ...
  radius: 5,
```

```
}

export async function createChicken(name: string, scene: Scene, param?: Param) {
  ...
  mesh.name = name;
  mesh.position = new Vector3(radius, 0, 0);
  ...
}
```

可以看到小雞與方塊有了距離。

▲ 圖 11-7　加入旋轉半徑

將小雞轉個方向，並讓距離遠一點。

➜ src\components\game-menu\chicken-fly\chicken.ts

```
...
const defaultParam: Required<Param> = {
  ...
  radius: 10,
}

export async function createChicken(name: string, scene: Scene, param?: Param) {
  ...
```

```
    mesh.position = new Vector3(radius, 0, 0);
    mesh.rotation = new Vector3(0, Tools.ToRadians(90), 0);
    ...
}
```

▲ 圖 11-8　旋轉小雞方向

 Tips

Tools.ToRadians 可以直接將角度轉換成徑度，

　　讓小雞開始旋轉吧。

　　可以指定一開始小雞旋轉平面的角度，接著小雞會繞著旋轉後的平面之 Y 軸進行旋轉，這個部分稍微複雜一點，需要使用 Quaternion。

Tips

想要深入了解的讀者可以參考以下連結：

◆ Rotation Quaternions：https://doc.babylonjs.com/features/featuresDeepDive/
mesh/transforms/center_origin/rotation_quaternions

鱈魚：「這次的動畫需真要 3 幀，也就是起始、中間與終點狀態。」

助教：「是不是因為前面灌水，所以現在要努力一點？('◉ ⌣ ◉`)」

鱈魚：「才不是哩，不要亂挑撥讀者的信任啊。Σ(´Д`;)」

這是因為轉一圈需要從 0 度轉到 360 度，但是在 Quaternion 中一次沒辦法旋轉超過 180 度，超過 180 度的話會變成其輻角，也就是說正轉 190 度會變成逆轉 170 度。

所以這裡我們需要轉 2 次 180 度，從 0 到 180，再從 180 到 360，總共 3 幀。

→ src\components\game-menu\chicken-fly\chicken.ts

```ts
import {
  Vector3, Scene, SceneLoader,
  AnimationGroup, Color3,
  TransformNode, BezierCurveEase,
  Animation, Quaternion, Axis, Tools,
} from "@babylonjs/core";
...
export async function createChicken(name: string, scene: Scene, param?: Param) {
  ...

  rootNode.position = position;

  function initRootNodeAnimation() {
    /** 計算起點於終點
     *
     ↵ 旋轉後的 Y 軸，以原點為中心，繞 Y 軸旋轉
     */
```

```
    const fromValue = rotation.toQuaternion();
    const midValue = fromValue.multiply(
      Quaternion.RotationAxis(Axis.Y, Tools.ToRadians(180))
    );
    const toValue = midValue.multiply(
      Quaternion.RotationAxis(Axis.Y, Tools.ToRadians(180))
    );

    const frameRate = 60;
    const animation = new Animation('circular-motion', 'rotationQuaternion',
frameRate,
      Animation.ANIMATIONTYPE_QUATERNION,
      Animation.ANIMATIONLOOPMODE_CYCLE
    );

    const keys = [
      {
        frame: 0, value: fromValue,
      },
      {
        frame: frameRate / 2, value: midValue,
      },
      {
        frame: frameRate, value: toValue,
      },
    ];
    animation.setKeys(keys);
    rootNode.animations = [animation];

    // 開始動畫
    scene.beginAnimation(rootNode, 0, frameRate, true, 0.3);
  }
  initRootNodeAnimation();

  ...
}
```

沒有意外的話，會看到小雞開始繞著方塊飛了！ ༼ つ◕‿◕ つ ༽

▲ 圖 11-9　小雞圓周運動

現在讓我們回到島嶼物件，產生多一點小雞吧。

我們讓每隻小雞錯開，實現方式就是讓每隻小雞的 Y 軸都轉一點角度。

➜ src\components\game-menu\chicken-fly\index.ts

```
import {
  MeshBuilder,
  Scene, StandardMaterial, Tools, TransformNode, Vector3
} from '@babylonjs/core';
import { defaults, range } from 'lodash-es';
...
export async function createChickenFlyIsland(scene: Scene, param: Param):
Promise<ChickenFlyIsland> {
  ...

  async function createChickens() {
    /**
     * 從 createChicken 函數獲取第三個參數的類型（索引為 2，因為索引從 0 開始）
     *
     * 從 0 到 360 度取 6 等分，產生 6 隻小雞
     */
    const divisions = 6;
```

```
const list: Parameters<typeof createChicken>[2][] =
  range(0, divisions)
    .map((n) => n * (360 / divisions))
    .map((degree) => ({
      rotation: new Vector3(0, Tools.ToRadians(degree), 0),
    }));

  ...
}
...
}
```

得到一個小雞飛行隊！∠(ᐛ 」∠)_

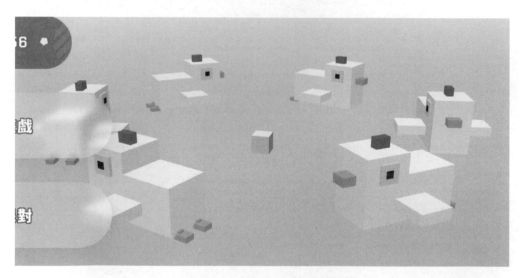

▲ 圖 11-10　產生多隻小雞

讓小雞轉起來吧！

加入 **delay** 參數，讓每隻小雞的旋轉錯開。

➔ src\components\game-menu\chicken-fly\chicken.ts

```
import {
  Vector3, Scene, SceneLoader,
```

```
  AnimationGroup, Color3,
  TransformNode, BezierCurveEase,
  Animation, Quaternion, Axis, Tools,
} from "@babylonjs/core";
import { defaults } from "lodash-es";
import { createAnimation } from "../../../common/utils";

interface Param {
  ...
  delay?: number;
}

const defaultParam: Required<Param> = {
  ...
  delay: 0,
}

export async function createChicken(name: string, scene: Scene, param?: Param) {
  const {
    scaling, position, rotation, radius, delay
  } = defaults(param, defaultParam);

  ...

  function initChickenAnimation() {
    const { animation, frameRate } = createAnimation(mesh, 'rotation',
      new Vector3(Tools.ToRadians(-360), Tools.ToRadians(90), 0), {
      // easingFunction: new BezierCurveEase(),
    });
    animation.loopMode = Animation.ANIMATIONLOOPMODE_CYCLE;
    scene.beginDirectAnimation(mesh, [animation], 0, frameRate, true, 0.6);
  }
  setTimeout(initChickenAnimation, delay);

  ...
}
```

加入每隻小雞的延遲時間。

→ src\components\game-menu\chicken-fly\index.ts

```
...
export async function createChickenFlyIsland(scene: Scene, param: Param):
Promise<ChickenFlyIsland> {
  ...
  async function createChickens() {
    ...
    const divisions = 6;
    const list: Parameters<typeof createChicken>[2][] =
      range(0, divisions)
        .map((n) => n * (360 / divisions))
        .map((degree, i) => ({
          rotation: new Vector3(0, Tools.ToRadians(degree), 0),
          delay: i * 600,
        }));

    ...
  }
  ...
}
```

現在會看到每隻小雞開始旋轉了。ᕕ(ﾟ∀ﾟ)ᕗ

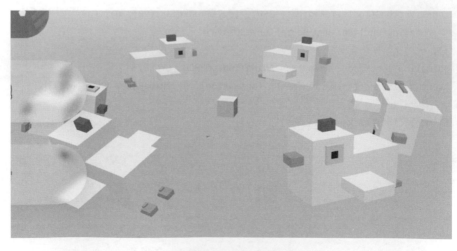

▲ 圖 11-11 旋轉小雞

現在讓我們把標題文字放進來吧。

前往 GitLab 中的 public\chicken-fly\title.glb 位置下載標題模型，並放置專案資料夾相同目錄位置。

接著與企鵝島的標題概念相同，新增 title.ts。

➔ src\components\game-menu\chicken-fly\title.ts

```
import {
  Scene, Vector3, SceneLoader,
  StandardMaterial, Color3, TransformNode,
} from "@babylonjs/core";
import { defaults } from "lodash-es";

interface Option {
  scaling?: number;
  position?: Vector3;
  rotation?: Vector3;
}

const defaultOption: Required<Option> = {
  scaling: 1,
  position: Vector3.Zero(),
  rotation: Vector3.Zero(),
}

export async function createTitle(name: string, scene: Scene, option?: Option) {
  const {
    scaling, position, rotation
  } = defaults(option, defaultOption);

  const rootNode = new TransformNode('chicken-fly-island-title');
  const result = await SceneLoader.ImportMeshAsync('', '/chicken-fly/', 'title.glb',
scene);

  /** 這次的模型已經內建材質，所以取 root 就行 */
  const mesh = result.meshes[0];
  mesh.name = name;
  mesh.position = position;
```

```
  mesh.rotation = rotation;
  mesh.scaling = new Vector3(scaling, scaling, scaling);
  mesh.parent = rootNode;

  function initMaterial() {
    const material = new StandardMaterial('chicken-fly-title-material', scene);
    material.diffuseColor = Color3.White();
    material.emissiveColor = Color3.FromHexString('#101519');
    material.specularColor = Color3.Black();
    mesh.material = material;
    return material;
  }
  initMaterial();

  return {
    rootNode,
    setParent: rootNode.setParent.bind(rootNode),
  }
}
```

來建立標題文字吧，同時刪除觀察用的方塊。

➡ src\components\game-menu\chicken-fly\title.ts

```
...
import { createTitle } from './title';
...

export async function createChickenFlyIsland(scene: Scene, param: Param):
Promise<ChickenFlyIsland> {
  ...
  async function createText() {
    const title = await createTitle('chicken-fly-title', scene);
    title.setParent(rootNode);
  }
  await createText();
  ...
}
```

標題文字出現了。

▲ 圖 11-12 小雞快飛標題

現在讓文字就定位。

➔ src\components\game-menu\chicken-fly\index.ts

```
...
export async function createChickenFlyIsland(scene: Scene, param: Param):
Promise<ChickenFlyIsland> {
  ...
  async function createText() {
    const title = await createTitle('chicken-fly-title', scene, {
      rotation: new Vector3(Tools.ToRadians(118.2), Tools.ToRadians(273.6), 0),
      scaling: 3,
    });
    title.setParent(rootNode);
  }
  await createText();
  ...
}
```

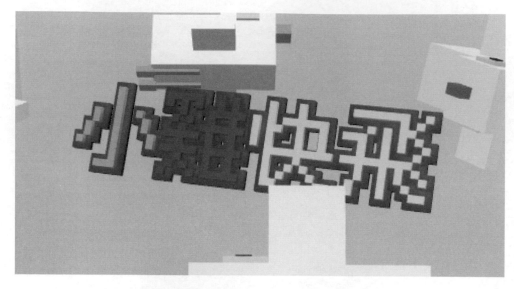

▲ 圖 11-13- 小雞快飛標題就位

幫標題加個漂浮動畫。

➜ src\components\game-menu\chicken-fly\title.ts

```
import {
  Scene, Vector3, SceneLoader,
  StandardMaterial, Color3, TransformNode,
  Animation,
  SineEase,
  EasingFunction,
} from "@babylonjs/core";
...
export async function createTitle(name: string, scene: Scene, option?: Option) {
  ...

  // 建立動畫
  function initAnimation() {
    const frameRate = 20;
    const floatAnimation = new Animation(
      'float', 'position', frameRate / 3,
      Animation.ANIMATIONTYPE_VECTOR3,
      Animation.ANIMATIONLOOPMODE_CYCLE
```

```
  );

  const offset = 0.2;
  floatAnimation.setKeys([
    {
      frame: 0,
      value: position.add(new Vector3(0, -offset, 0)),
    },
    {
      frame: frameRate / 2,
      value: position.add(new Vector3(0, offset, 0)),
    },
    {
      frame: frameRate,
      value: position.add(new Vector3(0, -offset, 0)),
    },
  ]);
  mesh.animations.push(floatAnimation);

  const easingFunction = new SineEase();
  easingFunction.setEasingMode(EasingFunction.EASINGMODE_EASEINOUT);
  floatAnimation.setEasingFunction(easingFunction);

  scene.beginAnimation(mesh, 0, frameRate, true);
  }
  initAnimation();
  ...
}
```

最後調整一下小島和鏡頭位置，不要和選單重疊。

→ src\components\game-menu-background.vue

```
...
<script setup lang="ts">
...
const shots: Shot[] = [
  ...
  {
    name: 'chicken-fly',
```

```
    camera: {
      target: new Vector3(0.9, 14, 29.6),
      alpha: 0.5,
      beta: 1.2,
      radius: 31,
    }
  },
];
...
async function createIslands(scene: Scene) {
  const results = await Promise.allSettled([
    ...
    createChickenFlyIsland(scene, {
      position: new Vector3(-2, 16, 37),
    }),
  ]);
  ...
}
...
</script>
```

▲ 圖 11-14 小雞島完成

鱈魚：「現在讓我們試試切換遊戲吧！ ℘ (´∀`℘)」

助教：「島嶼的 active 狀態效果呢？(´„•ω•„)」

鱈魚：「 ℘ (⊙ ↝ ⊙'℘) ...」

助教：「 ? (´„•ω•„)」

鱈魚：「那個部份我們就讓讀者們自由發揮吧！ ℘('ε'℘)」

助教：「還有這樣的嗎？ ⌐(°A,°)¬」

鱈魚：「這是為了鼓勵創作好嗎，絕對不是因為我就沒點子。∠(ᐛ 」∠)_ 」

我們先在 App.vue 綁定數字鍵盤 4 和 6，測試看看 game-menu 切換遊戲效果。

我們首先將 game-menu 的 prevGame 和 nextGame 都用 throttle 處理一下，以免連續切換。

➜ src\components\game-menu.vue

```
...
<script setup lang="ts">
...
import { debounce, throttle } from 'lodash-es';
...
const prevGame = throttle(() => {
  currentIndex.value--;
  if (currentIndex.value < 0) {
    currentIndex.value += games.length;
  }
}, 2000, {
  leading: true,
  trailing: false,
});
const nextGame = throttle(() => {
  currentIndex.value++;
```

```
  currentIndex.value %= games.length;
}, 2000, {
  leading: true,
  trailing: false,
});
...
</script>
...
```

Tips

throttle 可以將連續觸發改為指定間隔觸發，例如原本按按鈕的速度是每秒 5 次，使用 throttle 設定為每秒 1 次後，就可以確保按鈕每秒只觸發一次，可以有效避免使用者按太快導致異常。

想要深入了解的讀者可以參考以下連結：

◆ https://ithelp.ithome.com.tw/articles/10222749

接著在 App.vue 取得 game-menu 元件 ref。

➜ src\App.vue

```
<template>
  ...
  <game-menu
    ref="menu"
    class=" absolute w-full h-full"
  />
</template>

<script setup lang="ts">
import { ref } from 'vue';

import LoadingOverlay from './components/loading-overlay.vue';
import GameMenu from './components/game-menu.vue';

const menu = ref<InstanceType<typeof GameMenu>>();
```

```
</script>
...
```

　　註冊鍵盤事件並呼叫 prevGame 和 nextGame。

➔ src\App.vue

```
...
<script setup lang="ts">
...
window.addEventListener('keydown', (event) => {
  switch (event.key) {
    case '4': {
      menu.value?.prevGame();
      break;
    }

    case '6': {
      menu.value?.nextGame();
      break;
    }
  }
})
</script>
...
```

　　鱈魚：「然後讀者們會發現完全沒有反應！ ᕕ(ﾟ∀ﾟ)ᕗ」

　　助教：「你來亂的喔。 ┌(°ﾍﾟ,°)┐ 」

　　鱈魚：「沒有啦，因為我們只改了數值，沒有加上數值變化要怎麼甚麼的邏輯。ᕙ(˘ω˘ᕙ) 」

可以在 devtool 中看到 selectedGame 數值會發生變化。

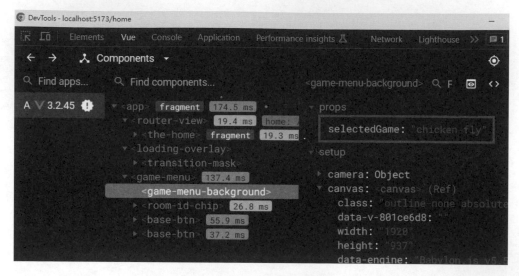

▲ 圖 11-15　選擇遊戲

現在我們要在遊戲選項變化時，同時移動鏡頭到每個對應的島嶼。

我們希望不同時機點移動鏡頭的方式會不太一樣，改寫一下鏡頭移動的功能。

- 鏡頭移動速度改成從參數輸入。

- easingFunction 改成從參數輸入。

➡ src\components\game-menu-background.vue

```ts
...
<script setup lang="ts">
...
const defaultEase = new BackEase();
defaultEase.setEasingMode(EasingFunction.EASINGMODE_EASEINOUT);
/** 鏡頭移動預設值 */
const moveCameraDefaultParams: {
  speed: number;
  easingFunction: EasingFunction;
} = {
```

```
  speed: 0.15,
  easingFunction: defaultEase,
}
async function moveCamera(params?: typeof moveCameraDefaultParams) {
  const shot = shots.find(({ name }) => name === props.selectedGame);
  if (!shot) return;

  const { speed, easingFunction } = defaults(params, moveCameraDefaultParams);

  const results = [
    createAnimation(camera, 'target', shot.camera.target, { easingFunction }),
    createAnimation(camera, 'alpha', shot.camera.alpha, { easingFunction }),
    createAnimation(camera, 'beta', shot.camera.beta, { easingFunction }),
    createAnimation(camera, 'radius', shot.camera.radius),
  ]

  /** 直接覆蓋鏡頭動畫，以免再次呼叫 moveCamera 的新舊動畫互相影響 */
  camera.animations = results.map(({ animation }) => animation);
      ...
  const ref = scene.beginAnimation(camera, 0, results[0].frameRate, false, speed);
  ...
}
...
onMounted(async () => {
  ...
  moveCamera();
});
...
</script>
...
```

接著將預設遊戲改回企鵝。

➡ src\components\game-menu.vue

```
...
<script setup lang="ts">
...
const currentIndex = ref(0);
```

11-35

```
...
</script>
...
```

現在讓我們實作切換鏡頭的部分吧。

➜ src\components\game-menu-background.vue

```
...
<script setup lang="ts">
...
watch(() => props.selectedGame, () => {
  const easingFunction = new BezierCurveEase(0.800, -0.155, 0.000, 1.000);

  moveCamera({
    speed: 0.3,
    easingFunction,
  });
});
...
</script>
...
```

現在切換遊戲，鏡頭也會移動到對應島嶼了！

▲ 圖 11-16 切換遊戲鏡頭移動

11.2 建立遊戲場景

此章節程式碼可以在以下連結取得。

- https://gitlab.com/deepmind1/animal-party/web/-/tree/feat/game-chicken-fly-scene

現在讓我們準備開發遊戲，預期小雞會背對螢幕，畫面前方會持續有障礙物飛來，玩家要控制小雞上下左右避開障礙物，碰撞障礙 3 次後小雞會墜機，存活到最後者獲勝。

讓我們建立遊戲場景元件吧。

➜ src\games\chicken-fly\game-scene.vue

```ts
<template>
</template>

<script setup lang="ts">
import { ref } from 'vue';

const emit = defineEmits<{
  (e: 'init'): void;
}>();
</script>

<style scoped lang="sass">
</style>
```

接著將 App.vue 目前的 game-menu 刪除，換成小雞遊戲的場景元件。

➜ src\App.vue

```
<template>
  <router-view />
  <loading-overlay />

  <game-scene class=" absolute w-full h-full" />
</template>

<script setup lang="ts">
import LoadingOverlay from './components/loading-overlay.vue';
import GameScene from './games/chicken-fly/game-scene.vue';
</script>
...
```

第一步一樣讓我們建立 babylon 基本場景。

➜ src\games\chicken-fly\game-scene.vue

```
<template>
  <div class="overflow-hidden">
    <canvas
      ref="canvas"
      class="outline-none w-full h-full"
    />
  </div>
</template>

<script setup lang="ts">
import { onBeforeUnmount, onMounted, ref } from 'vue';
import {
  ArcRotateCamera,
  Engine, HemisphericLight, MeshBuilder, Scene, Vector3,
} from '@babylonjs/core';
import '@babylonjs/core/Debug/debugLayer';
import '@babylonjs/inspector';

const emit = defineEmits<{
```

```
  (e: 'init'): void;
}>();

const canvas = ref<HTMLCanvasElement>();

let engine: Engine;
let scene: Scene;

function createEngine(canvas: HTMLCanvasElement) {
  const engine = new Engine(canvas, true);
  return engine;
}
function createScene(engine: Engine) {
  const scene = new Scene(engine);
  return scene;
}
function createCamera(scene: Scene) {
  const camera = new ArcRotateCamera(
    'camera',
    Math.PI / 2,
    Math.PI / 4,
    34,
    new Vector3(0, 0, 2),

    scene
  );

  return camera;
}

async function init() {
  if (!canvas.value) {
    console.error(' 無法取得 canvas DOM');
    return;
  }

  engine = createEngine(canvas.value);
  scene = createScene(engine);
  createCamera(scene);
```

```
/** 反覆渲染場景，這樣畫面才會持續變化 */
engine.runRenderLoop(() => {
  scene.render();
});

window.addEventListener('keydown', (ev) => {
  // 按下 Shift+I 可以切換視窗
  if (ev.shiftKey && ev.keyCode === 73) {
    if (scene.debugLayer.isVisible()) {
      scene.debugLayer.hide();
    } else {
      scene.debugLayer.show();
    }
  }
});
}

onMounted(async () => {
  await init();
  window.addEventListener('resize', handleResize);

  emit('init');
});

onBeforeUnmount(() => {
  engine.dispose();
  window.removeEventListener('resize', handleResize);
});

function handleResize() {
  engine.resize();
}
</script>
```

鱈魚：「內容基本上和企鵝遊戲場景一樣，讓我們複製貼上吧！ ᕦ(ﾟ∀ﾟ)ᕤ」

助教：「所以每新增一個遊戲場景，就要一直複製貼上喔？ ╭(ﾟA,ﾟ)╮」

鱈魚：「程式碼重複兩次以上才需要開始考慮復用啦！ ᕦ(ﾟ∀ﾟ)ᕤ」

助教：「已經重複 3 次了。(°c＿°`)」

鱈魚：「哪 ...... 真的欸。(⊙ ⁄⁄ ⊙')」

助教：「...... (°c＿°`)」

鱈魚：「這時候讓我們出動 Vue 的 Composition API 吧。(˚∀˚)✧」

建立檔案並加入基礎內容。

➜ src\composables\use-babylon-scene.ts

```
import { defaults } from 'lodash-es';



export function useBabylonScene() {
  return {}
}
```

現在讓我們把要複用的功能依序加入吧，首先加入變數。

➜ src\composables\use-babylon-scene.ts

```
import {
  ArcRotateCamera,
  Engine, Scene, Vector3
} from '@babylonjs/core';
import { onBeforeUnmount, onMounted, ref } from 'vue';

export function useBabylonScene () {
  const canvas = ref<HTMLCanvasElement>();

  let engine: Engine | undefined;
  let scene: Scene | undefined;
  let camera: ArcRotateCamera | undefined;

  onMounted(async () => {
    window.addEventListener('resize', handleResize);
  });
```

```
onBeforeUnmount(() => {
  engine?.dispose();
  window.removeEventListener('resize', handleResize);
});

function handleResize() {
  engine?.resize();
}

return {
  canvas,
  engine,
  scene,
  camera,
}
}
```

　　接著我們新增建立各類物件的 function，除了預設 function 外，使用者也可以透過參數加入建立物件的 function。

➜ src\composables\use-babylon-scene.ts

```
import { defaults } from 'lodash-es';
import {
  ArcRotateCamera,
  Engine, HemisphericLight, Scene, Vector3,
} from '@babylonjs/core';
import { onBeforeUnmount, onMounted, ref } from 'vue';

interface UseBabylonSceneParams {
  createEngine?: (canvas: HTMLCanvasElement) => Engine;
  createScene?: (engine: Engine) => Scene;
  createCamera?: (scene: Scene) => ArcRotateCamera;
  init?: (params: {
    canvas: HTMLCanvasElement;
    engine: Engine;
    scene: Scene;
    camera: ArcRotateCamera;
  }) => Promise<void>;}
```

```
const defaultParams: Required<UseBabylonSceneParams> = {
  createEngine(canvas) {
    return new Engine(canvas, true);
  },
  createScene(engine) {
    const scene = new Scene(engine);
    /** 使用預設光源 */
    scene.createDefaultLight();
    const defaultLight = scene.lights.at(-1);
    if (defaultLight instanceof HemisphericLight) {
      defaultLight.direction = new Vector3(0.5, 1, 0);
    }

    return scene;
  },
  createCamera(scene: Scene) {
    const camera = new ArcRotateCamera(
      'camera',
      Math.PI / 2,
      Math.PI / 4,
      34,
      new Vector3(0, 0, 2),
      scene
    );

    return camera;
  },
  init: () => Promise.resolve(),
}

export function useBabylonScene(params?: UseBabylonSceneParams) {
  ...
}
```

最後來呼叫參數吧。

src\composables\use-babylon-scene.ts

```
...
export function useBabylonScene(params?: UseBabylonSceneParams) {
```

```
...

const {
  createEngine, createScene, createCamera, init
} = defaults(params, defaultParams);

onMounted(async () => {
  if (!canvas.value) {
    console.error(' 無法取得 canvas DOM');
    return;
  }
  engine = createEngine(canvas.value);
  scene = createScene(engine);
  camera = createCamera(scene);

  window.addEventListener('resize', handleResize);

  /** 反覆渲染場景，這樣畫面才會持續變化 */
  engine.runRenderLoop(() => {
    scene?.render();
  });

  await init({
    canvas: canvas.value, engine, scene, camera,
  }); });

onBeforeUnmount(() => {
  engine?.dispose();
  window.removeEventListener('resize', handleResize);
});

function handleResize() {
  engine?.resize();
}

return {
  canvas,
  engine,
  scene,
  camera,
```

```
  }
}
```

現在讓我們回到遊戲場景元件，引用 use-babylon-scene 吧。

➜ src\composables\use-babylon-scene.ts

```
...
<script setup lang="ts">
import { onBeforeUnmount, onMounted, ref } from 'vue';
import {
  ArcRotateCamera,
  Engine, HemisphericLight, MeshBuilder, Scene, Vector3,
} from '@babylonjs/core';
import { SkyMaterial } from '@babylonjs/materials';
import '@babylonjs/core/Debug/debugLayer';
import '@babylonjs/inspector';

import { useBabylonScene } from '../../composables/use-babylon-scene';

const emit = defineEmits<{
  (e: 'init'): void;
}>();

const { canvas } = useBabylonScene({
  async init({ scene }) {
    window.addEventListener('keydown', (ev) => {
      // 按下 Shift+I 可以切換視窗
      if (ev.shiftKey && ev.keyCode === 73) {
        if (scene.debugLayer.isVisible()) {
          scene.debugLayer.hide();
        } else {
          scene.debugLayer.show();
        }
      }
    });

    emit('init');
  }
});
```

```
</script>
```

會發現程式碼變得相當精簡，感覺真不錯。(´▽`)╯

現在讓我們建立天空，建立一個 skybox，同時加入 createCamera，調整一下鏡頭位置。

➔ src\games\chicken-fly\game-scene.vue

```
...
<script setup lang="ts">
const { canvas } = useBabylonScene({
  createCamera(scene) {
    const camera = new ArcRotateCamera(
      'camera',
      Math.PI / 2,
      Math.PI / 2,
      0,
      new Vector3(0, 0, 0),
      scene,
    );

    camera.attachControl(canvas.value, true);
    camera.wheelDeltaPercentage = 0.1;

    return camera;
  },

  async init({ scene }) {
    createSky(scene);
    ...
  }
});

function createSky(scene: Scene) {
  const skyMaterial = new SkyMaterial("skyMaterial", scene);
  skyMaterial.backFaceCulling = false;
  skyMaterial.turbidity = 0.2;
  skyMaterial.luminance = 0.1;
```

```
    skyMaterial.rayleigh = 0.2;

    /** 太陽位置 */
    skyMaterial.useSunPosition = true;
    skyMaterial.sunPosition = new Vector3(1, 0.9, -2);

    const skybox = MeshBuilder.CreateBox("skyBox", { size: 100 }, scene);
    skybox.material = skyMaterial;

    return skybox;
}</script>
```

▲ 圖 11-17 產生天空

這裡我們建立一個 Box 加上 skyMaterial 產生天空材質。

所謂的 skybox 就真的是一個 box，只要把鏡頭塞在 box 內，就會有一種在天空內的感覺，所以如果把 box 的尺寸設小一點，將鏡頭移動到 box 外，就可以看到整個 box 的樣子。

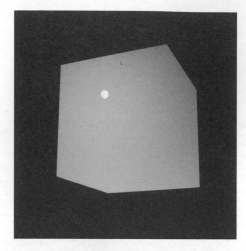

▲ 圖 11-18　天空盒

 Tips

想要深入了解的讀者可以參考以下連結：

◆ https://doc.babylonjs.com/toolsAndResources/assetLibraries/materialsLibrary/
skyMat

11.3 **起飛吧！小雞！**

此章節程式碼可以在以下連結取得。

■ https://gitlab.com/deepmind1/animal-party/web/-/tree/feat/game-chicken-
fly-character

現在讓我們新增小雞的 Class 吧，基本上和企鵝 Class 概念相同。

➔ src\games\chicken-fly\chicken.ts

```ts
import { Vector3, Color3, AbstractMesh, Scene, SceneLoader } from "@babylonjs/core";
/** 引入 loaders，這樣才能載入 glb 檔案 */
import '@babylonjs/loaders';
import { defaultsDeep } from "lodash-es";

export interface Params {
  /** 起始位置 */
  position?: Vector3;
  ownerId: string;
  /** 玩家顏色 */
  color?: Color3;
}

export class Chicken {
  mesh?: AbstractMesh;
  name: string;
  scene: Scene;
  params: Required<Params> = {
    position: new Vector3(0, 0, 0),
    ownerId: '',
    color: new Color3(0.9, 0.9, 0.9),
  };

  constructor(name: string, scene: Scene, params?: Params) {
    this.name = name;
    this.scene = scene;
    this.params = defaultsDeep(params, this.params);
  }

  async init() {
    const { position } = this.params;

    const result = await SceneLoader.ImportMeshAsync('', '/chicken-fly/', 'flying-
chicken.glb', this.scene);

    const chicken = result.meshes[0];
```

```
    chicken.position = position;
    chicken.scaling = new Vector3(0.15, 0.15, 0.15);
    chicken.rotation = new Vector3(0, Math.PI / 2, 0);

    return this;
  }
}
```

現在讓我們在場景引入小雞並調整一下鏡頭吧。

➔ src\games\chicken-fly\game-scene.vue

```
...
<script setup lang="ts">
...
import { Chicken } from './chicken';
...
const { canvas } = useBabylonScene({
  createCamera(scene) {
    const camera = new ArcRotateCamera(
      ...
      6,
      new Vector3(0, 0, 0),
      scene,
    );
    ...
  },

  async init({ scene }) {
    createSky(scene);
    createChickens(scene);
    ...
  }
});
...
async function createChickens(scene: Scene) {
  const item = await new Chicken('1', scene, {
    ownerId: '',
    position: new Vector3(0, 0, 0),
  }).init();
```

```
}
</script>
```

現在可以在畫面看到小雞了！

▲ 圖 11-19 小雞起飛

現在讓我們加入小雞的 hitbox，這裡我們用兩個方塊組成身體和翅膀的 hitbox。

➜ src\games\chicken-fly\chicken.ts

```
...
export class Chicken {
  ...
  private createHitBox() {
    const visibility = 0.5;

    /** 建立 body、wing 的 hit box */
    const body = MeshBuilder.CreateBox(`${this.name}-body-hit-box`, {
      width: 0.3, depth: 0.76, height: 0.48
    });
    body.visibility = visibility;

    const wing = MeshBuilder.CreateBox(`${this.name}-wing-hit-box`, {
```

```
      width: 0.8,
      depth: 0.23,
      height: 0.1
    });
    wing.position = new Vector3(0, -0.1, -0.02);
    wing.visibility = visibility;
    wing.setParent(body);

    /** 將 body 移至定位 */
    body.position = this.params.position;

    /** 產生物理效果 */
    wing.physicsImpostor = new PhysicsImpostor(
      wing,
      PhysicsImpostor.BoxImpostor,
      {
        mass: 0.1,
      },
      this.scene
    );
    body.physicsImpostor = new PhysicsImpostor(
      body,
      PhysicsImpostor.BoxImpostor,
      {
        mass: 1,
        restitution: 0.5
      },
      this.scene
    );

    return body;
  }

  async init() {
    const result = await SceneLoader.ImportMeshAsync('', '/chicken-fly/', 'flying-
chicken.glb', this.scene);

    const hitBox = this.createHitBox();
    this.mesh = hitBox;
```

```
    const chicken = result.meshes[0];
    chicken.setParent(hitBox);
    chicken.scaling = new Vector3(0.15, 0.15, 0.15);
    chicken.rotation = new Vector3(0, Math.PI / 2, 0);
    chicken.position = new Vector3(0, -0.1, -0.2);

    return this;
  }
}
```

現在會看到小雞被半透明的方塊包住。

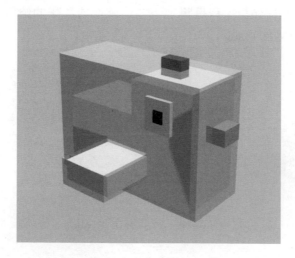

▲ 圖 11-20 小雞 hit box

確認範圍正確後，我們把 hit box 隱藏吧。

➜ src\games\chicken-fly\chicken.ts

```
...
export class Chicken {
  ...
    private createHitBox() {
      const visibility = 0;
      ...
  }
```

```
    ...
}
```

現在讓我們試試調整一下小雞的飛行姿態。

鱈魚：「實際上飛機的飛行姿態很複雜，讓我們簡化一下飛行控制。」

助教：「確定不是偷懶？(´一`)」

鱈魚：「才不是勒，我們又不是要做飛行模擬遊戲，當然是越簡單越好啊，你這樣會損壞我的形象捏。(⊙ ⌒⌒ ⊙')」

概念與類比搖桿類似，只是將上下左右轉換成小雞的俯仰角（pitch）、翻滾角（roll）與偏航角（yaw）。

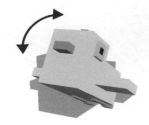

▲ 圖 11-21 俯仰角（pitch）、翻滾角（roll）、偏航角（yaw）

接著準備新增控制飛行姿態的 method，這裡我們一樣要使用四元數旋轉（Quaternion），先將 utils 的 createAnimation function 加上支援使用四元數。

→ src\common\utils.ts

```
...
QuarticEase, Quaternion, Vector3 } from '@babylonjs/core';
...
/** 建立從目前狀態至目標狀態的動畫
 *
 * @param target 目標物件
 * @param property 過度的目標屬性
 * @param to 目標狀態
```

```
  * @param option 參數
  * @returns
  */
export function createAnimation(
   ...
   to: number | Vector3 | Color3 | Quaternion,
   ...
) {
   ...
   if (typeof to === 'number') {
     ...
   } else if (to instanceof Vector3) {
     ...
   } else if (to instanceof Color3) {
     ...
   } else if (to instanceof Quaternion) {
     animationType = Animation.ANIMATIONTYPE_QUATERNION
   }
   ...
}
...
```

小雞加上控制姿態的 method。

➡ src\games\chicken-fly\chicken.ts

```
...
export class Chicken {
  /** 最大角度 */
  private maxAngle = Tools.ToRadians(45);

  ...
  /** 設定飛行姿態
   *
   * @param pitch 俯仰角
   * @param roll 翻滾角
   */
  setAttitude(pitch: number, roll: number) {
    if (!this.mesh?.physicsImpostor) return;
```

```
    /** 俯仰角（pitch）對應 -x 軸 */
    const newPitch = clamp(-pitch, -this.maxAngle, this.maxAngle);
    /** 翻滾角（roll）對應 -z 軸 */
    const newRoll = clamp(-roll, -this.maxAngle, this.maxAngle);
    /** 讓偏航角（yaw）跟隨翻滾角（roll）旋轉，讓小雞的姿態看起來豐富一點 */
    const yaw = newRoll / 2;

    const quaternion = Quaternion.RotationYawPitchRoll(yaw, newPitch, newRoll);

    /** 產生動畫 */
    const {
      animation, frameRate
    } = createAnimation(this.mesh, 'rotationQuaternion', quaternion, {
      speedRatio: 3
    });

    this.scene.beginDirectAnimation(this.mesh, [animation], 0, frameRate);  }
}
```

接著在場景註冊鍵盤事件，讓我們試試看效果。

➔ src\games\chicken-fly\game-scene.vue

```
...
<script setup lang="ts">
...
const currentAngle = {
  pitch: 0,
  roll: 0,
}
async function createChickens(scene: Scene) {
  const chicken = await new Chicken('1', scene, {
    ownerId: '',
    position: new Vector3(0, 0, 0),
  }).init();

  window.addEventListener('keydown', (e) => {
    switch (e.key) {
      case '8': {
```

```
        currentAngle.pitch += 0.1;
        break;
      }
      case '2': {
        currentAngle.pitch -= 0.1;
        break;
      }
      case '4': {
        currentAngle.roll += 0.1;
        break;
      }
      case '6': {
        currentAngle.roll -= 0.1;
        break;
      }
    }

    const { pitch, roll } = currentAngle;
    chicken.setAttitude(pitch, roll);
  });
}
</script>
```

現在可以試試用數字鍵盤的 8、2、4、6 控制小雞姿態。

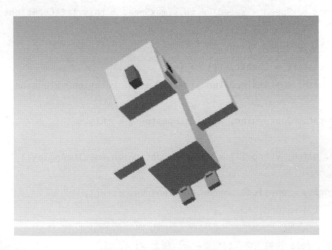

▲　圖 11-22　控制小雞飛行姿態

小雞準備好飛向更自由的世界了！✧*。٩(´ʊ`*)و✧*。

讓我們根據小雞的飛行姿態產生移動加速度吧！

這裡希望小雞的旋轉角度越斜，在其對應方向的加速度，也就是升力就越大，但是速度越接近最大速度時，升力就會等比縮小，最終到 0。

➔ src\games\chicken-fly\chicken.ts

```
...
export class Chicken {
  ...
  private maxAngle = Tools.ToRadians(45);
  private maxSpeed = 1;
  /** 用來放大升力 */
  private liftCoefficient = 3;
  ...
  /** 根據目前速度取得升力比率，速度越大，比率越小 */
  private getLiftRatio(speed: number) {
    // 如果速度到達上限，則加速度為 0
    if (Math.abs(speed) >= Math.abs(this.maxSpeed)) {
      return 0;
    }

    // 將速度映射到加速度，速度越高，加速度越小，最後乘上升力係數放大數值。
    return (this.maxSpeed - speed) / this.maxSpeed * this.liftCoefficient;
  }
  /** 根據目前姿態產生加速度 */
  private processLift = throttle(() => {
    if (!this.mesh?.physicsImpostor || !this.mesh?.rotationQuaternion) return;

    const { rotationQuaternion, physicsImpostor } = this.mesh;

    const { x: pitch, y: roll } = rotationQuaternion.toEulerAngles();

    const velocity = physicsImpostor.getLinearVelocity();
    if (!velocity) return;

    /** X 座標與 roll 旋轉方向剛好相反，所以乘上 -1 */
    const xLift = this.getLiftRatio(velocity.x) * -roll;
```

```
      const yLift = this.getLiftRatio(velocity.y) * pitch;

      const force = new Vector3(xLift, yLift, 0);
      physicsImpostor.applyForce(force, Vector3.Zero());
    }, 30)

  async init() {
    ...
    // 持續在每個 frame render 之前呼叫
    this.scene.registerBeforeRender(() => {
      this.processLift();
    });

    return this;
  }

  ...
}
```

接著在場景開啟物理效果。

→ src\games\chicken-fly\game-scene.vue

```
...
<script setup lang="ts">
import * as CANNON from 'cannon-es';
import {
  ArcRotateCamera,
  CannonJSPlugin,
  Engine, HemisphericLight, MeshBuilder, Scene, Vector3,
} from '@babylonjs/core';
...
const { canvas } = useBabylonScene({
  ...
  async init({ scene }) {
    const physicsPlugin = new CannonJSPlugin(true, 8, CANNON);
    scene.enablePhysics(new Vector3(0, 0, 0), physicsPlugin);

    ...
  }
```

```
});
...
</script>
```

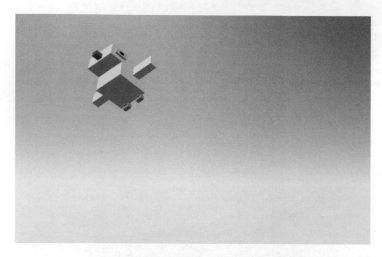

▲ 圖 11-23　根據姿態產生升力

　　現在調整姿態後，小雞終於可以飛行 ......， 啊，啊啊啊啊啊啊啊啊！飛出
畫面外啦！ ᕦ(´ロ`ᕦ)

　　現在讓我們限制一下小雞的飛行範圍，避免小雞前後移動和瘋狂亂轉。

➜　src\games\chicken-fly\chicken.ts

```
import { Vector3, Color3, AbstractMesh, Scene, SceneLoader, MeshBuilder,
PhysicsImpostor, Mesh, Tools } from "@babylonjs/core";
/** 引入 loaders，這樣才能載入 glb 檔案 */
import '@babylonjs/loaders';
import { clamp, defaultsDeep, throttle } from "lodash-es";
import { createAnimation } from "../../common/utils";

export interface PenguinParams {
  /** 起始位置 */
  position?: Vector3;
  ownerId: string;
  /** 玩家顏色 */
```

```
  color?: Color3;
}

export class Chicken {
  ...
  /** 限制小雞的移動範圍 */
  private limitPosition() {
    if (!this.mesh) return;

    /** 避免小雞前後移動，所以 z 固定為 0 */
    this.mesh.position.z = 0;

    /** 角速度歸零，讓角色停止亂轉 */
    this.mesh?.physicsImpostor?.setAngularVelocity(new Vector3(0, 0, 0));
  }

  async init() {
    ...

    // 持續在每個 frame render 之前呼叫
    this.scene.registerBeforeRender(() => {
      this.processLift();
      this.limitPosition();
    });

    return this;
  }
...
}
```

接著在場景中新增空氣牆。

➜ src\games\chicken-fly\game-scene.vue

```
...
<script setup lang="ts">
...
/** 遊戲場景邊界 */
const sceneBoundary = {
```

```
    x: 5,
    y: 2.5,
  }
...
const { canvas } = useBabylonScene({
    ...

    async init({ scene }) {
      ...
      createChickens(scene);
      createBoundary(scene);
      ...
    }
});
...
function createBoundary(scene: Scene) {
  const list = [
    {
      name: 'boundary-top',
      option: { height: 0.5, width: 10, depth: 0.5 },
      position: new Vector3(0, sceneBoundary.y, 0),
    },
    {
      name: 'boundary-bottom',
      option: { height: 0.5, width: 10, depth: 0.5 },
      position: new Vector3(0, -sceneBoundary.y, 0),

    },
    {
      name: 'boundary-left',
      option: { height: 10, width: 0.5, depth: 0.5 },
      position: new Vector3(-sceneBoundary.x, 0, 0),

    },
    {
      name: 'boundary-right',
      option: { height: 10, width: 0.5, depth: 0.5 },
      position: new Vector3(sceneBoundary.x, 0, 0),
    },
```

```
  ]

  list.forEach(({ name, option, position }) => {
    const wall = MeshBuilder.CreateBox(name, option, scene);
    wall.position = position;
    wall.physicsImpostor = new PhysicsImpostor(wall, PhysicsImpostor.BoxImpostor,
      { mass: 0, friction: 0, restitution: 0 }, scene
    );
  });
}
...
</script>
```

可以看到周圍被圈起來了。

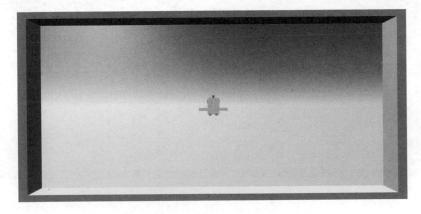

▲ 圖 11-24 建立空氣牆

現在讓我們新增一堆小雞試試看牆有沒有效果吧。

→ src\games\chicken-fly\game-scene.vue

```
...
<script setup lang="ts">
...
async function createChickens(scene: Scene) {
  const chicken = await new Chicken('1', scene, {
    ...
  }).init();
```

```
  for (let i = 0; i < 10; i++) {
    await new Chicken(`${i + 2}`, scene, {
      ownerId: '',
      position: new Vector3(1, 1, 0),
    }).init();
  }

  ...
}</script>
```

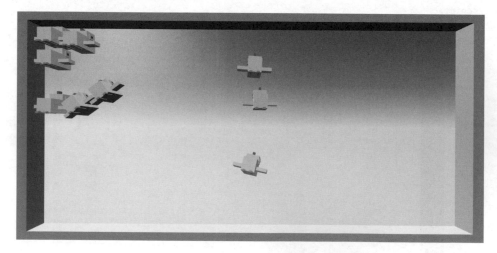

▲ 圖 11-25　多個小雞

讓我們把牆隱藏起來吧。

➜ src\games\chicken-fly\game-scene.vue

```
...
<script setup lang="ts">
...
function createBoundary(scene: Scene) {
  ...

  list.forEach(({ name, option, position }) => {
    ...
```

```
    wall.isVisible = false;
  });
}

...
}</script>
```

最後讓我們把小雞染色吧，這樣才可以讓玩家知道自己的小雞是哪一個。

➜ src\games\chicken-fly\chicken.ts

```
import { Vector3, Color3, AbstractMesh, Scene, SceneLoader, MeshBuilder,
PhysicsImpostor, Mesh, Tools, Quaternion, PBRMaterial } from "@babylonjs/core";
...
export class Chicken {
  ...
  async init() {
    ...
    /** 找到 body mesh 將材質改顏色 */
    const bodyMesh = result.meshes.find(
      ({ name }) => name === 'body'
    );
    /** 確認模型材質為 PBRMaterial */
    if (bodyMesh?.material instanceof PBRMaterial) {
      bodyMesh.material.albedoColor = this.params.color;
    }
    ...
  }
  ...
}
```

試試看把小雞設為紅色。

➜ src\games\chicken-fly\game-scene.vue

```
...

<script setup lang="ts">
...
```

```
async function createChickens(scene: Scene) {
  const chicken = await new Chicken('1', scene, {
    ...
    color: new Color3(1, 0, 0),
  }).init();
  ...
}
</script>
```

可以看到可以控制的小雞變成紅色。

▲ 圖 11-26　小雞染色

11.4 完成天空

此章節程式碼可以在以下連結取得。

- https://gitlab.com/deepmind1/animal-party/web/-/tree/feat/game-chicken-fly-sky

現在讓我們加入其他天空會有的元素吧，首先讓我們加入雲朵，這裡我們使用 Solid Particle System 實現。

Solid Particle System 主要用於產生雲朵、小行星帶等等，這類有大量重複形狀的物體，比起直接建立同等數量的物體來說，更加節省記憶體與計算負擔。

Tips

想要深入了解的讀者可以參考以下連結：

◆ https://doc.babylonjs.com/features/featuresDeepDive/particles/solid_particle_system

新增建立雲朵的 function。

➜ src\games\chicken-fly\game-scene.vue

```
...

<script setup lang="ts">
...
import {
  ArcRotateCamera,
  CannonJSPlugin,
  Color3,
  MeshBuilder, PhysicsImpostor,
  Scene, SolidParticleSystem, Vector3,
} from '@babylonjs/core';
...
const { canvas } = useBabylonScene({
  ...
  async init({ scene }) {
    ...
    createBoundary(scene);
    createClouds(scene);
    ...
  }
});
```

```
...
function createClouds(scene: Scene) {
  const clouds = new SolidParticleSystem('clouds', scene);

  /** 建立雲朵 */
  const cloud = MeshBuilder.CreateBox('cloud', {
    width: 1.5, height: 0.5, depth: 2,
  }, scene);

  /** 建立 50 個一樣的 mesh */
  clouds.addShape(cloud, 50);
  /** 新增至 SolidParticleSystem 後原本的 cloud 就可以停用了 */
  cloud.dispose();

  /** 實際建立 mesh */
  const mesh = clouds.buildMesh();

  return clouds;
}...
</script>
```

▲ 圖 11-27 建立雲朵

鱈魚：「雲朵出現了！ ໒(´∀`໒)」

助教：「你哪隻魚眼看這個像雲朵啊！(⊙ 益 ⊙')」

鱈魚：「別急別急，這是因為雲朵擠在一起啦。ヽ(ˊ▽ˋ)ㄏ」

讓我們把雲分散吧。

➜ src\games\chicken-fly\game-scene.vue

```
...
<script setup lang="ts">
...
function createClouds(scene: Scene) {
  ...

  /** 實際建立 mesh */
  const mesh = clouds.buildMesh();
  /** 開啟透明效果 */
  mesh.hasVertexAlpha = true;

  /** 建立初始化 function */
  clouds.initParticles = () => {
    clouds.particles.forEach((particle) => {
      /** 隨機分布 */
      particle.position.x = Scalar.RandomRange(-sceneBoundary.x * 2, sceneBoundary.x *
2);
      particle.position.y = Scalar.RandomRange(-sceneBoundary.y, -sceneBoundary.y *
0.9);
      particle.position.z = Scalar.RandomRange(0, -150);

      /** 隨機尺寸 */
      particle.scaling = new Vector3(
        Scalar.RandomRange(1, 3),
        Scalar.RandomRange(1, 2),
        Scalar.RandomRange(1, 3)
      );
      /** 半透明白色 */
      particle.color = new Color4(1, 1, 1, 0.5);    });
  };

  /** 初始化 */
  clouds.initParticles();
  clouds.setParticles();
```

```
  return clouds;
}...
</script>
```

雲海出現啦！ヾ(๑'ᵕ')ノ˚

▲ 圖 11-28　多個雲朵

現在我們讓雲動起來，這裡希望時間遊戲越久，物體的運動速度就越快。

我們新增一個隨著時間增加的計數器，再將累積的數值依照需求轉換成速度即可。

→ src\games\chicken-fly\game-scene.vue

```
...
<script setup lang='ts'>
...
import { useInterval } from '@vueuse/core';

/** 每秒 +1 */
const counter = useInterval(1000);
...
function createClouds(scene: Scene) {
  ...

  /** 更新邏輯 */
  clouds.updateParticle = (particle) => {
    /** 雲朵移動速度 */
```

```
    particle.velocity.z = 0.1 + counter.value * 0.005;
    particle.position.addInPlace(particle.velocity);

    return particle;
  }

  /** 持續呼叫，粒子才會渲染 updateParticle 後的結果 */
  scene.onAfterRenderObservable.add(() => {
    clouds.setParticles();
  })

  return clouds;
}
...
</script>
```

鱈魚：「雲開始飄動，小雞看起來有向前飛的感覺了！ヾ(๑'ʊ`๑)ﾉ゙」

助教：「可是雲一下子就飄沒了。(´･ω･`)」

鱈魚：「讓我們把雲回收再利用吧。(ﾐ｀ω´ﾐ)✧」

讓飄出畫面的雲回到初始點，我們就不用一直產生新的雲，這樣可以有效提升性能。

➜ src\games\chicken-fly\game-scene.vue

```
...
<script setup lang="ts">
...
function createClouds(scene: Scene) {
  ...

  const cloudDepth = 2;
  /** 建立雲朵 */
  const cloud = MeshBuilder.CreateBox('cloud', {
    width: 1.5, height: 0.5, depth: cloudDepth,
  }, scene),
```

```
  ...
  /** 更新邏輯 */
  clouds.updateParticle = (particle) => {
    /** 雲的尺寸是 cloudDepth，最大倍數是 3 倍，
     * 所以位置大於 cloudDepth * 3 一定超出畫面了。
     */
    if (particle.position.z > cloudDepth * 3) {
      particle.position.z = -150;
    }
    ...
  }
  ...
}
...
</script>
```

現在雲好了，讓我們加個速度線特效吧。

我們讓速度線包裹光暈，所以先讓我們啟動光暈圖層。

➔ src\games\chicken-fly\game-scene.vue

```
...
<script setup lang="ts">
import {
  ...
  GlowLayer,
} from '@babylonjs/core';
...
const { canvas } = useBabylonScene({
  ...
  async init({ scene }) {
    const physicsPlugin = new CannonJSPlugin(true, 8, CANNON);
    scene.enablePhysics(new Vector3(0, 0, 0), physicsPlugin);

    const glowLayer = new GlowLayer('glow', scene, {
      blurKernelSize: 16
    });
    ...
```

```
  }
});
...
</script>
```

blurKernelSize 表示光暈的解析度，越高越耗費效能。

現在讓我們新增速度線的 SolidParticleSystem，概念與雲朵一模一樣，只差在有材質。

➜ src\games\chicken-fly\game-scene.vue

```
...
<script setup lang="ts">
...
function createSpeedLines(scene: Scene) {
  const lines = new SolidParticleSystem('speed-lines', scene);
  const depth = 1;

  const line = MeshBuilder.CreateBox('line', {
    width: 0.01, height: 0.01, depth,
  }, scene);

  /** 建立材質，並加入光暈 */
  const material = new StandardMaterial('speed-line', scene);
  material.emissiveColor = Color3.White();
  line.material = material;

  lines.addShape(line, 100);
  line.dispose();

  const mesh = lines.buildMesh();
  mesh.hasVertexAlpha = true;

  /** 將材質加入粒子系統中 */
  lines.setMultiMaterial([material]);

  lines.initParticles = () => {
    lines.particles.forEach((particle) => {
```

```
      particle.position.x = Scalar.RandomRange(-sceneBoundary.x, sceneBoundary.x);
      particle.position.y = Scalar.RandomRange(-sceneBoundary.y, sceneBoundary.y);
      particle.position.z = Scalar.RandomRange(0, -50);

      particle.color = new Color4(1, 1, 1, 0.05);
    });
  };

  lines.initParticles();
  lines.setParticles();

  lines.updateParticle = (particle) => {
    if (particle.position.z > depth * 2) {
      particle.position.z = -50;
    }

    particle.velocity.z = 0.5 + counter.value * 0.005;
    particle.position.addInPlace(particle.velocity);

    return particle;
  }

  scene.onAfterRenderObservable.add(() => {
    lines.setParticles();
  })

  return lines;
}
...
</script>
```

現在可以在畫面看到速度線特效了。

▲ 圖 11-29 加入速度線

最後讓我們加入會攔截小雞的壞壞雞吧，預期會有複雜的飛行路線，讓玩家更有挑戰性，讓我們新增 Class。

概念與小雞相同，只是簡化很多。

- hit box 只用一個方塊

- 人物固定為黑色

➡ src\games\chicken-fly\bad-chicken.ts

```
import { Vector3, Color3, AbstractMesh, Scene, SceneLoader, MeshBuilder,
PhysicsImpostor, Mesh, Tools, Quaternion, PBRMaterial } from "@babylonjs/core";
/** 引入 loaders，這樣才能載入 glb 檔案 */
import '@babylonjs/loaders';
import { clamp, defaultsDeep, throttle } from "lodash-es";

export interface Params {
  /** 起始位置 */
  position?: Vector3;
}

export class BadChicken {
  mesh?: AbstractMesh;
  name: string;
```

```
  scene: Scene;
  params: Required<Params> = {
    position: new Vector3(0, 0, 0),
  };


  constructor(name: string, scene: Scene, params?: Params) {
    this.name = name;
    this.scene = scene;
    this.params = defaultsDeep(params, this.params);
  }

  private createHitBox() {
    const hitBox = MeshBuilder.CreateBox(`${this.name}-hit-box`, {
      width: 1, depth: 1.8, height: 1.2
    });
    hitBox.visibility = 0.5;
    hitBox.position = this.params.position;

    return hitBox;
  }

  async init() {
    const result = await SceneLoader.ImportMeshAsync('', '/chicken-fly/', 'flying-
chicken.glb', this.scene);

    const hitBox = this.createHitBox();
    this.mesh = hitBox;

    const bodyMesh = result.meshes.find(
      ({ name }) => name === 'body'
    );
    if (bodyMesh?.material instanceof PBRMaterial) {
      bodyMesh.material.albedoColor = Color3.Black();
    }

    const chicken = result.meshes[0];
    chicken.setParent(hitBox);
    chicken.scaling = new Vector3(0.4, 0.4, 0.4);
```

```
    chicken.rotation = new Vector3(0, -Math.PI / 2, 0);
    chicken.position = new Vector3(0, -0.25, 0.5);

    return this;
  }

}
```

接著在場景新增壞壞雞，同時調整一下所有場景與人物的過程。

➔ src\games\chicken-fly\game-scene.vue

```
...
<script setup lang='ts'>
...
import { BadChicken } from './bad-chicken';
...

const { canvas } = useBabylonScene({
  ...

  async init({ scene }) {
    ...

    createSky(scene);
    createBoundary(scene);
    createClouds(scene);
    createSpeedLines(scene);

    await createChickens(scene);
    await createBadChickens(scene);

    ...
  }
});
...
async function createBadChickens(scene: Scene) {
  await new BadChicken('bad-chicken', scene).init();
}
...
```

```
</script>
...
```

▲ 圖 11-30 壞壞雞登場

確認 hit box 範圍沒問題，讓我們隱藏 hit box。

➜ src\games\chicken-fly\bad-chicken.ts

```
...
export class BadChicken {
  ...
  private createHitBox() {
    ...
    hitBox.visibility = 0;
    ...
  }
  ...
}
```

接著讓我們每隔 20 公尺產生一隻壞壞雞，總共產生 5 隻。

➜ src\games\chicken-fly\game-scene.vue

```
...
<script setup lang='ts'>
...
```

```
import { range } from 'lodash-es';
...
async function createBadChickens(scene: Scene) {
  /** 小雞間距 */
  const gap = 50;

  const tasks = range(3).map(
    (value) => new BadChicken(`bad-chicken-${value}`, scene, {
      position: new Vector3(0, 0, (value + 1) * -gap),
    }).init()
  );
  const chickens = await Promise.all(tasks);
  return chickens;
}...
</script>
...
```

現在可以看到畫面前方出現黑色壞壞雞，看起來只有 1 隻是因為其他隻超出畫面了。

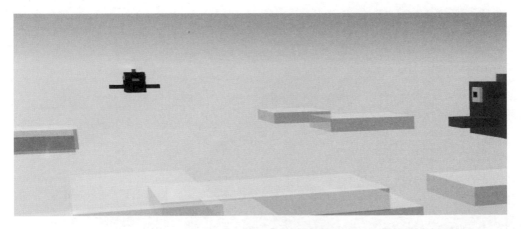

▲ 圖 11-31　多隻壞壞雞

現在我們讓壞壞雞動起來並讓壞壞雞超過指定位置時自動回收。

新增參數與設定速度用的 method。

➜ src\games\chicken-fly\bad-chicken.ts

```
...
export interface Params {
  /** 起始位置 */
  position?: Vector3;

  /** 物體 z 軸超過此位子會被回收 */
  recyclePosition?: number;
  /** 被回收後的新起點 */
  recycleStartPosition?: number;
}

export class BadChicken {
  ...
  params: Required<Params> = {
    ...
    recyclePosition: 0,
    recycleStartPosition: -100,
  };

  /** 速度基準值 */
  private speed = 0;

  constructor(...) {
    ...
  }

  ...

  async init() {
    ...
    this.scene.registerBeforeRender(() => {
      hitBox.position.addInPlace(new Vector3(0, 0, this.speed));

      if (hitBox.position.z > this.params.recyclePosition) {
        hitBox.position.z = this.params.recycleStartPosition;
      }
    });
```

```
  return this;
}

/** 設定速度基準值 */
setSpeed(speed: number) {
  this.speed = speed;
}
}
```

現在壞壞雞也會飛了！ヽ(●｀∀´●)/

▲ 圖 11-32　多隻壞壞雞

壞壞雞只會直直飛感覺很沒挑戰性，讓我們加點變化吧。

我們利用三角函數中，cos 會從 1 到 -1 的特性，來實作壞壞雞偏移飛行的效果。

➜ src\games\chicken-fly\bad-chicken.ts

```
...
export interface Params {
  ...
```

```typescript
  sceneBoundary?: {
    x: number;
    y: number;
  }
}

export class BadChicken {
  ...
  params: Required<Params> = {
    ...
    sceneBoundary: {
      x: 5,
      y: 2
    }
  };
  ...
  /** 目前相位 */
  private phase = {
    x: 0,
    y: 0,
  };
  /** 變換頻率 */
  private circularFrequency = {
    x: Scalar.RandomRange(-Math.PI / 100, Math.PI / 100),
    y: Scalar.RandomRange(-Math.PI / 100, Math.PI / 100),
  };

  ...

  async init() {
    ...

    /** 移動人物 */
    this.scene.registerBeforeRender(() => {
      const {
        x: xMax, y: yMax
      } = this.params.sceneBoundary;

      const {
```

```
    phase: {
      x: xPhase,
      y: yPhase
    },
    circularFrequency: {
      x: xCircularFrequency,
      y: yCircularFrequency
    },
  } = this;

  /** 計算位移 */
  const x = xMax * Math.cos(xPhase);
  const y = yMax * Math.cos(yPhase);
  const z = hitBox.position.z + this.speed;

  hitBox.position = new Vector3(x, y, z);

  /** 累加相位 */
  this.phase = {
    x: xPhase + xCircularFrequency,
    y: yPhase + yCircularFrequency,
  }


  /** 檢查是否需要回收 */
  if (hitBox.position.z > this.params.recyclePosition) {
    hitBox.position.z = this.params.recycleStartPosition;
  }
});

  return this;
}
...
}
```

現在壞壞雞會各種花式飛行了。

▲ 圖 11-33　壞壞雞花式飛行

我們新增一個專們用於回收的 method，讓壞壞雞在每次回收花式飛行加速。

➜ src\games\chicken-fly\bad-chicken.ts

```
...
export class BadChicken {
  ...
  private recycle() {
    if (!this.mesh) return;

    const xMax = Math.PI / 100 + this.speed / 5;
    const yMax = Math.PI / 100 + this.speed / 5;
    this.circularFrequency = {
      x: Scalar.RandomRange(-xMax, xMax),
      y: Scalar.RandomRange(-yMax, yMax),
    };

    this.mesh.position.z = this.params.recycleStartPosition;
  }

  async init() {
    ...
```

```
  /** 移動人物 */
  this.scene.registerBeforeRender(() => {
    ...
    /** 檢查是否需要回收 */
    if (hitBox.position.z > this.params.recyclePosition) {
      this.recycle();
    }
  });

  return this;
}
...
}
```

11.5 撞雞啦！

此章節程式碼可以在以下連結取得。

- https://gitlab.com/deepmind1/animal-party/web/-/tree/feat/game-chicken-fly-state

人物、場景和障礙物都就位了，現在讓我們加入小雞被壞壞雞碰撞的邏輯吧。

概念和企鵝遊戲的碰撞偵測一模一樣，調整一下 createChickens，只留下紅色小雞並新增 detectCollideEvents 用於碰撞偵測。

➜ src\games\chicken-fly\game-scene.vue

```
...
<script setup lang='ts'>
...
const { canvas } = useBabylonScene({
  ...
  async init({ scene }) {
    ...
    const chickens = await createChickens(scene);
    const badChickens = await createBadChickens(scene);
    ...
    scene.registerBeforeRender(() => {
      detectCollideEvents(chickens, badChickens);
    });

    emit('init');
  }
});
...
// 偵測碰撞事件
function detectCollideEvents(chickens: Chicken[], badChickens: BadChicken[]) {
  /** 依據檢查壞雞 */
  badChickens.forEach(({ mesh: badChickenMesh }) => {
    if (!badChickenMesh) return;

    /** 不可能碰到小雞，跳過 */
    if (badChickenMesh.position.z < -1 ||
      badChickenMesh.position.z > 1) return;

    /** 依序檢查小雞 */
    chickens.forEach((chicken) => {
      if (!chicken.mesh) return;

      /** 已停用，跳過 */
      if (chicken.mesh.isDisposed()) return;

      if (badChickenMesh.intersectsMesh(chicken.mesh)) {
        console.log(`${chicken.name} 撞雞`);
      }
```

```
    });
  });
}
...
async function createChickens(scene: Scene) {
  ...
  return [chicken];
}
</script>
```

現在發生碰撞時會在 console 印出「1 撞雞」的文字，接著讓我們實作碰撞效果。

先來梳理一下小雞的狀態變化：

- 一般狀態可以自由移動。

- 被碰撞時，產生翻滾效果且生命值減 1，期間不會再被碰撞。

- 被碰撞狀態持續 2 秒後回復一般狀態，其他無法進行姿態控制。

- 總共 3 點生命值，歸零時，小雞墜機，持續下降且無法控制。

鱈魚：「那就依據建立每個狀態之間切換的功能吧。」

助教：「你說像企鵝那樣嗎？有沒有更好理解的方法啊？(ˋ´ω`ˍ)」

鱈魚：「那我們就用有限狀態機來實現吧。ヽ(●ˋ∀´●)ﾉ」

助教：「狀態雞？這個章節怎麼一堆雞啊......('◉ㅅ◉`)」

所謂的有限狀態機，顧名思義就是指狀態有限的狀態機，狀態機在任意時刻中，一定只能處於一種狀態，觸發事件後可以轉移至特定狀態，

鱈魚：「也就是說如果有狀態 A 和 B，要嘛就只能是狀態 A，不然就是狀態 B。(•ω•)✧」

助教：「講人話好嗎？讀者要跑光啦。ლ(ˇ口ˇ ლ)」

鱈魚：「簡單來就，紅綠燈就是一種狀態機啦。根據時間觸發，只會在紅黃綠之間切換。ㄔ('ε'ㄔ)」

助教：「一開始就這樣解釋不就得了。('●ω●')」

讓我們借用一下 XState 官網的圖片。

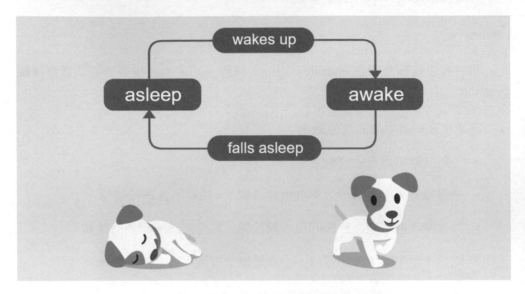

▲ 圖 11-34　狀態與事件

（圖片來源：https://xstate.js.org/docs/guides/introduction-to-state-machines-and-statecharts/#transitions-and-events）

狗狗有 asleep 和 awake 共 2 種狀態：

- asleep 狀態時，觸發 wakes up 事件後會轉移至 awake 狀態

- awake 狀態時，觸發 falls asleep 事件後會轉移至 asleep 狀態

以上就是狀態機的基本概念。

現在讓我們使用 XState 這個套件，我們只要定義有哪些狀態、狀態之間如何切換即可，套件內已經幫我們處理好所有的細節了！、(≧∀≦)、

在命令提示視窗執行以下命令安裝套件。

```
npm i xstate
```

現在讓我們引入 XState 並列舉一下人物的狀態。

➔ src\games\chicken-fly\chicken.ts

```
...
import { createMachine, interpret } from "xstate";

enum State {
  /** 可以自由移動 */
  IDLE = 'idle',
  /** 被攻擊後，人物會翻滾，持續硬直 2s */
  ATTACKED = 'attacked',
  /** 人物死亡，無法控制 */
  DEAD = 'dead',
}
...
```

createMachine 用於建立狀態機，interpret 用於使用狀態機。

接著讓我們在小雞的 Class 中建立狀態機並加入生命值變數吧。

➔ src\games\chicken-fly\chicken.ts

```
...
export class Chicken {
  ...
  /** 儲存被解釋後的狀態機服務 */
  private stateService;

  constructor(name: string, scene: Scene, params?: Params) {
    ...

    const stateMachine = createMachine(
      {
        id: 'chicken',
```

```
      initial: State.IDLE,
      /** 儲存狀態機內部變數 */
      context: {
        health: 3,
      },
      states: {},
      /** 官方建議啟用
       * https://xstate.js.org/docs/guides/actions.html#actions
       */
      predictableActionArguments: true,
    },
    {
      actions: {},
    }
  );

  this.stateService = interpret(stateMachine);
  /** 啟動狀態機服務 */
  this.stateService.start();
} ...
}
```

其中最重要的 2 個參數是 states 與 actions。

states 用來定義每個狀態進出入的動作與可被轉移至何種狀態。

actions 就是定義可被各個 state 呼叫的動作。

讓我們依據剛剛整理的條件慢慢建立 states 與 actions：

- 一般狀態可以自由移動。

讓我們在 states 中新增 idle 狀態。

➔ src\games\chicken-fly\chicken.ts

```
...
export class Chicken {
  ...
  constructor(name: string, scene: Scene, params?: Params) {
```

```
    ...
    const stateMachine = createMachine(
      {
        ...
        states: {
          [State.IDLE]: {
            on: {},
          },
        },
        ...
      },
      ...,
      }
    );
    ...
  }
  ...
}
```

接著是：

■ 被碰撞時，產生翻滾效果且生命值減 1，期間不會再被碰撞。

這就表示 idle 可以觸發 attacked 事件後，轉移至 attacked 狀態，且 actions
需要翻滾與減少生命值的動作。

➡ src\games\chicken-fly\chicken.ts

```
...
export class Chicken {
  ...
  constructor(name: string, scene: Scene, params?: Params) {
    ...
    const stateMachine = createMachine(
      {
        ...
        states: {
          [State.IDLE]: {
            on: {
```

```
            /** 表示呼叫 attacked 事件後會轉移至 attacked 狀態 */
            [State.ATTACKED]: State.ATTACKED,
          },
        },
        [State.ATTACKED]: {
          entry: ['minusHealth', 'tumbling'],
        },
      },
      ...
    },
    {
      actions: {
        minusHealth: (context) => {
          context.health -= 1;
        },
        tumbling: (context) => {
          console.log('tumbling : ', context.health);
        },
      },
    }
  );
}
...
/** 觸發 attacked 事件 */
attacked() {
  const snapshot = this.stateService.getSnapshot();
  if (snapshot.done) return;

  this.stateService.send({ type: State.ATTACKED });
}
...
}
```

entry 表示進入此狀態時需要被執行的動作清單；minusHealth 是減少生命；tumbling 是翻滾動畫，動畫效果讓我們完成狀態機後再來實作。

接下來是：

- 被碰撞狀態持續 2 秒後回復一般狀態，其他無法進行姿態控制。

　　表示 attacked 狀態無法直接轉移至其他狀態，會在 2 秒後自行轉移至 idle
狀態，而且期間無法控制姿態。

➜ src\games\chicken-fly\chicken.ts

```
...
export class Chicken {
  ...
  constructor(name: string, scene: Scene, params?: Params) {
    ...
    const stateMachine = createMachine(
      {
        ...
        states: {
          ...
          [State.ATTACKED]: {
            ...
            after: [
              {
                delay: 2000,
                target: State.IDLE,
              },
            ],
          },
        },
        ...
      },
      ...
    );
  }
  ...
  setAttitude(pitch: number, roll: number) {
    const snapshot = this.stateService.getSnapshot();

    /** attacked 狀態無法控制姿態 */
    if (snapshot.value === State.ATTACKED) return;
    ...
  }
```

```
    ...
}
```

after 用於定義此狀態之後會轉移至何種狀態，可以指定時間與目標。

最後是：

- 總共 3 點生命值，歸零時，小雞墜機，持續下降且無法控制。

表示每次 idle 狀態觸發 attacked 事件時，應該先檢查 health 是否為 0，為 0 則轉移至 dead 狀態，否則轉移至 attacked 狀態。

→ src\games\chicken-fly\chicken.ts

```
...
export class Chicken {
  ...
  constructor(name: string, scene: Scene, params?: Params) {
    ...

    const stateMachine = createMachine(
      {
        ...
        states: {
          [State.IDLE]: {
            on: {
              [State.ATTACKED]: [
                {
                  cond: (context) => context.health > 1,
                  target: State.ATTACKED,
                },
                {
                  target: State.DEAD,
                }
              ],
            },
          },
          ...
          [State.DEAD]: {
```

```
            entry: ['tumbling', 'dead'],
            type: 'final',
          },
        },
        ...
      },
      {
        actions: {
          ...
          dead: () => {
            console.log('dead');
          },
        },
      }
    );
  }
  ...
  setAttitude(pitch: number, roll: number) {
    const snapshot = this.stateService.getSnapshot();

    /** attacked 狀態或最終狀態，皆無法控制姿態 */
    if (snapshot.value === State.ATTACKED || snapshot.done) return;
    ...
  }
  ...
}
```

cond 表示條件判斷，與程式語言中的 if 概念相同，這裡表示 health 大於 1 時，才會轉移至 attacked 狀態，否則前往 dead 狀態。

dead 狀態中 type 為 final，表示此狀態為最後狀態，不會再發生任何轉移。

現在讓我們試試用鍵盤事件直接呼叫小雞的 attacked method。

設定數字鍵 0 可以觸發。

➜ src\games\chicken-fly\game-scene.vue

```
...
<script setup lang='ts'>
```

```
...
async function createChickens(scene: Scene) {
  ...
  window.addEventListener('keydown', (e) => {
    switch (e.key) {
      ...
      case '0': {
        chicken.attacked();
        break;
      }
    }

    ...
  });
  ...
}
</script>
```

現在按下數字鍵 0 後，應該會在 console 中看到以下訊息。

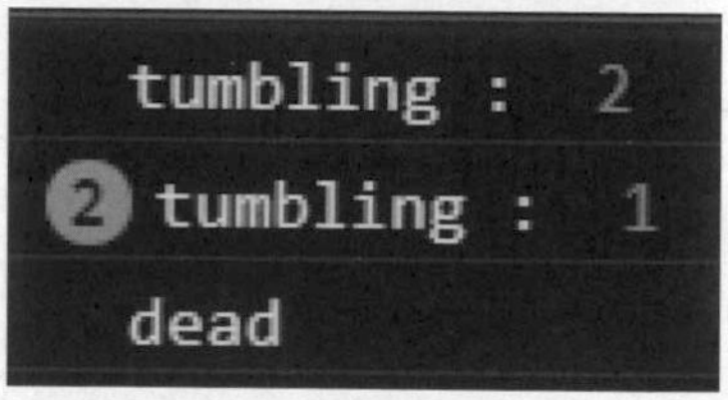

▲ 圖 11-35 狀態轉移訊息

可以發現在按下 3 次數字鍵 0 後，狀態就進入 dead，小雞這時候就會失去控制了。

狀態機用起來是不是更清楚明瞭呢？官方還有視覺化工具喔！ヽ(*´︶`*)ノ

下圖就是我們目前小雞的狀態機。

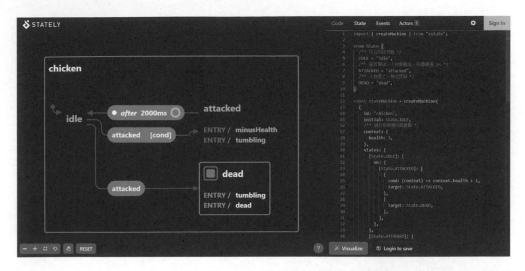

▲ 圖 11-36 狀態機視覺化

 Tips

想要玩玩看的讀者可以參考以下連結：

◆ https://stately.ai/viz

鱈魚：「還可以點擊事件，會動畫呈現轉移效果。(ㅍ_>ㅍ)◊」

助教：「既然狀態機這麼好用，為甚麼企鵝遊戲不用這個？ ㎝(‧´ɪ‧ ㎝)」

鱈魚：「有比較才會有傷害嘛，讀者們可以試試看使用狀態機重構企鵝遊戲的企鵝喔。 ㄜ(´∀`ㄜ)」

現在讓我們實現碰撞和死亡動畫效果吧。

首先調整一下場景中的 counter，在數值到達 150 時停止增加。

➜ src\games\chicken-fly\game-scene.vue

```
...
<script setup lang='ts'>
```

```
...
/** 每秒 +1 */
const {
  counter, pause
} = useInterval(1000, { controls: true });
whenever(
  () => counter.value >= 150,
  () => pause(),
)
...
</script>
```

接著新增 tumbling method，產生旋轉效果，並在 stateMachine actions 的 tumbling 呼叫此 method。

➜ src\games\chicken-fly\chicken.ts

```
...
export class Chicken {
  ...
  constructor(name: string, scene: Scene, params?: Params) {
    ...
    const stateMachine = createMachine(
      ...
      {
        actions: {
          ...
          tumbling: () => {
            this.tumbling();
          },
          ...
        },
      }
    );
    ...
  }

  ...
```

```
/** 限制小雞的移動範圍 */
private limitPosition() {
  ...
  /** 角速度歸零，讓角色停止亂轉 */
  if (snapshot.value === State.IDLE) {
    this.mesh?.physicsImpostor?.setAngularVelocity(new Vector3(0, 0, 0));
  }
}

private tumbling() {
  if (!this.mesh?.physicsImpostor) return;

  /** 加入角速度，小雞就會開始旋轉了 */
  this.mesh.physicsImpostor.setAngularVelocity(
    new Vector3(
      Scalar.RandomRange(-10, 10),
      10,
      Scalar.RandomRange(-10, 10),
    ),
  );
}

async init() {...}
...
}
```

現在按下數字鍵 0，小雞會瘋狂旋轉了。ᕕ(゜∀゜)ᕗ

最後讓我們完成 dead 效果吧。

➜ src\games\chicken-fly\chicken.ts

```
...
export class Chicken {
  ...
  constructor(name: string, scene: Scene, params?: Params) {
    ...
    const stateMachine = createMachine(
      ...
```

```
      {
        actions: {
          ...
          dead: () => {
            this.dead();
          },
        },
      }
    );
    ...
  }


  ...
  private async dead() {
    if (!this.mesh) return;

    /** 移出畫面 */
    const {
      animation, frameRate
    } = createAnimation(
      this.mesh,
      'position',
      this.mesh.position.add(new Vector3(0, -5, 0))
    )

    const animatable = this.scene.beginDirectAnimation(this.mesh, [animation], 0,
frameRate);
    await animatable.waitAsync();

    /** 動畫結束後，停用 mesh */
    this.mesh.dispose();
  }
  async init() {...}
  ...
}
```

現在小雞會在進入 dead 狀態時往下掉，墜雞啦！ ᕙ(˚∀˚。)ᕗ

接下來讓我們準備開始打造搖桿吧。

11.6 Web API 授權

此章節程式碼可以在以下連結取得。

- https://gitlab.com/deepmind1/animal-party/web/-/tree/feat/game-chicken-fly-permission

這個遊戲較為特別的地方在於玩家不是使用虛擬搖桿控制，而是使用手機的陀螺儀，根據手機的姿態控制小雞的飛行姿態。

還記得一開始定義 Player 型別的時候有預留一個 permissions 資料欄位嗎？現在讓我們完成這個部分吧。

permissions 用來表示目前玩家的 Web API 權限狀態，要使用陀螺儀的話我們需要 gyroscope 權限。

讓我們調整一下 Player 型別定義。

➜ src\types\game.type.ts

```
...
/** PermissionState 是 Web 內建定義 */
export type PlayerPermissionState = PermissionState | 'not-support';

export interface PlayerPermission {
  gyroscope: PlayerPermissionState;
}
```

```
/** 玩家 */
export interface Player {
  /** 唯一 ID */
  readonly clientId: string;
  /** 表示玩家手機端允許的 API 清單 */
  permission?: PlayerPermission;
}
...
```

Tips

PermissionState 是 Web API 內健行別，想要深入了解的讀者可以參考以下連結：

◆ https://developer.mozilla.org/en-US/docs/Web/API/PermissionStatus

接下來在我們開發小雞遊戲的搖桿頁面之前，我們必須先幫網頁的 dev server 加上 SSL 證書，變成 https 連線才行，否則無法使用 gyroscope API。

還記得先前開啟 dev server 時，終端機顯示的內容嗎？

▲ 圖 11-37 開啟 dev server

可以注意到網址是 http，現在讓我們把他變成 https 吧。

輸入以下命令安裝必要套件。

```
npm i -D @vitejs/plugin-basic-ssl
```

接著調整 vite 設定。

➔ vite.config.ts

```
...
import basicSsl from '@vitejs/plugin-basic-ssl';

// https://vitejs.dev/config/
export default defineConfig(({ command, mode }) => {
  return {
    plugins: [
      ...

      basicSsl(),
    ],
    server: {
      https: true,
      ...
    }
  }
})
```

現在使用 npm run dev 重新啟動 dev server。

這個時候不同作業系統可能會出現以下訊息：

Windows 出現警告視窗。

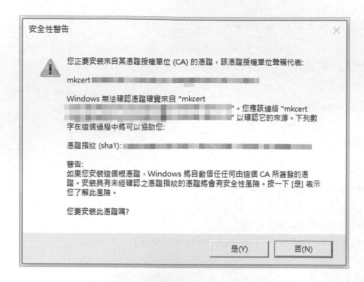

▲ 圖 11-38 Window 安全性警告

Mac 需要 sudo 權限產生 SSL 證書,所以要輸入密碼

▲ 圖 11-39 Mac 需要 sudo 權限

按下「是」或輸入密碼後繼續就可以了,最後會發現終端機的內容變成。

▲ 圖 11-40 開啟 https dev server

https 出現了!(´„•ω•„)

現在要改用 https 的網址打開網頁,但是大家可能會發現網頁出現如下圖警告。

▲ 圖 11-41　連線警告

這是因為瀏覽器不認得這個 SSL 證書，所以警告使用者要注意，但是不用擔心，不認得很正常，因為證書是我們自己產生的嘛。(´▽`)∕

按「進階」後，點擊「繼續前往 192.168.0.XXX 網站」就對了！(˶′•ω•˶)

現在讓我們新增小雞遊戲的搖桿頁面吧！

鱈魚：「基本上就是把企鵝的搖桿複製......」

助教：「...('◉ ◞ ◉')」

鱈魚：「你是又有甚麼意見了？(ᴖ ᵕ ᴖ')」

助教：「程式碼又重複了欸。('◉ ◞ ◉')」

鱈魚：「好啦，讓我們調整一下吧。ヽ(●ﾟ∀ﾟ●)ﾉ 」

仔細觀察後可以發現，搖桿的結構、螢幕轉向警告等等其實都可以重複使用，讓我們將這些重複使用的東西提取出來，變成可重複使用的元件吧。

→ src\components\player-gamepad-container.vue

```vue
<template>
  <div
    class="w-full h-full flex text-white select-none overflow-hidden"
    :class="bgClass"
    @touchmove="(e) => e.preventDefault()"
  >
    <slot />

    <div class="code-name">
      {{ codeName }}
    </div>

    <q-dialog
      v-model="isWrongOrientation"
      persistent
    >
      <q-card class="p-8">
        <q-card-section class="flex flex-col items-center gap-6">
          <q-spinner-box
            color="primary"
            size="10rem"
          />
          <div class="text-4xl">
            請將手機轉為 {{ targetOrientation }}
          </div>
          <div class="text-base">
            轉為 {{ targetOrientation }} 後，此視窗會自動關閉
          </div>
        </q-card-section>
      </q-card>
    </q-dialog>
  </div>
</template>
```

```
<script setup lang="ts">
import { computed } from 'vue';

import { useClientPlayer } from '../composables/use-client-player';
import { useScreenOrientation } from '@vueuse/core';

interface Props {
  /** 指定螢幕方向，為空表示不限制 */
  orientation?: 'landscape' | 'portrait';
}
const props = withDefaults(defineProps<Props>(), {
  orientation: undefined,
});

const { orientation } = useScreenOrientation();
const { codeName, colorName } = useClientPlayer();

const isWrongOrientation = computed(() => {
  if (!props.orientation) {
    return false;
  }

  return orientation.value?.includes(props.orientation) ?? false;
});
const bgClass = computed(() => `bg-${colorName.value}`);

const targetOrientation = computed(() =>
  props.orientation === 'landscape' ? '直向' : '橫向'
);
</script>

<style scoped lang="sass">
.code-name
  position: absolute
  top: 0
  left: 50%
  transform: translateX(-50%)
  display: flex
```

```
    justify-content: center
    padding: 0.1rem
    font-size: 10rem
    text-shadow: 0px 0px 2rem rgba(#000, 0.5)
</style>
```

可以看到原本放按鈕的地方我們保留了 slot，這樣各個遊戲就可以依據具體需求加入按扭了。

接著新增小雞遊戲搖桿，先加入一個確認按鈕。

➜ src\views\the-player-gamepad-chicken-fly.vue

```
<template>
  <player-gamepad-container>
    <gamepad-btn
      class="absolute top-10 right-20 opacity-90"
      size="3rem"
      icon="done"
      @trigger="(status) => handleBtnTrigger('confirm', status)"
    />
  </player-gamepad-container>
</template>

<script setup lang="ts">
import { computed, onMounted } from 'vue';
import { KeyName } from '../types';

import GamepadBtn from '../components/gamepad-btn.vue';
import PlayerGamepadContainer from '../components/player-gamepad-container.vue';

import { useLoading } from '../composables/use-loading';
import { useClientPlayer } from '../composables/use-client-player';

const loading = useLoading();
const { emitGamepadData } = useClientPlayer();

function handleBtnTrigger(keyName: `${KeyName}`, status: boolean) {
  emitGamepadData([{
    name: keyName,
```

```
    value: status,
  }]);
}

onMounted(() => {
  loading.hide();
});
</script>
```

是不是清爽很多呢？這個就是元件化的優點。(ㅍ‿ㅍ)◇

開始開發小雞搖桿之前，我們需要在大廳搖桿中取得感測器相關的權限。

現在讓我們回到大廳搖桿，第一步讓我們用 player-gamepad-container 重構一下吧。

➜ src\views\the-player-gamepad-lobby.vue

```
<template>
  <player-gamepad-container orientation="portrait">
    <gamepad-d-pad
      class="absolute bottom-5 left-8 opacity-90"
      @trigger="({ keyName, status }) => handleBtnTrigger(keyName, status)"
    />
    <gamepad-btn
      class="absolute bottom-10 right-20 opacity-90"
      size="6rem"
      icon="done"
      @trigger="(status) => handleBtnTrigger('confirm', status)"
    />
  </player-gamepad-container>
</template>

<script setup lang="ts">
import { onMounted } from 'vue';
import { KeyName } from '../types';

import PlayerGamepadContainer from '../components/player-gamepad-container.vue';
import GamepadBtn from '../components/gamepad-btn.vue';
import GamepadDPad from '../components/gamepad-d-pad.vue';
```

```
import { useLoading } from '../composables/use-loading';
import { useClientPlayer } from '../composables/use-client-player';

const loading = useLoading();
const { emitGamepadData } = useClientPlayer();

function handleBtnTrigger(keyName: `${KeyName}`, status: boolean) {
  emitGamepadData([{
    name: keyName,
    value: status,
  }]);
}

onMounted(() => {
  loading.hide();
});
</script>
```

極度清爽！(·ω·)ノ ヽ(·ω·)

現在讓我們新增一個元件，用來取得或確認 Web API 授權。

➜ src\components\permission-card.vue

```
<template>
  <q-card>
    permission-card
  </q-card>
</template>

<script setup lang="ts">
import { cloneDeep } from 'lodash-es';
import { ref } from 'vue';
import { PlayerPermission } from '../types';

const emit = defineEmits<{
  (e: 'update', data: PlayerPermission): void;
}>();
</script>
```

```
<style scoped lang="sass">
</style>
```

　　將 permission-card 放到 the-player-gamepad-lobby 中並用 dialog 包裹，這樣就可以用 dialog 的方式打開 permission-card。

➜ src\views\the-player-gamepad-lobby.vue

```
<template>
  <player-gamepad-container orientation="portrait">
    ...
    <q-dialog v-model="permissionCardVisible">
      <permission-card />
    </q-dialog>
  </player-gamepad-container>
</template>

<script setup lang="ts">
...
const permissionCardVisible = ref(false);
...
</script>
```

　　現在讓我把 App.vue 中的遊戲場景刪除，改成放 the-player-gamepad-lobby，方便開發。

➜ src\App.vue

```
<template>
  ...
  <the-player-gamepad-lobby class="  w-full h-full z-10" />
</template>

<script setup lang="ts">
...
import ThePlayerGamepadLobby from './views/the-player-gamepad-lobby.vue';
</script>

...
```

讓我們在大廳搖桿加入開啟 dialog 的按鈕。

➡ src\views\the-player-gamepad-lobby.vue

```
<template>
  <player-gamepad-container orientation="portrait">
    ...
    <gamepad-btn
      class="absolute top-10 right-20 opacity-90"
      size="4rem"
      icon="api"
      @click="openPermissionCard()"
    />
    ...
  </player-gamepad-container>
</template>

<script setup lang="ts">
...
const permissionCardVisible = ref(false);
function openPermissionCard() {
  permissionCardVisible.value = true;
}
...
</script>
```

▲ 圖 11-42 加入開啟 dialog 按鈕

接著讓我們慢慢完成 permission-card 內容。

我們使用 VueUse 提供的 usePermission 取得授權相關功能。

➡ src\components\permission-card.vue

```
...
<script setup lang="ts">
import { usePermission, refDefault } from '@vueuse/core';
import { ref, watch } from 'vue';
import { PlayerPermission } from '../types';

const emit = defineEmits<{
  (e: 'update', data: PlayerPermission): void;
}>();

const gyroscopePermission = usePermission('gyroscope', { controls: true });
const gyroscopeState = refDefault(gyroscopePermission.state, 'not-support');
</script>
...
```

Tips

想要深入了解的讀者可以參考以下連結：

- https://vueuse.org/core/usePermission/#usepermission

接著列舉狀態詳細資訊並提供取得資訊 function。

➡ src\components\permission-card.vue

```
...
<script setup lang="ts">
...
/** 狀態詳細資訊 */
const stateInfoMap: Record<PlayerPermissionState, {
  icon: string;
  color: string;
```

```
    description: string;
}> = {
  'granted': {
    icon: 'check_circle',
    color: 'green',
    description: '已同意',
  },
  'denied': {
    icon: 'cancel',
    color: 'deep-orange',
    description: '授權被拒絕',
  },
  'prompt': {
    icon: 'info',
    color: 'cyan',
    description: '尚未進行授權',
  },
  'not-support': {
    icon: 'help',
    color: 'grey',
    description: '不支援此 API',
  },
}
function getStateInfo(value: PlayerPermissionState) {
  return stateInfoMap?.[value] ?? stateInfoMap['not-support'];
}
</script>
...
```

　　並用 computed 列出權限清單。

➜ src\components\permission-card.vue

```
...
<script setup lang="ts">
...
const permissions = computed<{
  key: keyof PlayerPermission,
  icon: string;
```

```
  label: string;
  caption: string;
  state: PlayerPermissionState;
  stateInfo: ReturnType<typeof getStateInfo>;
  onClick: UsePermissionReturnWithControls['query'];
}[]>(() => ([
  {
    key: 'gyroscope',
    icon: 'screen_rotation_alt',
    label: ' 陀螺儀 ',
    caption: ' 可以偵測手機旋轉角度，通常用於體感遊戲 ',
    state: gyroscopeState.value,
    stateInfo: getStateInfo(gyroscopeState.value),
    onClick: gyroscopePermission.query,
  },
]));
</script>
...
```

最後用 watch 偵測 permissions，變更時使用 emit 發出資料。

➔ src\components\permission-card.vue

```
...
<script setup lang="ts">
...
watch(permissions, (data) => {
  const result = data.reduce((acc, item) => {
    acc[item.key] = item.state;
    return acc;
  }, {} as PlayerPermission);

  emit('update', result);
}, {
  deep: true,
});</script>
...
```

以上我們就完成程式邏輯的部分了，最後讓我們完成 template 內容，列出
所有 API 授權狀態，並用 base-polygon 裝飾一下。

➜ src\components\permission-card.vue

```
<template>
  <q-card>
    <q-card-section class=" relative bg-teal-5 text-white overflow-hidden">
      <div class=" text-4xl  font-bold">
        Web API 授權清單
      </div>
      <div class="text-2xl flex flex-col gap-2 mt-4">
        <div>
          當狀態為 <q-icon name="info" /> 時，點擊對應項目進行授權
        </div>
        <div>
          若狀態為 <q-icon name="cancel" /> 時，請在瀏覽器設定中允許對應 API 權限
        </div>
      </div>

      <base-polygon
        opacity="0.3"
        size="20rem"
        class=" absolute -top-[13rem] -right-[3rem]"
        rotate="60deg"
        shape="pentagon"
      />
    </q-card-section>

    <q-card-section class=" relative overflow-hidden">
      <q-list>
        <q-item
          v-for="permission in permissions"
          :key="permission.key"
          v-ripple
          clickable
          @click="permission.onClick"
        >
          <q-item-section
```

```
          avatar
          top
      >
        <q-avatar
          :icon="permission.icon"
          color="grey"
          text-color="white"
        />

      </q-item-section>

      <q-item-section>
        <q-item-label class=" text-3xl">
          {{ permission.label }}
        </q-item-label>
        <q-item-label class=" text-2xl">
          {{ permission.caption }}
        </q-item-label>
      </q-item-section>

      <q-item-section side>
        <q-icon
          :name="permission.stateInfo.icon"
          :color="permission.stateInfo.color"
        />
      </q-item-section>
    </q-item>
  </q-list>

  <base-polygon
    opacity="0.1"
    size="22rem"
    class=" absolute -bottom-[13rem] -left-[3rem]"
    rotate="60deg"
    color="#AAA"
    shape="round"
    fill="spot"
  />
 </q-card-section>
</q-card>
```

```
</template>

<script setup lang="ts">
...
import BasePolygon from './base-polygon.vue';
...
</script>
```

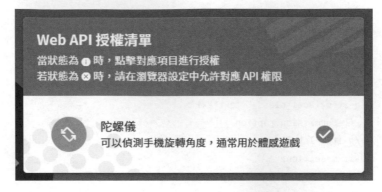

▲ 圖 11-43　Web API 授權清單

現在讓我們回到大廳搖桿，來接收一下 permission-card 提供的資料。

- 新增 handlePermission 用於接收資料變更。

- permissionCardVisible 預設為 true

現在 dialog 出現時，console 會跑出如下圖資訊。

▲ 圖 11-44　permission 資料

接下來我們需要將 permission 資料傳送至 server，新增發送玩家資料事件。

➜ src\types\socket.type.ts

```
...

interface OnEvents {
  ...
}

interface EmitEvents {
  ...
  'player:profile': (data: Player) => void;


  ...
}
...
```

接著在 use-client-player 新增發送資料用的 function。

➜ src\composables\use-client-player.ts

```
...
export function useClientPlayer() {
  ...
  async function emitProfile(data: Omit<Player, 'clientId'>) {
    if (!mainStore.client?.connected) {
      return Promise.reject('client 尚未連線 ');
    }

    mainStore.client.emit('player:profile', {
      clientId: mainStore.clientId,
      ...data,
    });
  }

  return {
    ...
    /** 發送玩家自身資料 */
    emitProfile,
  }
}
```

最後在大廳搖桿中呼叫 emitProfile。

以上完成發送的部分了，讓我們前往久違的 server 專案吧。

11.6.1 伺服器同步 permission 資料

此章節程式碼可以在以下連結取得。

- https://gitlab.com/deepmind1/animal-party/server/-/tree/feat/update-player-permission

第一步當然是調整型別定義，首先是 Player 部分。

➔ src\game-console\game-console.type.ts

```
...
/** PermissionState 是 Web 內建定義 */
export type PlayerPermissionState = PermissionState | 'not-support';

export interface PlayerPermission {
  gyroscope: PlayerPermissionState;
}

/** 玩家 */
export interface Player {
  /** 唯一 ID */
  readonly clientId: string;
  /** 表示玩家手機端允許的 API 清單 */
  permission?: PlayerPermission;
}
...
```

接著新增事件定義，client 端是 emit 事件，server 端當然是 on 事件囉。
(ェ_>ェ)✧

➜ types\socket.type.ts

```
...
export interface OnEvents {
  ...
  'player:profile': (data: Player) => void;

  ...
}
...
```

server 端實現很單純，其實就只是將玩家搖桿資訊轉送至遊戲機網頁即可，
所以我們新增一個事件表示從 server 發送玩家資料至遊戲機。

➜ types\socket.type.ts

```
...
export interface EmitEvents {
  ...
  'game-console:profile-update': (data: Player) => void;
}
...
```

新增處理此事件的 method 吧。

➜ src\game-console\game-console.gateway.ts

```
...
import {
  GameConsoleState,
  Player,
  UpdateGameConsoleState,
} from './game-console.type';
...
export class GameConsoleGateway {
  ...
```

```
@SubscribeMessage<keyof OnEvents>('player:profile')
async handlePlayerProfile(socket: ClientSocket, data: Player) {
  const client = this.wsClientService.getClient({ socketId: socket.id });
  if (!client) {
    const result: SocketResponse = {
      status: 'err',
      message: '此 socket 不存在 client',
    };
    return result;
  }

  const room = this.roomService.getRoom({ playerId: client.id });
  if (!room) {
    const result: SocketResponse = {
      status: 'err',
      message: 'client 未加入任何房間 ',
    };
    return result;
  }

  const founderClient = this.wsClientService.getClient({
    clientId: room.founderId,
  });
  if (!founderClient) {
    const result: SocketResponse = {
      status: 'err',
      message: '此 socket 不存在 client',
    };
    return result;
  }

  /** 發送至遊戲機網頁 */
  const targetSocket = this.server.sockets.sockets.get(
    founderClient.socketId,
  );
  if (!targetSocket) {
    const result: SocketResponse = {
      status: 'err',
```

```
      message: '不存在 room founder 對應之 Client',
    };
    return result;
  }
  targetSocket.emit('game-console:profile-update', data);

  const result: SocketResponse = {
    status: 'suc',
    message: '傳輸資料成功',
    data: undefined,
  };
  return result;
  }
}
```

鱈魚：「server 部分完成了！(´▽`)ﾉ」

助教：「這麼快嗎？╭(°A,°`)╮ 」

鱈魚：「因為重點不是 server 嘛。ヽ(●`∀´●)ﾉ」

11.7 遊戲機接收資料

現在讓我們回到遊戲機網頁，準備接收玩家傳送過來的資料吧。

首先還是新增一下新事件定義。

➔ src\types\socket.type.ts

```
...
interface OnEvents {
  ...
  'game-console:profile-update': (data: Player) => void;
}
...
```

現在讓我們在 use-client-game-console 新增接收 profile 的 hook 吧。

➜ src\composables\use-client-game-console.ts

```
...
export function useClientGameConsole() {
  ...
  const profileUpdateHook = createEventHook<Player>();

  onBeforeUnmount(() => {
   ...
    mainStore.client?.removeListener('game-console:profile-update', profileUpdateHook.
trigger);
  });

  return {
   ...

    /** 玩家資料事件，例如 Web API 權限更新等等 */
    onProfileUpdate: (fn: Parameters<typeof profileUpdateHook['on']>[0]) => {
      mainStore.client?.on('game-console:profile-update', profileUpdateHook.trigger);
      return profileUpdateHook.on(fn);
    },

    ...
  }
}
```

現在我們必須調整一下 game-console.store 中的 updateState 內部邏輯，因為目前的作法是直接覆蓋玩家清單。

➜ src\stores\game-console.store.ts

```
...
export const useGameConsoleStore = defineStore('game-console', () => {
  ...

  function updateState(state: UpdateStateParams) {
    ...
```

```
    if (state.players) {
      /** 將原本玩家清單的資料儲存至新的清單中 */
      players.value = state.players.map(({ clientId }) => {
        const target = players.value.find(
          (player) => player.clientId === clientId
        );

        return {
          clientId,
          ...target,
        }
      });
    }
  }

  ...
})
...
```

接著我們需要新增一個 function，用來將新的玩家資料更新或加入現有的玩
家清單中。

➡ src\stores\game-console.store.ts

```
...
export const useGameConsoleStore = defineStore('game-console', () => {
  ...

  function updateProfile(data: Player) {
    /** 檢查是否已存在 */
    const index = players.value.findIndex(({ clientId }) =>
      data.clientId === clientId
    );

    /** 不存在，新增 */
    if (index < 0) {
      players.value.push(data);
      return;
```

```
    }

    /** 更新資料 */
    players.value[index] = data;
  }

  return {
    ...
    updateProfile,
  }
})
...
```

最後讓我們到負責註冊玩家相關事件的 the-game-console 元件內，接收資料吧。

➜ src\views\the-game-console.vue

```
...

<script setup lang="ts">
...
function init() {
  ...

  gameConsole.onPlayerUpdate(...);
  gameConsole.onProfileUpdate((player) => {
    gameConsoleStore.updateProfile(player);
  });
  ...
}
...
</script>
...
```

現在讓我們把 App.vue 中的大廳搖桿刪除，完整的開一次派對，看看資料有沒有傳輸成功。

讓我們在 devtool 看看。

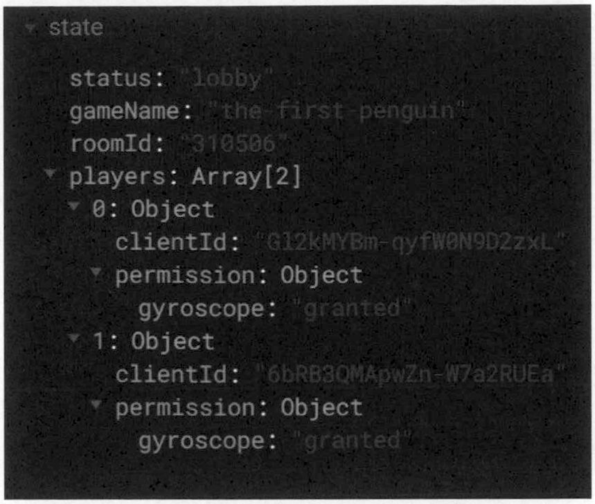

```
▼ state
    status: "lobby"
    gameName: "the-first-penguin"
    roomId: "310506"
  ▼ players: Array[2]
    ▼ 0: Object
        clientId: "G12kMYBm-qyfW0N9D2zxL"
      ▼ permission: Object
          gyroscope: "granted"
    ▼ 1: Object
        clientId: "6bRB3QMApwZn-W7a2RUEa"
      ▼ permission: Object
          gyroscope: "granted"
```

▲ 圖 11-45 接收玩家資料

現在我們可以收到玩家的 API 授權資料了！✧*。٩(´ㅇ`*)و✧*。

鱈魚：「在進入開始開發小雞搖桿之前，需要加入一個重要限制。」

助教：「這麼突然？甚麼限制啊？(*´･д･)?」

鱈魚：「玩遊戲要付費解鎖！ ੭ (´∀`੭)」

助教：（掏出球棒 (╪⊙д⊙)）

鱈魚：「說笑的啦，是遊戲遊玩條件啦。 ੭('ε'੭)」

小雞快飛遊戲一定要陀螺儀控制，也就是所有參加玩家都必須先取得陀螺儀權限才行，否則不能開始遊戲，讓我們來實作這個邏輯吧。

首先定義一下遊玩條件並在 emit 追加錯誤事件。

➜ src\components\game-menu.vue

```
...
<script setup lang="ts">
...
```

```
interface GameInfo {
  name: `${GameName}`;
  routeName: `${RouteName}`;
  /** 遊戲開始條件，如果任一項不符合會發出錯誤 */
  condition: {
    minPlayers: number;
    maxPlayers?: number;
    requiredPermissions: (keyof PlayerPermission)[]
  },
}

const emit = defineEmits<{
  ...
  (e: 'error', message: string): void;
}>();
...
</script>
...
```

接著將現有的遊戲依據定義加入內容。

➜ src\components\game-menu.vue

```
...
<script setup lang="ts">
...
const games: GameInfo[] = [
  {
    name: 'the-first-penguin',
    routeName: 'game-console-the-first-penguin',
    condition: {
      minPlayers: 2,
      /** 企鵝太多浮冰會塞不下 XD */
      maxPlayers: 6,
      requiredPermissions: [],
    }
  },
  {
    name: 'chicken-fly',
    routeName: 'game-console-chicken-fly',
```

```
    condition: {
      minPlayers: 2,
      maxPlayers: 30,
      requiredPermissions: [
        'gyroscope',
      ],
    }
  },
];
...
</script>
...
```

接著列舉權限相關文字訊息，這樣提示的時候玩家才看得懂。

➔ src\types\game.type.ts

```
...

export interface PlayerPermission {
  gyroscope: PlayerPermissionState;
}

export const permissionInfoMap: Record<keyof PlayerPermission, {
  label: string;
}> = {
  'gyroscope': {
    label: '陀螺儀',
  },
}
...
```

新增 checkGameCondition function 用來檢查是否有任何問題，並在 start Game 呼叫。

➔ src\components\game-menu.vue

```
...
<script setup lang="ts">
...
```

```
function checkGameCondition(condition: GameInfo['condition'], players: Player[]) {
  const {
    minPlayers, maxPlayers, requiredPermissions,
  } = condition;

  if (players.length < minPlayers) {
    return ` 人數太少惹，不能小於 ${minPlayers} 個小夥伴。`;
  }

  if (maxPlayers && players.length > maxPlayers) {
    return ` 太多人啦，不能超過 ${maxPlayers} 個人，請狠下心減少人數。`;
  }

  /** 檢查是否有玩家不符合資格 */
  for (const player of players) {
    for (const name of requiredPermissions) {
      if (player.permission?.[name] === 'granted') continue;

      const codeName = gameConsole.getPlayerCodeName(player.clientId);
      const permissionLabel = permissionInfoMap[name].label;
      return `${codeName} 玩家缺少「${permissionLabel}」權限，請點擊搖桿畫面右上角按鈕確認
權限狀態。`;
    }
  }

  return undefined;
}

const startGame = debounce(async () => {
  const game = selectedGame.value;
  const players = gameConsole.players.value;

  const error = checkGameCondition(game.condition, players);
  if (error) {
    emit('error', error);
    return;
  }

  gameConsole.setStatus('playing');
  gameConsole.setGameName(game.name);
```

```
  ...
}, 3000, {
  ...
});
...
</script>
...
```

最後在 the-game-console-lobby 接收錯誤訊息並用 notify 顯示吧。

➡ src\views\the-game-console-lobby.vue

```
<template>
  <game-menu
    ...
    @error="handleError"
  />

  ...
</template>

<script setup lang="ts">
...
function handleError(message: string) {
  $q.notify({
    type: 'negative',
    message
  });
}
...
</script>
```

現在故意將陀螺儀權限關閉，在小雞快飛遊戲按下開始，就會跑出錯誤訊息了！

▲ 圖 11-46 缺少必要權限

鱈魚：「現在讓我們來打造陀螺儀......」

助教：「等等啊！遊戲選單還沒辦法用搖桿控制要選哪個遊戲欸！('๑╯╰๑') 」

鱈魚：「對吼。ヾ(๑'ᵕ'๑)ﾉﾞ 」

讓我們把搖桿左右按鍵呼叫切換遊戲功能，同時重構一下。

➜ src\views\the-game-console-lobby.vue

```ts
<template>
  <game-menu
    ...
    @error="handleError"
  />

  ...
</template>

<script setup lang="ts">
...
/** 按鍵事件表，僅用到部分 KeyName */
type GamepadEventMap = {
  [K in KeyName]?: () => void;
};
const gamepadEventMap: GamepadEventMap = {
  'up': () => menu.value?.prev(),
  'down': () => menu.value?.next(),
  'left': () => menu.value?.prevGame(),
  'right': () => menu.value?.nextGame(),
  'confirm': () => menu.value?.click(),
}

function init() {
  gameConsole.setStatus('lobby');
  loading.hide();

  gameConsole.onGamepadData((data) => {
```

```
    const datum = data.keys.at(-1);
    if (!datum) return;

    const { name, value } = datum;
    if (value) return;

    gamepadEventMap[name]?.();
  });
}
</script>
```

同時讓我們在 useClientPlayer 追加 onPlayerUpdate，因為玩家搖桿也要知道目前玩家是否更新。

→ src\composables\use-client-player.ts

```
...
export function useClientPlayer() {
  ...
  const stateUpdateHook = createEventHook<GameConsoleState>();
  const playerUpdateHook = createEventHook<Player[]>();

  /** 元件解除安裝前，移除 Listener 以免記憶體洩漏 */
  onBeforeUnmount(() => {
    ...
    mainStore.client?.removeListener('game-console:player-update', playerUpdateHook.
trigger);
  });

  ...
  return {
    ...
    /** 玩家變更事件，例如玩家加入或斷線等等 */
    onPlayerUpdate: (fn: Parameters<typeof playerUpdateHook['on']>[0]) => {
      mainStore.client?.on('game-console:player-update', playerUpdateHook.trigger);
      return playerUpdateHook.on(fn);
    },

    ...
```

```
    }
}
```

接著在 the-player-gamepad 註冊事件。

→ src\views\the-player-gamepad.vue

```
...
<script setup lang="ts">
...
function init() {
  ...

  player.onGameConsoleStateUpdate((state) => {
    ...
  });

  player.onPlayerUpdate((players) => {
    console.log(`[ onPlayerUpdate ] players : `, players);

    gameConsoleStore.updateState({ players });
  });

  ...
}
init();
</script>
```

接下來讓我們開發姿態控制搖桿吧！ ヽ(*´︶`*)ノ

建立姿態控制搖桿

此章節程式碼可以在以下連結取得。

- https://gitlab.com/deepmind1/animal-party/web/-/tree/feat/game-chicken-
  fly-gamepad

現在讓我們回到小雞搖桿，讓我們偵測陀螺儀的資料吧。

老樣子，為了方便開發，讓我們把小雞搖桿加到 App.vue 中吧。

➜ src\App.vue

```
<template>
  ...

  <the-player-gamepad-chicken-fly class=" z-10" />
</template>

<script setup lang="ts">
import LoadingOverlay from './components/loading-overlay.vue';
import ThePlayerGamepadChickenFly from './views/the-player-gamepad-chicken-fly.vue';
</script>

...
```

接著新增姿態控制元件。

→ src\components\gamepad-device-motion.vue

```
<template>
  <div
    class="rounded-full bg-grey-10"
    :style="padStyle"
  >

  </div>
</template>

<script setup lang="ts">
export interface Angle {
  x: number;
  y: number;
  z: number
}

interface Props {
  /** 尺寸，直徑 */
  size?: string
}
const props = withDefaults(defineProps<Props>(), {
  size: '30rem'
});

const emit = defineEmits<{
  (e: 'angle', angle: Angle): void;
}>();

const padStyle = computed(() => {
  return {
    width: props.size,
    height: props.size,
  }
});
```

```
</script>

<style scoped lang="sass">

</style>
```

在搖桿中引入陀螺儀元件並調整確認按鈕位置。

➜ src\views\the-player-gamepad-chicken-fly.vue

```
<template>
  <player-gamepad-container>
    <gamepad-btn
      class="absolute top-1/2 right-1/2 translate-x-1/2 opacity-90"
      ...
    />

    <gamepad-device-motion class="absolute bottom-16 left-1/2 -translate-x-1/2
opacity-90" />
  </player-gamepad-container>
</template>

<script setup lang="ts">
...
import GamepadDeviceMotion from '../components/gamepad-device-motion.vue';
...
</script>
```

目前看起來長這樣。

▲ 圖 11-47　小雞快飛搖桿

由於預期使用手機姿態控制，所以我們就不限制橫握手機，以免自動旋轉亂轉。

預期會長得有點像類比搖桿，差在中間的圓點位置不是觸控，而是依照手機姿態移動，而且整體會有一點 3D 偏移的效果。

說起來很抽象，還是讓我們直接一步一步實現吧。(´▽`)/

第一步是將陀螺儀的數值取出來。

➔ src\components\gamepad-device-motion.vue

```
...
<script setup lang="ts">
import { computed, ref } from 'vue';

import { useDeviceMotion } from '@vueuse/core';
```

```
interface Angle {
  x: number;
  y: number;
  z: number
}

interface Props { ... }
...
const emit = defineEmits<{...}>();

const deviceMotion = useDeviceMotion();

function toFixed(number: number, fractionDigits = 3) {
  return Number(number.toFixed(fractionDigits));
}

const rotationRate = computed(() => {
  if (!deviceMotion.rotationRate.value) {
    return {
      alpha: 0,
      beta: 0,
      gamma: 0,
    }
  }

  const {
    alpha, beta, gamma
  } = deviceMotion.rotationRate.value;

  return {
    alpha: toFixed(alpha ?? 0),
    beta: toFixed(beta ?? 0),
    gamma: toFixed(gamma ?? 0),
  }
});
</script>
```

現在我們有角速度了，所以我們到底要怎麼知道速度呢？角速度積分後就是角度，所以我們只要把角速度積分就可以啦。ㄟ(•ω•)ㄏ

這裡我們使用 requestAnimationFrame 實現，requestAnimationFrame 可以要求瀏覽器在下次重繪畫面前呼叫特定函數，這樣積分的過程就會同步於畫面更新速度。

Tips

想要深入了解的讀者可以參考以下連結：

◆ https://developer.mozilla.org/zh-TW/docs/Web/API/window/requestAnimationFrame

➜ src\components\gamepad-device-motion.vue

```
...
<script setup lang="ts">
...
const angle = ref<Angle>({
  x: 0,
  y: 0,
  z: 0,
});

/** 紀錄上次時間，才有辦法計算時間變化量 */
let previousTimestamp = 0;
const integrateRotationRate: Parameters<typeof requestAnimationFrame>[0] = (timestamp)
 => {
  if (previousTimestamp) {
    const deltaTime = timestamp - previousTimestamp;

    const {
      alpha, beta, gamma,
    } = rotationRate.value;

    /** 積分就是目前值加上速度乘以時間變化量
```

```
    *
    * 除以 1000 是因為角速度單位是 degree/s，而 requestAnimationFrame 提供的時間是毫秒
（ms），
    * 所以要除以 1000
    */
   angle.value.z += alpha * deltaTime / 1000;
   angle.value.x += beta * deltaTime / 1000;
   angle.value.y += gamma * deltaTime / 1000;

   angle.value.x = toFixed(angle.value.x);
   angle.value.y = toFixed(angle.value.y);
   angle.value.z = toFixed(angle.value.z);
 }

 previousTimestamp = timestamp;
 requestAnimationFrame(integrateRotationRate);
}
requestAnimationFrame(integrateRotationRate);

</script>
```

現在我們把 angle 放到 template 中，實際看看效果。

➜ src\components\gamepad-device-motion.vue

```html
<template>
  <div
    class="rounded-full bg-grey-10"
    :style="padStyle"
  >
    <div class=" mt-5">
      <div class="text-4xl">
        angle:
      </div>
      <div>
        x: {{ angle.x }}
      </div>
      <div>
        y: {{ angle.y }}
```

```
      </div>
      <div>
        z: {{ angle.z }}
      </div>
    </div>
  </div>
</template>
...
```

現在用手機打開網頁看看，順利的話會出現如下圖效果。

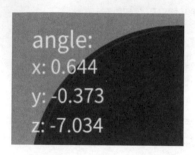

▲ 圖 11-48　角度變化

我們成功取得手機的角度變化了。ᕦ(ﾟ∀ﾟ)ᕤ

 Tips

可以注意到我們取得的角度其實並非手機真正的姿態角度，只是角度變化量累積得到的相對角度。

想要更準確的角度，其實要使用加速度計資料加上角速度進行融合濾波，得到的角度才是真正的手機姿態角度，不過作為遊戲控制搖桿，目前的設計不但簡單且非常夠用了。(´▽`)/

　　但是由於感測器誤差等等因素，時間一久，會因為誤差累積導致角度發生偏移。

所以讓我們加入歸零的功能吧。

➡ src\components\gamepad-device-motion.vue

```
<template>
  <div
    v-ripple
    ...
    @click="resetAngle"
  >
  </div>
</template>

<script setup lang="ts">
...
function resetAngle() {
  angle.value = {
    x: 0,
    y: 0,
    z: 0
  }
}
...
</script>
...
```

現在點擊黑色圓圈，角度就會歸零，接著讓我們加上中間圓點，概念與類比搖桿相同。

➡ src\components\gamepad-device-motion.vue

```
<template>
  <div
    class="rounded-full bg-grey-10 flex justify-center items-center"
  >
    <div
      class="thumb"
    />
  </div>
```

```
</template>

<script setup lang="ts">
...
</script>

<style scoped lang="sass">
.thumb
  width: 40%
  height: 40%
  background: white
  border-radius: 9999px
  opacity: 0.8
</style>
```

▲ 圖 11-49　加入圓點

　　接著根據角度計算偏移量，我們預設最大角度是 45 度，以 x 方向為例，偏移範圍要從 +-45 度映射至 +- pad 寬度的一半。

　　先在 utils 新增一個用來映射的 function。

➜ src\common\utils.ts

```
...
/** 將 value 從 a 範圍映射至 b 範圍 */
```

```ts
export function mapRange(
  value: number,
  aMin: number, aMax: number,
  bMin: number, bMax: number
) {
  return bMin + ((value - aMin) * (bMax - bMin)) / (aMax - aMin);
}
```

➜ src\components\gamepad-device-motion.vue

```
<template>
  <div
    ref="pad"
    ...
  >
    <div
      class="thumb"
      :style="thumbStyle"
    />
  </div>
</template>

<script setup lang="ts">
...

const pad = ref<HTMLElement>();
const { width, height } = useElementSize(pad);

const thumbStyle = ref({
  transform: `translate(0px, 0px)`,
});

watch(angle, (newAngle) => {
  const maxAngle = props.maxAngle;
  const maxX = width.value / 2;
  const maxY = height.value / 2;

  /** 螢幕的 x 方向對應 x 軸旋轉量 */
  const x = mapRange(
```

```
      clamp(newAngle.x, -maxAngle, maxAngle),
      -maxAngle, maxAngle, -maxX, maxX,
    );
    /** 螢幕的 y 方向對應 z 軸旋轉量 */
    const y = mapRange(
      clamp(newAngle.z, -maxAngle, maxAngle),
      -maxAngle, maxAngle, -maxY, maxY
    );

    thumbStyle.value = {
      transform: `translate3d(${x}px, ${y}px, 10px)`,
    }}, {
    deep: true
});
</script>
...
```

現在圓點會隨著手機姿態移動了！✧*。٩(ˊᗜˋ*)و✧*。

讓我們加點樣式和懸浮效果。

➜ src\components\gamepad-device-motion.vue

```
<template>
  <div class="container w-auto">
    <div
      ...
      :style="padStyle"
      ...
    >
      <div class="pad-ring top-0 left-0 absolute border w-full h-full rounded-full" />

      ...
    </div>
  </div>
</template>

<script setup lang="ts">
...
```

```
const padStyle = computed(() => {
  const { x, z } = angle.value;

  const transform = `rotateX(${-z}deg) rotateY(${x}deg)`;

  return {
    ...
    transform,
  }
});
...
</script>

<style scoped lang="sass">
.container
  perspective: 1000px

.pad
  transform-style: preserve-3d
  background: rgba(#111, 0.8)
.pad-ring
  transform: translateZ(-20px)
...
</style>
```

最後讓我們來傳輸資料吧，首先在傳輸資料追加 z 軸旋轉定義。

➜ src\types\player.type.ts

```
/** 按鍵類型 */
export enum KeyName {
  ...
  Z_AXIS = 'z-axis',
}
...
```

接著利用 throttle 調整資料發送時間間隔吧。

→ src\components\gamepad-device-motion.vue

```
...
<script setup lang="ts">
import { clamp, throttle } from 'lodash-es';
...
watch(angle, (newAngle) => {
  ...
  emitAngle(newAngle);
}, {
  deep: true
});


const emitAngle = throttle((angle: Angle) => {
  emit('angle', angle);
}, 50);</script>
...
```

現在讓我們在搖桿元件中接收姿態訊號並發送出去吧。

→ src\views\the-player-gamepad-chicken-fly.vue

```
<template>
  <player-gamepad-container>
    ...
    <gamepad-device-motion
      ...
      @angle="handleDeviceMotion"
    />
  </player-gamepad-container>
</template>

<script setup lang="ts">
...
import GamepadDeviceMotion, { Angle } from '../components/gamepad-device-motion.vue';
...
function handleDeviceMotion(angle: Angle) {
```

```
emitGamepadData([
  {
    name: 'x-axis',
    value: angle.x,
  },
  {
    name: 'y-axis',
    value: angle.y,
  },
  {
    name: 'z-axis',
    value: angle.z,
  },
]);
}
...
</script>
```

以上我們準備好姿態控制搖桿了！、(≧∀≦)、

11.9 小雞快飛！

此章節程式碼可以在以下連結取得。

- https://gitlab.com/deepmind1/animal-party/web/-/tree/feat/game-chicken-fly-finish

現在讓我們把小雞遊戲完成吧，第一步就是將遊戲場景加入對應的頁面中。

➜ src\views\the-game-console-chicken-fly.vue

```
<template>
  <game-scene
    class=" w-full h-full"
    @init="handleInit"
  />
</template>

<script setup lang="ts">
import GameScene from '../games/chicken-fly/game-scene.vue';

import { useLoading } from '../composables/use-loading';

const loading = useLoading();

function handleInit() {
  loading.hide();
}
</script>
```

同時將 App.vue 的 the-player-gamepad-chicken-fly 刪除，現在回到遊戲場景中，讓我們完成剩下的遊戲邏輯吧。

和企鵝遊戲一樣，所有玩家的小雞登場時，需要排列，但是我們在企鵝遊戲設計的 getSquareMatrixPositions，預設只能在 xz 平面上排列，而小雞遊戲需要在 xy 平面上排列，所以讓我們新增一個參數，可以指定要在哪個平面排列。

➜ src\common\utils.ts

```
...
/** 取得指定數量方形矩陣，可用於排列登場角色
 *
 * @param gap 間距
 * @param length 數量
 * @param origin 座標原點，預設 new Vector3(0, 0, 0)
 * @param plane 排列平面，預設 xz 平面
```

```
 */
export function getSquareMatrixPositions(
  gap: number,
  length: number,
  origin = new Vector3(0, 0, 0),
  plane: 'xy' | 'xz' | 'yz' = 'xz'
) {
  ...
  const result: Vector3[] = flow([
    ...
    /** 依序計算每個座標位置 */
    (array: number[]) => {
      return array.map((value, index) => {
        ...
        // 根據指定的平面計算座標
        switch (plane) {
          case 'xy':
            return new Vector3(col * gap, row * gap, 0);
          case 'yz':
            return new Vector3(0, col * gap, row * gap);
          case 'xz': default:
            return new Vector3(col * gap, 0, row * gap);
        }
      });
    },

    ...
  ])();
  return result;
}
...
```

接著在遊戲場景中新增 createChicken 並調整 createChickens，改為使用真正的玩家清單建立小雞。

➜ src\games\chicken-fly\game-scene.vue

```
<template>
  <div class="overflow-hidden">
```

```
    <canvas
      ref="canvas"
      class="outline-none w-full h-full"
    />
  </div>
</template>

<script setup lang='ts'>
...
import { colors } from 'quasar';
import { getPlayerColor, getSquareMatrixPositions } from '../../common/utils';
import { Player } from '../../types';
...
import { useClientGameConsole } from '../../composables/use-client-game-console';

const { getPaletteColor, textToRgb } = colors;
...
const gameConsole = useClientGameConsole();
...

const { canvas } = useBabylonScene({
  ...

  async init({ scene }) {
    ...

    const chickens = await createChickens(gameConsole.players.value, scene);
    ...
  }
});
...
async function createChicken(id: string, position: Vector3, scene: Scene) {
  /** 依照玩家 ID 取得對應顏色名稱並轉換成 rgb */
  const codeName = gameConsole.getPlayerCodeName(id);
  const color = getPlayerColor({ codeName });
  const hex = getPaletteColor(color);
  const rgb = textToRgb(hex);

  const chicken = await new Chicken(`chicken-${id}`, scene, {
```

```
    ownerId: id,
    position,
    color: new Color3(rgb.r / 255, rgb.g / 255, rgb.b / 255),
  }).init();

  return chicken;
}
async function createChickens(players: Player[], scene: Scene) {
  const positions = getSquareMatrixPositions(
    0.5, players.length, undefined, 'xy'
  );

  const tasks = players.map(
    (player, i) => createChicken(
      player.clientId, positions[i], scene,
    )
  );

  const chickens = await Promise.all(tasks);

  return chickens;
}
</script>
```

現在進入遊戲可以看到玩家對應的小雞登場了！

▲ 圖 11-50 玩家小雞登場

再來讓我們加入控制功能吧，但是要先讓手機前往對應的搖桿畫面才行。

➔ src\views\the-player-gamepad.vue

```
...

<script setup lang="ts">
...
/** 遊戲與搖桿頁面對應資料 */
const gamepadMap: Record<GameName, {
  routeName: RouteName,
}> = {
  'the-first-penguin': {
    routeName: RouteName.PLAYER_GAMEPAD_THE_FIRST_PENGUIN,
  },
  'chicken-fly': {
    routeName: RouteName.PLAYER_GAMEPAD_CHICKEN_FLY,
  },
}
function init() {
  ...
  player.onGameConsoleStateUpdate((state) => {
    ...

    const target = gamepadMap[gameName];
    router.push({ name: target.routeName });
  });
  ...
}
init();
</script>
```

新增控制小雞相關 function。

➔ src\games\chicken-fly\game-scene.vue

```
...

<script setup lang="ts">
...
```

```
const { canvas } = useBabylonScene({
  ...
  async init({ scene }) {
    ...
    const chickens = await createChickens(gameConsole.players.value, scene);
    initGamepadEvent(chickens);
    ...
  }
});

...

function initGamepadEvent(chickens: Chicken[]) {
  gameConsole.onGamepadData((data) => {
    const { playerId } = data;

    /** 找到對應的小雞 */
    const target = chickens.find(
      ({ params }) => params.ownerId === playerId
    );
    if (!target) return;

    ctrlChicken(target, data);
  });
}

/** 根據 key 取得資料 */
const findSingleData = curry((keys: SignalData[], name: `${KeyName}`): SignalData |
undefined =>
  keys.find((key) => key.name === name)
);

/** 控制指定小雞 */
function ctrlChicken(chicken: Chicken, data: GamepadData) {
  const { keys } = data;
  const findData = findSingleData(keys);

  // 移動按鍵
  const xData = findData('x-axis');
```

```
const zData = findData('z-axis');

const x = xData?.value ?? 0;
const z = zData?.value ?? 0;

if (typeof x === 'number' && typeof z === 'number') {
  /** 搖桿傳送過來的角度單位是 Degrees，需要轉換成小雞姿態角度單位（Radians），
   *
   * 手機的 +z 軸是遊戲座標的 -x、+x 軸是遊戲座標的 -z
   */
  chicken.setAttitude(
    Tools.ToRadians(-z),
    Tools.ToRadians(-x)
  );
}
}
</script>
```

　　再來我們需要在碰撞偵測中實際呼叫小雞被碰撞，同時我們在小雞內新增變數，用來記錄死掉時間，方便最後排名玩家。

➡ src\games\chicken-fly\game-scene.vue

```
...
<script setup lang="ts">
...
// 偵測碰撞事件
function detectCollideEvents(chickens: Chicken[], badChickens: BadChicken[]) {
  /** 依據檢查壞雞 */
  badChickens.forEach(({ mesh: badChickenMesh }) => {
    ...
    /** 依序檢查小雞 */
    chickens.forEach((chicken) => {
      ...
      if (badChickenMesh.intersectsMesh(chicken.mesh)) {
        chicken.attacked();
      }
    });
  });
```

```
}
...
</script>
```

➡️ src\games\chicken-fly\chicken.ts

```
...
export class Chicken {
  ...
  params: Required<Params> = {...};

  diedAt = 0;

  ...
  constructor(name: string, scene: Scene, params?: Params) {
    ...
  }
  ...
  private async dead() {
    if (!this.mesh) return;

    /** 紀錄死亡時間 */
    this.diedAt = new Date().getTime();
    ...
  }
  ...
}
```

最後讓我們偵測是否剩下一隻小雞，是的話表示遊戲結束，顯示排名。

可以想像顯示排名的畫面可以在各個遊戲之間重複使用，所以讓我們把它獨立為單一元件吧。

➡️ src\components\player-leaderboard.vue

```
<template></template>

<script setup lang="ts">
import { ref } from 'vue';
```

```
interface Props {
  title?: string;
  titleClass?: string;
  /** 順序表示排名 */
  idList: string[];
}
const props = withDefaults(defineProps<Props>(), {
  title: '玩家排行榜',
  titleClass: '',
});
</script>


<style scoped lang="sass">
</style>
```

接著老樣子先把元件放到 App.vue 方便開發並用 dialog 包起來。

➜　src\App.vue

```
<template>
  ...

  <q-dialog
    model-value
    persistent
  >
    <player-leaderboard
      class="w-[40rem]"
      :id-list="['1', '2', '3']"
    />
  </q-dialog>
</template>

<script setup lang="ts">
...
import PlayerLeaderboard from './components/player-leaderboard.vue';
</script>
...
```

第一步先新增標題。

➡ src\components\player-leaderboard.vue

```
<template>
  <q-card class=" !rounded-3xl min-w-[20rem]">
    <slot name="title">
      <q-card-section
        class=" relative overflow-hidden text-4xl bg-orange-400 text-white text-center"
        :class="props.titleClass"
      >
        {{ props.title }}

        <base-polygon
          class=" absolute top-0 left-0 -translate-x-1/3 -translate-y-2/3"
          fill="spot"
          opacity="0.3"
        />

        <base-polygon
          class=" absolute bottom-0 right-0 translate-x-1/3 translate-y-2/3"
          shape="pentagon"
          rotate="10deg"
          opacity="0.3"
        />
      </q-card-section>
    </slot>
  </q-card>
</template>

<script setup lang="ts">
import { ref } from 'vue';

import BasePolygon from './base-polygon.vue';
...
</script>
...
```

▲ 圖 11-51　排行榜標題

接著將排名列出來吧，因為沒有真的玩家，所以需要自行產生假資料。

➔ src\components\player-leaderboard.vue

```
<template>
  <q-card class=" !rounded-3xl min-w-[20rem]">
    ...
    <slot name="board">
      <q-card-section
        class="bg-orange-100 py-8 px-10 flex flex-col gap-2"
        :class="props.boardClass"
      >
        <div
          v-for="item, i in playerList"
          :key="i"
          class="player-item flex gap-6 text-white"
        >
          <div class="flex items-center text-3xl font-black text-gray-500">
            {{ i + 1 }}
          </div>

          <div
            class=" relative p-4  text-3xl rounded-full flex-1 text-center overflow-
hidden"
            :class="item.class"
          >
            {{ item.codeName }}
          </div>
        </div>
      </q-card-section>
    </slot>
```

```
    </q-card>
</template>

<script setup lang="ts">
...
import { computed, ref } from 'vue';
import { getPlayerColor } from '../common/utils';
import { useClientGameConsole } from '../composables/use-client-game-console';
...

interface Props {
  ...
  boardClass?: string;
  /** 順序表示排名 */
  ...
}
const props = withDefaults(defineProps<Props>(), {
  ...
  boardClass: '',
});

/** 從 id list 轉換成 player code name */
const playerList = computed(() =>
  range(1, 5.map((index) => {
    const codeName = `${index}P`;
    const color = getPlayerColor({ codeName });

    return {
      codeName,
      class: `bg-${color}`,
    }
  })
);
</script>
...
```

▲ 圖 11-52 排行榜排名

一樣利用 base-polygon 妝點一下。

➜ src\components\player-leaderboard.vue

```
<template>
  <q-card class=" !rounded-3xl min-w-[20rem]">
    ...

    <slot name="board">
      <q-card-section ... >
        <div ... >
          ...

          <div ... >
            {{ item.codeName }}

            <base-polygon
              ...
              v-bind="item.polygon"
```

```
              />
          </div>
        </div>
      </q-card-section>
    </slot>
  </q-card>
</template>

<script setup lang="ts">
import { random, range, sample } from 'lodash-es';
...
import BasePolygon, { ShapeType, FillType } from './base-polygon.vue';
...

/** 從 id list 轉換成 player code name */
const playerList = computed(() =>
  range(1, 5).map((index) => {
    ...
    const polygon = {
      fill: sample(Object.values(FillType)),
      shape: sample(Object.values(ShapeType)),
      rotate: `${random(0, 45)}deg`,
    }

    return {
      codeName,
      class: `bg-${color}`,
      polygon,
    }
  })
);
</script>
...
```

▲ 圖 11-53　排行榜玩家點綴

接著調整一下名次文字。

→ src\components\player-leaderboard.vue

```
<template>
  <q-card class=" !rounded-3xl min-w-[20rem]">
    ...
    <slot name="board">
      <q-card-section ... >
        <div
          ...
          class="player-item flex gap-6 text-white"
        >
          <div class="ranking flex items-center text-3xl font-black">
            {{ i + 1 }}
          </div>

          ...
        </div>
      </q-card-section>
```

```
    </slot>
  </q-card>
</template>

<script setup lang="ts">
...
</script>

<style scoped lang="sass">
.player-item
  .ranking
    padding: 0rem 1.5rem
    background: rgba(#000, 0.1)
    border-radius: 999px
    text-shadow: 0 0 8px rgba(#000, 0.4)
    color: white

  &:nth-child(1)
    .ranking
      color: white
      text-shadow: 0 0 4px rgba(#000, 0.4)
      background: linear-gradient(30deg, #f5b402, #fcfc4c)
      border-radius: 0px
      clip-path: polygon(0% 15%, 15% 15%, 15% 0%, 85% 0%, 85% 15%, 100% 15%, 100% 85%,
85% 85%, 85% 100%, 15% 100%, 15% 85%, 0% 85%)

  &:nth-child(2)
    .ranking
      color: white
      text-shadow: 0 0 4px rgba(#000, 0.4)
      background: linear-gradient(30deg, #c7bfb5, #FFF)
      border-radius: 0px
      clip-path: polygon(30% 0%, 70% 0%, 100% 30%, 100% 70%, 70% 100%, 30% 100%, 0%
70%, 0% 30%)

  &:nth-child(3)
    .ranking
      color: white
      text-shadow: 0 0 4px rgba(#000, 0.4)
      background: linear-gradient(30deg, #b83a00, #ed8d15)
```

```
</style>
```

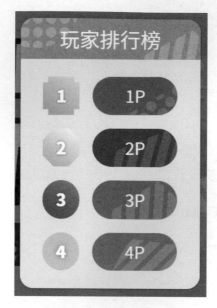

▲　圖 11-54　排行榜名次點綴

最後讓我們加點動畫吧。ᕕ(ﾟ ∀ ﾟ)ᕗ

使用 transform 配合 animation 產生各種動態。

➜　src\components\player-leaderboard.vue

```
<template>
  <q-card class=" !rounded-3xl min-w-[20rem]">
    ...
    <slot name="board">
      <q-card-section ... >
        <div
          v-for="item, i in playerList"
          ...
        >
          ...
          <div
```

```
                class="code-name ..."
                :class="item.class"
            >
                ...
            </div>
        </div>
      </q-card-section>
    </slot>
  </q-card>
</template>

<script setup lang="ts">
...
</script>

<style scoped lang="sass">
.player-item
  ...
  &:nth-child(1)
    .code-name
      animation: jump 2s infinite
      transform-origin: 50% 100%

    ...

  &:nth-child(2)
    .code-name
      animation: cheerful 2s infinite
      transform-origin: 50% 100%

    ...

  &:nth-child(3)
    .code-name
      animation: joy-bounce 2s infinite

    ...

$jump-in: cubic-bezier(0.755, 0.050, 0.855, 0.060)
$jump-out: cubic-bezier(0.230, 1.000, 0.320, 1.000)
@keyframes jump
  0%, 100%
```

```
    transform: scale( 1 )
    animation-timing-function: $jump-out
  20%
    transform: scale( 1.05, 0.9 )
    animation-timing-function: $jump-out
  40%
    transform: scale( 0.7, 1.3 ) translateY(-40%)
    animation-timing-function: $jump-in
  60%
    transform: scale( 0.9, 1.1 )
    animation-timing-function: $jump-out
  80%
    transform: scale( 1.1, 0.9 )


$cheerful-timing: cubic-bezier(1.000, 0.000, 0.160, 1.000)
@keyframes cheerful
  0%, 100%
    transform: skew(10deg) scaleY(1.1)
    animation-timing-function: $cheerful-timing
  25%, 75%
    transform: skew(0deg) scaleY(0.8)
    animation-timing-function: $cheerful-timing
  50%
    transform: skew(-10deg) scaleY(1.1)
    animation-timing-function: $cheerful-timing


$joy-bounce-timing: cubic-bezier(0.910, -0.010, 0.285, 1.350)
@keyframes joy-bounce
  0%, 100%
    transform: scale( 1 )
    animation-timing-function: $joy-bounce-timing
  50%
    transform: scale( 1.05, 0.9 )
    animation-timing-function: $joy-bounce-timing


</style>
```

▲ 圖 11-55　排行榜玩家動畫

最後讓我們加入 default slot，並將玩家改成實際玩家。

➜ src\components\player-leaderboard.vue

```
<template>
  <q-card class=" !rounded-3xl min-w-[20rem]">
    ...
    <slot name="default" />
  </q-card>
</template>

<script setup lang="ts">
...
import { useClientGameConsole } from '../composables/use-client-game-console';
...
const gameConsole = useClientGameConsole();

/** 從 id list 轉換成 player code name */
const playerList = computed(() =>
  props.idList.map((id) => {
    const codeName = gameConsole.getPlayerCodeName(id);
```

```
    const color = getPlayerColor({ codeName });

    const polygon = {
      fill: sample(Object.values(FillType)),
      shape: sample(Object.values(ShapeType)),
      rotate: `${random(0, 45)}deg`,
    }

    return {
      codeName,
      class: `bg-${color}`,
      polygon,
    }
  })
);
</script>

<style scoped lang="sass">
...
</style>
```

現在會發現因為找不到玩家，所以都變成 unknown。

▲ 圖 11-56 排行榜使用實際玩家

不過沒關係，現在刪掉 App.vue 的 player-leaderboard 元件後，回到小雞遊戲！(´▽`)/。

讓我們在小雞遊戲中使用 player-leaderboard，並加入 isGameOver 變數。

➜ src\games\chicken-fly\game-scene.vue

```
<template>
  <div class="overflow-hidden">
    ...
    <q-dialog
      v-model="isGameOver"
      persistent
    >
      <player-leaderboard :id-list="['1', '2', '3']">
        <div class="text-xl text-gray-400 p-5 text-center">
          按下 <q-icon name="done" /> 回到大廳
        </div>
      </player-leaderboard>
    </q-dialog>
  </div>
</template>

<script setup lang='ts'>
...

import PlayerLeaderboard from '../../components/player-leaderboard.vue';
...

const isGameOver = ref(false);
/** 遊戲場景邊界 */
...
</script>
```

接著加入以下內容：

- 新增小雞變數，儲存所有小雞

- 新增 getRankedIdList function，用來取得排名後的玩家 ID。

→ src\games\chicken-fly\game-scene.vue

```
<template>
  <div class="overflow-hidden">
    ...

    <q-dialog ... >
      <player-leaderboard :id-list="getRankedIdList(playerChickens)">
        ...
      </player-leaderboard>
    </q-dialog>
  </div>
</template>

<script setup lang='ts'>
...
const playerChickens: Chicken[] = [];

function getRankedIdList(chickens: Chicken[]) {
  const result = cloneDeep(chickens)
    .sort((a, b) => {
      /** diedAt 為 0 表示第一名，其他則從最大排到最小
       *
       * -1 表示將目前的 a 排在 b 前面
       * 1 表示將目前的 a 排在 b 後面
       */    if (a.diedAt === 0) return -1;
      if (b.diedAt === 0) return 1;

      return b.diedAt - a.diedAt;
    })
    .map(({ params }) => params.ownerId);

  return result;
}

const { canvas } = useBabylonScene({
  ...

  async init({ scene }) {
```

```
    ...

    const chickens = await createChickens(gameConsole.players.value, scene);
    initGamepadEvent(chickens);
    playerChickens.push(...chickens);

    ...
  }
});
...
</script>
```

在來新增判斷遊戲結束的 function。

➜ src\games\chicken-fly\game-scene.vue

```
<template>
...
</template>

<script setup lang='ts'>
...
const { canvas } = useBabylonScene({
  ...
  async init({ scene, engine }) {
    ...
    scene.registerBeforeRender(() => {
      detectCollideEvents(chickens, badChickens);
      detectGameOver(chickens, engine);
    });
    ...
  }
});
...
/** 偵測遊戲是否結束 */
function detectGameOver(chickens: Chicken[], engine: Engine) {
  const theLivingList = chickens.filter(({ mesh }) => !mesh?.isDisposed());

  /** 2 人以上表示遊戲還沒結束 */
```

```
  if (theLivingList.length >= 2) return;

  engine.stopRenderLoop();
  isGameOver.value = true;
}
...
</script>
```

最後加入返回大廳功能，就大功告成啦！✧*｡٩(´ ∀ ｀*)و✧*｡。

在企鵝遊戲的時候我們讓遊戲場景可以直接返回大廳頁面，其實這樣的設計不好，遊戲場景負責的工作應該專注於處理遊戲畫面，跳轉頁面等等工作應該使用 emit 通知父元件，也就是 the-game-console-chicken-fly 負責。

➜ src\games\chicken-fly\game-scene.vue

```
<template>
...
</template>

<script setup lang='ts'>
...
const emit = defineEmits<{
  (e: 'init'): void;
  (e: 'back-to-lobby'): void;
}>();
...
/** 控制指定小雞 */
function ctrlChicken(chicken: Chicken, data: GamepadData) {
  ...
  // 確認按鍵
  const confirmData = findData('confirm');
  if (confirmData && isGameOver.value) {
    emit('back-to-lobby');
    isGameOver.value = false;
    return;
  }
```

```
  ...
}
</script>
```

父元件處理返回大廳功能。

➜ src\views\the-game-console-chicken-fly.vue

```
<template>
  <game-scene
    ...
    @back-to-lobby="handleBackToLobby"
  />
</template>

<script setup lang="ts">
...
import { useRouter } from 'vue-router';
import { RouteName } from '../router/router';

const loading = useLoading();
const router = useRouter();
...
async function handleBackToLobby() {
  await loading.show();
  router.push({
    name: RouteName.GAME_CONSOLE_LOBBY
  });
}
</script>
```

現在我們可以完整遊玩一次了！✧*。٩(´▽`*)و✧*。

▲ 圖 11-57 完整小雞遊戲

助教：「實際玩起來，小雞有點難控制欸。(›´ω‹)」

鱈魚：「那是你技術差吧？(๏ ⩊ ๏')」

助教：「你行你試試看？(╬⊙д⊙)」

鱈魚：「嗯 ....，好像真有點難，欸嘿ヾ(๏'◡'๏)ノ゙」

讓我們調整一下小雞的控制邏輯，讓小雞更好控制一點。

➜ src\games\chicken-fly\chicken.ts

```
...
export class Chicken {
  ...
  private maxSpeed = 4;
  /** 用來放大升力 */
  private liftCoefficient = 15;
  ...
  constructor(name: string, scene: Scene, params?: Params) {
    ...
```

```
  }
  ...
  /** 根據目前速度取得升力比率，速度越大，比率越小 */
  private getLift(speed: number, angle: number) {
    const ratio = (this.maxSpeed - Math.abs(speed)) / this.maxSpeed;

    // 檢查速度和角度是否同向
    const sameDirection = (speed > 0 && angle > 0) || (speed < 0 && angle < 0);

    // 如果同向，為計算出的比率；如果反向，將比率放大，如此便可以快速轉向
    const result = sameDirection ? ratio : 1 / ratio;
    return angle * result * this.liftCoefficient;
  }
  /** 根據目前姿態產生加速度 */
  private processLift = throttle(() => {
    ...
    /** X 座標與 roll 旋轉方向剛好相反，所以乘上 -1 */
    const xLift = this.getLift(velocity.x, -roll);
    const yLift = this.getLift(velocity.y, pitch);

    ...
  }, 10)
  ...
}
```

以上我們完成小雞遊戲了！♪ㄟ(ﾟ∀ﾟ)ㄏ

　　不過目前發現一個小問題，就是我們從小雞遊戲回到大廳時，畫面卻會前往企鵝遊戲，這是因為選單預設沒有隨著上次選擇的遊戲變化，讓我們調整一下吧。

　　首先在 use-client-game-console 追加提供目前遊戲名稱。

➜ src\composables\use-client-game-console.ts

```
...
export function useClientGameConsole() {
...
  return {
```

```
  ...
  currentGame: computed(() => gameConsoleStore.gameName),
 }
}
```

接著調整遊戲選單 currentIndex 初始化邏輯。

➔ src\composables\use-client-game-console.ts

```
...
<script setup lang="ts">
...
const currentIndex = ref(
  games.findIndex(({ name }) => name === gameConsole.currentGame.value) ?? 0
);

...
</script>
```

現在回到大廳會正常前往上次選擇的遊戲了。ヽ(*´︶`*)ﾉ

⬧ 11.10 增強遊戲體驗

此章節程式碼可以在以下連結取得。

- https://gitlab.com/deepmind1/animal-party/web/-/tree/feat/game-chicken-fly-scene-mode

助教：「現在有個小問題。(´・ω・`)」

鱈魚：「一直輸是你自己的問題。(´−`)」

助教：「才不是，是遊戲沒有教學，大家都不知道怎麼玩啊。ლ(・´口`・ლ)」

鱈魚：「那是他們沒有慧 ......」

助教：（拿出球棒。(╬⊙д⊙)）

鱈魚：「讓我們來做個引導教學吧！ヾ(◍'ڡ`◍)/˚」

把流程改成遊戲開始後會先進到教學模式，當每個玩家都按下確認後才會正式開始吧。

目前計畫在遊戲場景新增 mode 參數，根據參數會有不同的呈現方式。

首先定義模式。

➔ src\types\game.type.ts

```
...
/** 遊戲模式 */
export enum GameSceneMode {
  /** 依照預設規則遊玩 */
  NORMAL = 'normal',
  /** 用來作為背景展示使用 */
  SHOWCASE = 'showcase',
  /** 方便用來讓玩家練習，會有人物不會死亡等等效果 */
  TRAINING = 'training',
}
```

小雞新增鎖血狀態，絕對不會死掉，方便玩家練習。

➔ src\games\chicken-fly\chicken.ts

```
...
export class Chicken {
  ...
  /** 為 true 時，人物血量不會降低 */
  healthLock = false;
  ...
```

```
  constructor(name: string, scene: Scene, params?: Params) {
    ...

    const stateMachine = createMachine(
      ...
      {
        actions: {
          minusHealth: (context) => {
            if (this.healthLock) return;
            context.health -= 1;
          },
          ...
        },
      }
    );
    ...
  }

  ...
}
```

接著在遊戲場景內實現各種模式的邏輯。

- 練習、展示模式時，場景不加速。

- 展示模式時，壞雞不移動。

- 練習模式時，小雞不會墜機。

➔ src\games\chicken-fly\game-scene.vue

```
...
<script setup lang='ts'>
...
interface Props {
  mode?: `${GameSceneMode}`;
}
const props = withDefaults(defineProps<Props>(), {
  mode: 'normal',
});
```

```
const emit = defineEmits<{...}>();

/** 每秒 +1 */
const {
  counter, pause, resume
} = useInterval(1000, { controls: true });
whenever(
  () => counter.value >= 150,
  () => pause(),
)
watch(() => props.mode, (mode) => {
  /** 練習、展示模式不加速 */
  if (['training', 'showcase'].includes(mode)) {
    pause();
  }

  /** 一般模式開始計時 */
  if (mode === 'normal') {
    resume();
  }
}, { immediate: true });
...
async function createBadChickens(scene: Scene) {
  ...
  watch(counter, (value) => {
    /** 展示模式壞雞不移動 */
    if (props.mode === 'showcase') return;

    ...
    });
  }, { immediate: true });

  return chickens;
}
...


async function createChicken(id: string, position: Vector3, scene: Scene) {
```

```
  ...
  if (props.mode === 'training') {
    chicken.healthLock = true;
  }

  return chicken;
}
...


</script>
```

然後在場景中追加一個將意外出界的小雞救回的 function。

➜ src\games\chicken-fly\game-scene.vue

```
...
<script setup lang='ts'>
...
const { canvas } = useBabylonScene({
  ...
  async init({ scene, engine }) {
    ...

    scene.registerBeforeRender(() => {
      ...
      detectOutOfBounds(chickens);
    });
    ...
  }
});
...
/** 救回出界的小雞 */
function detectOutOfBounds(chickens: Chicken[]) {
  chickens.forEach((chicken) => {
    if (!chicken.mesh || chicken.diedAt > 0) return;

    const { x, y } = chicken.mesh.position;

    if (Math.abs(x) > sceneBoundary.x ||
      Math.abs(y) > sceneBoundary.y) {
```

```
      chicken.mesh.position = new Vector3(0, 0, 0);

      chicken.mesh.physicsImpostor?.setLinearVelocity(
        new Vector3(0, 0, 0),
      )
    }
  });
}
...
</script>
```

接著調整小雞遊戲頁面，將遊戲場景分為練習用和正式上場。

➜ src\games\chicken-fly\game-scene.vue

```
<template>
  <div class="w-full h-full bg-white">
    <transition
      name="opacity"
      mode="out-in"
    >
      <!-- 練習 -->
      <div
        v-if="sceneMode === 'training'"
        class="w-full h-full"
      >
        <game-scene
          mode="training"
          class=" w-full h-full"
          @init="handleInit"
        />
        <div class=" absolute top-0 left-0 text-9xl">
          練習模式
        </div>
      </div>

      <!-- 正式 -->
      <div
        v-else
```

```
      class="w-full h-full"
    >
      <game-scene
        :mode="sceneMode"
        class=" w-full h-full"
        @back-to-lobby="handleBackToLobby"
      />
    </div>
  </transition>
 </div>
</template>

<script setup lang="ts">
import { ref } from 'vue';
import { RouteName } from '../router/router';
import { GameSceneMode } from '../types';
...
const sceneMode = ref<`${GameSceneMode}`>('training');

...
</script>
```

現在進入遊戲會發現畫面有個超大的練習模式。

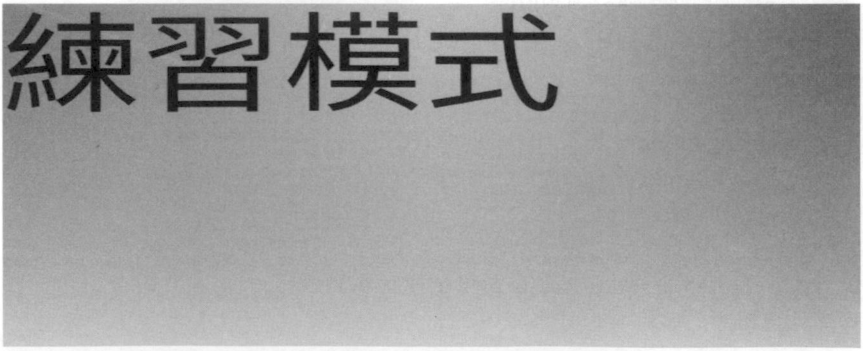

▲ 圖 11-58　小雞遊戲練習模式

這就表示我們成功了，現在讓我們完成練習模式的教學內容吧。

首先新增教學卡片。

→ src\games\chicken-fly\tutorial-card.vue

```
<template>
  <q-card class="flex flex-col rounded-[2rem] shadow-sm overflow-hidden text-sky-900">
    <q-card-section class=" relative text-white text-4xl text-center bg-sky-600 p-7
 !rounded-none">
      遊戲説明

      <base-polygon
        class=" absolute top-0 left-0 -translate-x-1/4 -translate-y-1/2"
        opacity="0.2"
        rotate="30deg"
      />
      <base-polygon
        class=" absolute bottom-0 right-0 translate-x-1/4 translate-y-1/2"
        opacity="0.2"
        fill="spot"
        shape="pentagon"
        rotate="30deg"
      />
    </q-card-section>

    <q-card-section class=" text-xl p-10 text-center">
      躲避黑色壞壞雞，碰撞 3 次會墜雞喔！
    </q-card-section>

    <q-card-section class="section">
      <div class="pad ">
        <div class="thumb thumb-action" />
      </div>
      <p class=" text-center flex-1">
        旋轉手機即可控制小雞姿態 <br>（建議關閉手機自動旋轉）
      </p>
    </q-card-section>

    <q-card-section class="section">
      <div class="pad click-pad">
        <div class="thumb" />
      </div>
      <p class=" text-center flex-1">
```

```
          按下圓圈即可將姿態歸零
      </p>
    </q-card-section>


    <q-card-section class="section justify-center flex-1">
      <p>
        練習完成後，按下
        <q-icon
          name="done"
          color="white"
          class="p-2 bg-[#000]/60 rounded-full"
        />
        準備完成
      </p>
    </q-card-section>
  </q-card>
</template>

<script setup lang="ts">
import BasePolygon from '../../components/base-polygon.vue';
</script>

<style scoped lang="sass">
.section
  @apply text-xl p-6 px-10 flex items-center flex-nowrap gap-6

.pad
  @apply rounded-full flex justify-center items-center
  width: 10rem
  height: 10rem
  background: rgba(#111, 0.6)
.thumb
  width: 40%
  height: 40%
  background: white
  border-radius: 9999px
  opacity: 0.8

.thumb-action
  animation: thumb-action 3s ease-in-out infinite
```

```
@keyframes thumb-action
  0%, 100%
    transform: translate(0%, 0%)
  25%
    transform: translate(0%, 50%)
  50%
    transform: translate(-50%, 0%)
  75%
    transform: translate(0%, -50%)

.click-pad
  animation: click-pad 1s ease-in-out infinite

@keyframes click-pad
  0%, 20%, 100%
    transform: scale(1)
  40%
    transform: scale(0.98)
    opacity: 0.8

</style>
```

在頁面中引入教學卡片。

➡️ src\views\the-game-console-chicken-fly.vue

```
<template>
  <div class="w-full h-full bg-white">
    <transition
      name="opacity"
      mode="out-in"
    >
      <!-- 練習 -->
      <div
        v-if="sceneMode === 'training'"
        class="w-full h-full flex bg-sky-100"
      >
        <div class="flex flex-col w-[70%]">
          <game-scene
            mode="training"
```

```
          class=" w-full h-[80%] rounded-br-3xl"
          @init="handleInit"
        />
      </div>


      <!-- 遊戲說明 -->
      <div class="p-4 flex-1 ">
        <tutorial-card class="w-full h-full" />
      </div>
    </div>


    <!-- 正式 -->
    ...
  </transition>
 </div>
</template>

<script setup lang="ts">
...
import TutorialCard from '../games/chicken-fly/tutorial-card.vue';
...
</script>
```

▲ 圖 11-59 加入教學卡片

看起來真不錯。ヽ(•ω•)ﾉ

接著讓我們引入玩家頭像，用來呈現準備完成的玩家。

➜ src\views\the-game-console-chicken-fly.vue

```
<template>
  <div ...>
    <transition
      ...
    >
      <!-- 練習 -->
      <div
        ...
      >
        <div class=" ...">
          <game-scene
            ...
          />

          <!-- 玩家頭像 -->
          <transition-group
            tag="div"
            class=" absolute bottom-0 w-full overflow-hidden px-10 pointer-events-none"
            name="avatar"
          >
            <player-list-avatar
              v-for="player in players"
              :key="player.clientId"
              :player="player"
              :code-name="player.codeName"
            />
          </transition-group>
        </div>

        <!-- 遊戲說明 -->
        ...
      </div>
```

```
      <!-- 正式 -->
      ...
    </transition>
  </div>
</template>

<script setup lang="ts">
...
import { useClientGameConsole } from '../composables/use-client-game-console';

const gameConsole = useClientGameConsole();
...

const players = computed(() => {
  return gameConsole.players.value.map((player) => {
    const codeName = gameConsole.getPlayerCodeName(player.clientId);

    return {
      ...player,
      codeName,
      ok: false,
    }
  });
});
</script>
```

現在可以看到玩家出現了。

▲ 圖 11-60 顯示目前玩家

　　最後讓我們偵測玩家是否按下確認，並在所有玩家都按下確認後切換 mode 吧。

　　新增變數與偵測搖桿資料事件。

➔ src\views\the-game-console-chicken-fly.vue

```ts
...
<script setup lang="ts">
...

/** 紀錄準備完成玩家 */
const readiedPlayerIdList = ref<string[]>([]);

const players = computed(() => {
  return gameConsole.players.value.map((player) => {
    const codeName = gameConsole.getPlayerCodeName(player.clientId);
    const ok = readiedPlayerIdList.value.includes(player.clientId);
```

```
    return {
      ...player,
      codeName,
      ok,
    }
  });
});

gameConsole.onGamepadData((data) => {
  const lastDatum = data.keys.at(-1);
  if (lastDatum?.name !== 'confirm') return;

  readiedPlayerIdList.value.push(data.playerId);
});
</script>
```

新增玩家準備完成樣式

→ src\views\the-game-console-chicken-fly.vue

```
<template>
  <div ...>
    <transition
      ...
    >
      <!-- 練習 -->
      <div
        ...
      >
        <div ...>
          ...
          <!-- 玩家頭像 -->
          <transition-group
            ...
          >
            <player-list-avatar
              ...
              class="player"
              :class="{ 'ready': player.ok }"
```

```
          />
        </transition-group>
      </div>

      <!-- 遊戲說明 -->
      ...
    </div>

    <!-- 正式 -->
    ...
  </transition>
  </div>
</template>
...
<style scoped lang="sass">
.player
  transform: translateY(0%)
  transition-duration: 0.4s
  &.ready
    opacity: 0.5
    transform: translateY(10%)
    &::after
      position: absolute
      content: 'OK'
      color: white
      text-shadow: 0 0 6px rgba(#000, 0.4)
      left: 50%
      bottom: 15%
      font-size: 2rem
      transform: translateX(-50%)
</style>
```

▲ 圖 11-61 2P 玩家準備完成

判斷是否所有玩家都準備完成，準備完成時將 mode 切換至 normal。

➜ src\views\the-game-console-chicken-fly.vue

```
...
<script setup lang="ts">
...
import { whenever } from '@vueuse/core';
...

const players = computed(() => {
  ...
});

whenever(
  () => players.value.every(({ ok }) => ok),
  () => {
    sceneMode.value = 'normal';
  }
);
...
</script>
```

現在可以從練習切換至正式開打了，只是忽然正式開始有點倉促，讓我們在正式開始之前倒數 321 吧。

新增倒數元件。

➜ src\components\countdown-overlay.vue

```
<template>
  <transition name="opacity">
    <!-- 小於 0 後元件自動消失 -->
    <div
      v-if="counter >= 0"
      class=" absolute w-full h-full flex justify-center items-center"
    >
      <transition
        name="countdown"
```

```
        mode="out-in"
        appear
      >
        <div
          class="text"
          :key="text"
        >
          {{ text }}
        </div>
      </transition>
    </div>
  </transition>
</template>

<script setup lang="ts">
import { useIntervalFn } from '@vueuse/core';
import { ref, watch } from 'vue';

interface Props {
  /** 預設從 3 開始倒數 */
  startValue?: number;
  startText?: string;
  fontSize?: string;
  immediate?: boolean;
}
const props = withDefaults(defineProps<Props>(), {
  startValue: 3,
  startText: 'Start',
  fontSize: '16rem',
  immediate: true,
});

const emit = defineEmits<{
  (e: 'count', value: number): void;
  (e: 'done'): void;
}>();

const counter = ref(props.startValue),
watch(counter, (value) => {
```

```
  if (value <= -1) {
    pause();
    emit('done');
    return;
  }

  emit('count', value);
});

/** 0 之前顯示數字，0 的時候顯示 startText */
const text = computed(() => {
  if (counter.value === 0) {
    return props.startText;
  }

  return counter.value;
});

const { resume, pause } = useIntervalFn(() => {
  counter.value -= 1;
}, 1500, {
  immediate: props.immediate,
})

defineExpose({
  start: resume,
})
</script>

<style scoped lang="sass">
.text
  font-family: 'Luckiest Guy'
  color: white
  font-size: v-bind('props.fontSize')

.countdown
  &-enter-active, &-leave-active
    transition-duration: 0.6s
    transition-timing-function: cubic-bezier(0.295, 1.650, 0.410, 0.995)
```

```
  &-enter-from, &-leave-to
    transform: scale(0)
    opacity: 0 !important
</style>
```

功能很單純，就是 321 start 這樣。(′„•ω•„)

讓我們放到 App.vue 看看效果。

➜ src\App.vue

```
<template>
  ...

  <countdown-overlay />
</template>

<script setup lang="ts">
...
import countdownOverlay from './components/countdown-overlay.vue';

</script>
...
```

▲ 圖 11-62 倒數計時

樣式看起來有點單調，讓顏色豐富一點好了。

- 如同 title-logo 元件一般，文字使用外框包裹。

- 新增 getClass function 取得隨機顏色，讓效果更活潑

➜ src\components\countdown-overlay.vue

```
<template>
  <transition name="opacity">
    ...
    <div
      ...
    >
      <transition
        ...
      >
        <div :key="text">
          <!-- 文字 -->
          <div
            ref="textRef"
            class="text"
            :class="getClass()"
          >
            {{ text }}
          </div>

          <!-- 文字外框 -->
          <div
            class="text absolute top-1/2 left-1/2 -translate-x-1/2 -translate-y-1/2
stroke-color"
            :style="strokeStyle"
            v-html="textRef?.innerHTML"
          />
        </div>       </transition>
    </div>
  </transition>

  <svg
    version="1.1"
    style="display: none;"
  >
    <defs>
      <filter :id="svgFilterId">
        <feMorphology
          operator="dilate"
```

```
        radius="6"
      />
      <feComposite
        operator="xor"
        in="SourceGraphic"
      />
    </filter>
  </defs>
</svg>
</template>

<script setup lang="ts">
...

const textRef = ref<HTMLDivElement>();

const svgFilterId = `svg-filter-${nanoid()}`;
const strokeStyle = computed(() => ({
  filter: `url(#${svgFilterId})`
}));

const colors = [
  'text-red-400',
  'text-lime-400',
  'text-sky-400',
  'text-purple-400',
  'text-amber-400',
  'text-teal-400',
  'text-pink-400',
  'text-fuchsia-400',
];
function getClass() {
  return sample(colors);
}

...
</script>
...
```

▲ 圖 11-63　倒數加入外框與顏色

　　現在看起來好多了，讓我們刪除 App.vue 中的倒數元件，實際加入小雞遊戲吧。

　　改成由倒數計時元件將 gameMode 改為 normal。

➔ src\views\the-game-console-chicken-fly.vue

```
<template>
  <div ...>
    <transition
      ...
    >
      ...
      <!-- 正式 -->
      <div
        ...
      >
        <game-scene
          :mode="sceneMode"
          class=" absolute w-full h-full"
          @back-to-lobby="handleBackToLobby"
        />

        <countdown-overlay @done="handleTimeout" />      </div>
    </transition>
  </div>
</template>


<script setup lang="ts">
```

```
...
import CountdownOverlay from '../components/countdown-overlay.vue';
...

function handleTimeout() {
  sceneMode.value = 'normal';
}
...

whenever(
  ...
  () => {
    sceneMode.value = 'showcase';
  }
);
...
</script>
...
```

▲ 圖 11-64　小雞遊戲完成

小雞遊戲大功告成！✧*｡٩(´ ロ ｀*)و✧*｡

第**12**章

狐狸與老鼠

現在讓我們進入下一個遊戲吧！ゝ(๏'ᴗ'๏)ﾉ

赤狐會在雪地上聆聽積雪下的小老鼠竄動的聲音，再抓準時機跳起，插入雪中抓老鼠，仔細感受雪地的震動，抓到最大隻的老鼠吧。

第一步老樣子先來打造遊戲選單用的小島。

12.1 建立狐狸島

此章節程式碼可以在以下連結取得。

- https://gitlab.com/deepmind1/animal-party/web/-/tree/feat/game-fox-and-mouse-island

第**12**章 狐狸與老鼠

第一步先在遊戲列舉新增老鼠與狐狸遊戲。

➔ src\types\game.type.ts

```
...
/** 遊戲名稱 */
export enum GameName {
  ...
  FOX_AND_MOUSE = 'fox-and-mouse',
}
...
```

接著在 src\components\game-menu 資料夾中新增 fox-and-mouse 資料夾，並新增 index.ts 檔案。

➔ src\components\game-menu\fox-and-mouse\index.ts

```
import {
  Scene, TransformNode
} from '@babylonjs/core';
import { ModelIsland } from '../../../types';

export type FoxAndMouseIsland = ModelIsland;

export async function createFoxAndMouseIsland(scene: Scene): Promise<FoxAndMouseIsland>
 {
  const rootNode = new TransformNode('fox-and-mouse-island');

  return {
    name: 'fox-and-mouse',
    rootNode,
    setActive() {
      //
    },
  }
}
```

讓我們慢慢完成小島內容，先把 game-menu 放到 App.vue 中，方便開發。

➜ src\App.vue

```
<template>
  ...

  <game-menu class=" absolute w-full h-full" />
</template>

<script setup lang="ts">
import LoadingOverlay from './components/loading-overlay.vue';
import GameMenu from './components/game-menu.vue';
</script>
...
```

首先新增 the-game-console-fox-and-mouse，作為狐狸遊戲頁面容器。

➜ src\views\the-game-console-fox-and-mouse.vue

```
<template></template>

<script setup lang="ts">
</script>
```

在 router 加入此頁面。

➜ src\router\router.ts

```
...
export enum RouteName {
  ...
  GAME_CONSOLE_FOX_AND_MOUSE = 'game-console-fox-and-mouse',
  ...
}

const routes: RouteRecordRaw[] = [
  ...
  {
    path: `/game-console`,
```

```
    ...
    children: [
      ...
      {
        path: `fox-and-mouse`,
        name: RouteName.GAME_CONSOLE_FOX_AND_MOUSE,
        component: () => import('../views/the-game-console-fox-and-mouse.vue')
      },
    ]
  },
  ...
]
...
```

接著新增狐狸島的鏡頭位置並將速度調快，跳過進場動畫。

修正一下 moveCamera 參數型別，應該是全部選填才對。

➜ src\components\game-menu-background.vue

```
...
<script setup lang="ts">
...
const shots: Shot[] = [
  ...
  {
    name: 'fox-and-mouse',
    camera: {
      target: new Vector3(10, 0.5, 40),
      alpha: 3,
      beta: 1,
      radius: 20,
    }
  },
];
...
async function moveCamera(params?: Partial<typeof moveCameraDefaultParams>) { ... }
...
onMounted(async () => {
```

```
  ...
  moveCamera({ speed: 10 });
});
...
</script>
```

在遊戲選單中新增狐狸遊戲。

➔ src\components\game-menu.vue

```
...
<script setup lang="ts">
...
const games: GameInfo[] = [
  ...
  {
    name: 'fox-and-mouse',
    camera: {
      target: new Vector3(8.4, -4.2, 42),
      alpha: 2.5,
      beta: 1.1,
      radius: 24.2,
    }
  },
];
...
</script>
```

最後把預設遊戲先改為 fox-and-mouse。

➔ src\stores\game-console.store.ts

```
...
export const useGameConsoleStore = defineStore('game-console', () => {
  ...
  const gameName = ref<`${GameName}`>('fox-and-mouse');
  ...
})
```

現在畫面會變成這樣。

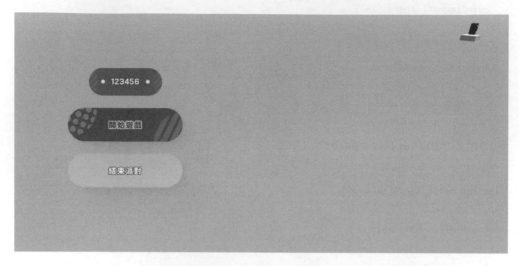

▲ 圖 12-1　調整進場鏡頭

現在鏡頭喬好了，讓我們來把小島做出來吧！

首先新增 public\fox-and-mouse 資料夾，前往 GitLab 同一個目錄下載 3D 模型並放入。

 Tips

讀者可以前往 GitLab 下載模型：

◆ https://gitlab.com/deepmind1/animal-party/web/-/tree/feat/game-fox-and-mouse-island/public/fox-and-mouse

接著在 game-menu-background 加入小島吧。

不過在加入小島前，讓我們用 use-babylon-scene 重構一下，讓程式變得更簡潔。(ㅍ‿ㅍ)✧

➜ src\components\game-menu-background.vue

```
...
<script setup lang="ts">
...
import { useBabylonScene } from '../composables/use-babylon-scene';
...
const { canvas, camera } = useBabylonScene({
  createScene(engine: Engine) { ... },
  createCamera(scene) { ... },
  async init({ scene, camera }) {
    createSea(scene);
    await createIslands(scene);

    window.addEventListener('keydown', (ev) => {
      // 按下 Shift+I 可以切換視窗
      if (ev.shiftKey && ev.keyCode === 73) {
        if (scene.debugLayer.isVisible()) {
          scene.debugLayer.hide();
        } else {
          scene.debugLayer.show();
        }
      }
    });

    /** 發出完成事件，表示畫面初始化完成 */
    emit('completed');

    moveCamera(camera, { speed: 10 });
  }
});
...
async function moveCamera(
  camera: ArcRotateCamera, params?: Partial<typeof moveCameraDefaultParams>
) {
  ...
  const ref = camera._scene.beginAnimation(camera, 0, results[0].frameRate, false,
 speed);
  ...
}
```

```
watch(() => props.selectedGame, () => {
  ...
  if (camera) {
    moveCamera(camera, {

      speed: 0.3,
      easingFunction,
    });
  }
});

function createSea(scene: Scene) { ... }

/** 將所有島嶼儲存至此 */
let islands: ModelIsland[] = [];
async function createIslands(scene: Scene) { ... }
</script>
```

整體程式變得更清爽了。(´▽`)/﹗

現在實際加入狐狸島。

→ src\components\game-menu-background.vue

```
...
<script setup lang="ts">
...
import { createFoxAndMouseIsland } from './game-menu/fox-and-mouse';
...
/** 將所有島嶼儲存至此 */
let islands: ModelIsland[] = [];
async function createIslands(scene: Scene) {
  const results = await Promise.allSettled([
    ...
    createFoxAndMouseIsland(scene),
  ]);

  ...
}
</script>
```

　　不過由於狐狸島目前甚麼都沒有，所以甚麼都看不到，讓我們加入島嶼模型並加入參數。

➜ src\components\game-menu\fox-and-mouse\index.ts

```ts
import {
  Scene, SceneLoader, TransformNode, Vector3
} from '@babylonjs/core';
import { ModelIsland } from '../../../types';
import { defaults } from 'lodash-es';

export type FoxAndMouseIsland = ModelIsland;

interface Param {
  position?: Vector3;
  rotation?: Vector3,
}
const defaultParam: Required<Param> = {
  position: new Vector3(0, 0, 0),
  rotation: new Vector3(0, 0, 0),
}

export async function createFoxAndMouseIsland(scene: Scene, param: Param):
Promise<FoxAndMouseIsland> {
  const { position, rotation } = defaults(param, defaultParam);

  const rootNode = new TransformNode('fox-and-mouse-island');

  async function createIsland() {
    const result = await SceneLoader.ImportMeshAsync('', '/fox-and-mouse/', 'fox-island.glb', scene);
    const mesh = result.meshes[0];
    mesh.parent = rootNode;

    return mesh;
  }
  await createIsland();

  rootNode.position = position;
```

```
  rootNode.rotation = rotation;

  return {
    name: 'fox-and-mouse',
    rootNode,
    setActive() {
      //
    },
  }
}
```

現在會注意到島跑到畫面右上角。

▲ 圖 12-2　引入狐狸島模型

這是因為預設在原點，現在讓我們將基本內容就位：

- 調整島嶼位置、旋轉與尺寸。

- 新增一個燈，以免狐狸島太暗。

→ src\components\game-menu-background.vue

```
...
<script setup lang="ts">
...
async function createIslands(scene: Scene) {
  const results = await Promise.allSettled([
    ...
    createFoxAndMouseIsland(scene, {
```

```
    position: new Vector3(-1, 0, 44),
  }),
]);

...
}
</script>
```

➜ src\components\game-menu\fox-and-mouse\index.ts

```
...

export async function createFoxAndMouseIsland(scene: Scene, param: Param):
Promise<FoxAndMouseIsland> {
  ...

  function createLight() {
    const light = new HemisphericLight(
      'fox-and-mouse-island-light',
      new Vector3(-5, 5, 5),
      scene,
    );
    light.intensity = 0.1;
    light.parent = rootNode;

    return light;
  }
  createLight();

  async function createIsland() {
    ...
    mesh.scaling = new Vector3(0.5, 0.5, 0.5);
    mesh.rotation = new Vector3(0, Tools.ToRadians(-90), 0);
    mesh.parent = rootNode;

    return mesh;
  }
  ...
}
```

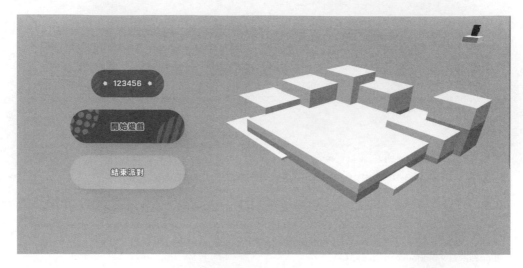

▲ 圖 12-3　島嶼模型就位

現在讓我們加入狐狸吧。(ˊ ω ˋ)/

與建立島嶼中的企鵝相同概念，新增狐狸檔案。

➜ src\components\game-menu\fox-and-mouse\fox.ts

```ts
import {
  Scene, Vector3, AnimationGroup, SceneLoader,
} from "@babylonjs/core";
import { defaults } from "lodash-es";

interface Option {
  scaling?: number;
  position?: Vector3;
  rotation?: Vector3;
  playAnimation?: 'idle' | 'walk' | 'dive' | '';
}

const defaultOption: Required<Option> = {
  scaling: 1,
  position: Vector3.Zero(),
  rotation: Vector3.Zero(),
  playAnimation: '',
}
```

```
export async function createFox(name: string, scene: Scene, option?: Option) {
  const {
    scaling, position, rotation
  } = defaults(option, defaultOption);

  const result = await SceneLoader.ImportMeshAsync('', '/fox-and-mouse/', 'fox.glb',
 scene);

  const mesh = result.meshes[0];
  mesh.name = name;
  mesh.position = position;
  mesh.rotation = rotation;
  mesh.scaling = new Vector3(scaling, scaling, scaling);

  function initAnimation(animationGroups: AnimationGroup[]) {
    animationGroups.forEach((animationGroup) => {
      animationGroup.stop();
    });

    /** 播放指定動畫 */
    if (option?.playAnimation) {
      const animation = animationGroups.find(({ name }) => name === option.
playAnimation);
      animation?.play(true);
    }
  }
  initAnimation(result.animationGroups);

  return { mesh }
}
```

接著引入狐狸們。

➔ src\components\game-menu\fox-and-mouse\index.ts

```
...
import { createFox } from './fox';
...
```

```
export async function createFoxAndMouseIsland(scene: Scene, param: Param):
Promise<FoxAndMouseIsland> {
  ...

  async function createFoxes() {
    /**
     * 從 createFox 函數獲取第三個參數的類型（索引為 2，因為索引從 0 開始）
     */
    const list: Parameters<typeof createFox>[2][] = [
      {
        scaling: 0.3,
        position: new Vector3(1, 1, 0),
        playAnimation: 'dive',
      },
      {
        scaling: 0.3,
        position: new Vector3(2, 1, 2),
        rotation: new Vector3(0, Tools.ToRadians(90), 0),
        playAnimation: 'dive',
      },
      {
        scaling: 0.3,
        position: new Vector3(-2, 1, 0),
        rotation: new Vector3(0, Tools.ToRadians(-90), 0),
        playAnimation: 'dive',
      },
      {
        scaling: 0.25,
        position: new Vector3(3.5, 2, -1.5),
        rotation: new Vector3(0, Tools.ToRadians(180), 0),
        playAnimation: 'dive',
      },
      {
        scaling: 0.3,
        position: new Vector3(-3.5, 2.4, -4),
        rotation: new Vector3(0, Tools.ToRadians(180), 0),
        playAnimation: 'dive',
      },
    ];
```

```
  /** 建立 createFox 的 Promise 矩陣 */
  const tasks = list.map((option, i) =>
    createFox(`fox-${i}`, scene, option)
  );
  /** 當所有的 Promise 都完成時，回傳所有 Promise 的結果。 */
  const results = await Promise.allSettled(tasks);

  /** 將成功建立的人物綁定至 rootNode */
  results.forEach((result) => {
    if (result.status === 'rejected') return;
    result.value.mesh.setParent(rootNode);
  });
}
await createFoxes();

...
}
```

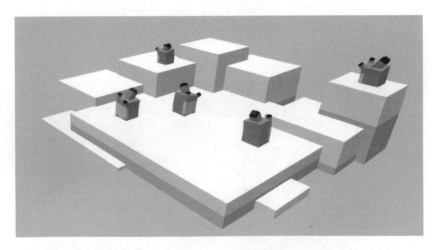

▲ 圖 12-4 狐狸登場

現在讓我們引入標題。

➜ src\components\game-menu\fox-and-mouse\index.ts

```
...
import { createFox } from './fox';
...
```

```
export async function createFoxAndMouseIsland(scene: Scene, param: Param):
Promise<FoxAndMouseIsland> {
  ...

  async function createTitle() {
    const result = await SceneLoader.ImportMeshAsync('', '/fox-and-mouse/', 'fox-
title.glb', scene);
    const mesh = result.meshes[0];
    mesh.name = 'fox-title';
    mesh.scaling = new Vector3(0.8, 0.8, 0.8);
    mesh.position = new Vector3(-1.5, 0.8, 2.7);
    mesh.parent = rootNode;

    return mesh;
  }
  await createTitle();

  ...
}
```

▲ 圖 12-5 加入標題

狐狸島完成！讓我們把預設遊戲改回企鵝並把鏡頭快轉取消吧。

→ src\components\game-menu-background.vue

```
...
<script setup lang="ts">
const { canvas, camera } = useBabylonScene({
  ...
  async init({ scene, camera }) {
    ...

    moveCamera(camera);
  }
});
...
</script>
```

→ src\stores\game-console.store.ts

```
...
export const useGameConsoleStore = defineStore('game-console', () => {
  ...
  const gameName = ref<`${GameName}`>('the-first-penguin');
  ...
})
...
```

把 App.vue 的遊戲選單刪除後，讓我們試試看選擇遊戲。

這裡讓我們修正一個小問題，就是 useBabylonScene 裡面的 engine、scene、camera，改成使用 shallowRef 包裝，以免外部使用的時候，Vue 的響應式系統漏接。

→ src\composables\use-babylon-scene.ts

```
...
export function useBabylonScene(params?: UseBabylonSceneParams) {
  ...
```

```
const engine = shallowRef<Engine>();
const scene = shallowRef<Scene>();
const camera = shallowRef<ArcRotateCamera>();

...

onMounted(async () => {
  ...
  engine.value = createEngine(canvas.value);
  scene.value = createScene(engine.value);
  camera.value = createCamera(scene.value);

  window.addEventListener('resize', handleResize);

  /** 反覆渲染場景，這樣畫面才會持續變化 */
  engine.value.runRenderLoop(() => {
    scene.value?.render();
  });

  await init({
    canvas: canvas.value,
    engine: engine.value,
    scene: scene.value,
    camera: camera.value,
  });
});

onBeforeUnmount(() => {
  engine.value?.dispose();
  window.removeEventListener('resize', handleResize);
});

function handleResize() {
  engine.value?.resize();
}

return {
  canvas,
  engine,
```

```
    scene,
    camera,
  }
}
```

 Tips

為甚麼要用 shallowRef 而非 ref，這是因為這類 babylonjs 提供的物件內部變化
與 Vue 沒有直接關係，而 babylon 物件內容又相當深層且龐大，使用 ref 響應
其內部所有內容只會浪費效能，所以這裡採用 shallowRef。

想要深入了解的讀者可以參考以下連結：

* https://cn.vuejs.org/guide/best-practices/performance.html#reduce-reactivity-overhead-for-large-immutable-structures

game-menu-background 元件內容也隨之調整。

➜ src\components\game-menu-background.vue

```
...
<script setup lang="ts">
...
watch(() => props.selectedGame, () => {
  const easingFunction = new BezierCurveEase(0.800, -0.155, 0.000, 1.000);

  if (camera.value) {
    moveCamera(camera.value, {
      speed: 0.3,
      easingFunction,
    });
  }
});
...
</script>
```

現在開啟派對後，我們有 3 個遊戲可以選了！ (≧∀≦)ﾉ

接著讓我們開發狐狸與老鼠的遊戲吧！(ㅍ‿ㅍ)◇

12.2 建立遊戲場景

此章節程式碼可以在以下連結取得。

- https://gitlab.com/deepmind1/animal-party/web/-/tree/feat/game-fox-and-mouse-sence

現在讓我們準備開發遊戲，預期有以下內容：

控制狐狸在雪地上移動，雪地下會產生比玩家數量多 3 個老鼠。

- 玩家的狐狸踩在老鼠上方的雪地時，手機會震動。

- 玩家按下抓老鼠按鍵後，狐狸會跳起並插在雪地中。

- 讓我們建立遊戲場景元件吧。

➜ src\games\fox-and-mouse\game-scene.vue

```ts
<template>
</template>

<script setup lang="ts">
import { ref } from 'vue';

const emit = defineEmits<{
  (e: 'init'): void;
}>();
</script>
```

```
<style scoped lang="sass">
</style>
```

接著在 App.vue 加入場景元件。

➜　src\App.vue

```
<template>
  <router-view />
  <loading-overlay />

  <game-scene class=" absolute w-full h-full" />
</template>

<script setup lang="ts">
import LoadingOverlay from './components/loading-overlay.vue';
import GameMenu from './components/game-menu.vue';

import GameScene from './games/fox-and-mouse/game-scene.vue';
</script>
...
```

使用 useBabylonScene 建立遊戲場景吧。

➜　src\games\fox-and-mouse\game-scene.vue

```
<template>
  <div class="overflow-hidden">
    <canvas
      ref="canvas"
      class="outline-none w-full h-full"
    />
  </div>
</template>

<script setup lang="ts">
import { onBeforeUnmount, onMounted, ref } from 'vue';
import {
  ArcRotateCamera,
```

```
  Engine, HemisphericLight, MeshBuilder, Scene, Vector3,
} from '@babylonjs/core';
import { SkyMaterial } from '@babylonjs/materials';
import '@babylonjs/core/Debug/debugLayer';
import '@babylonjs/inspector';

import { useBabylonScene } from '../../composables/use-babylon-scene';

const emit = defineEmits<{
  (e: 'init'): void;
}>();

const { canvas } = useBabylonScene({
  async init({ scene }) {
    window.addEventListener('keydown', (ev) => {
      // 按下 Shift+I 可以切換視窗
      if (ev.shiftKey && ev.keyCode === 73) {
        if (scene.debugLayer.isVisible()) {
          scene.debugLayer.hide();
        } else {
          scene.debugLayer.show();
        }
      }
    });

    emit('init');
  }
});
</script>
```

現在會看到畫面出現熟悉的一片深藍色了，接著讓我們新增雪地吧。

➜ src\games\fox-and-mouse\game-scene.vue

```
...
<script setup lang="ts">
...
const { canvas } = useBabylonScene({
  async init({ scene }) {
```

```
    createSnowfield(scene);

    ...
  }
});


function createSnowfield(scene: Scene) {
  const snowfield = MeshBuilder.CreateGround('snowfield ', { height: 150, width: 150 });

  const material = new BackgroundMaterial("snowfield-material", scene);
  material.useRGBColor = false;
  material.primaryColor = new Color3(0.93, 0.95, 0.95);

  snowfield.material = material;

  return snowfield;
}
</script>
```

現在畫面變成一片白亮亮，啥都看不到。

讓我們加一些小石頭，讓畫面不要那麼單調，同時調整一下鏡頭角度。

➜ src\games\fox-and-mouse\game-scene.vue

```
...
<script setup lang="ts">
...
/** 遊戲場景邊界 */
const sceneBoundary = {
  x: 11,
  z: 7,
}

const { canvas } = useBabylonScene({
  async init({ scene, camera }) {
    camera.beta = 0.5;
```

```
    createSnowfield(scene);
    createStones(scene);

    ...
  }
});
...

function createStones(scene: Scene) {
  const stones = new SolidParticleSystem('stones', scene, {
    useModelMaterial: true
  });

  const material = new StandardMaterial('stone-material', scene);
  material.diffuseColor = new Color3(0.5, 0.5, 0.5);
  material.emissiveColor = new Color3(0.5, 0.5, 0.5);

  const stone = MeshBuilder.CreateBox('stone', {
    width: 0.5,
    depth: 0.5,
    height: 0.2,
  });
  stone.material = material;

  stones.addShape(stone, 5);
  /** 新增至 SolidParticleSystem 後原本的 cloud 就可以停用了 */
  stone.dispose();

  /** 實際建立 mesh */
  const mesh = stones.buildMesh();

  /** 建立初始化 function */
  stones.initParticles = () => {
    stones.particles.forEach((particle) => {
      /** 隨機分布 */
      particle.position.x = Scalar.RandomRange(-sceneBoundary.x * 2, sceneBoundary.x *
2);

      particle.position.z = Scalar.RandomRange(-sceneBoundary.z * 2, sceneBoundary.z *
2);
```

```
    const scaling = Scalar.RandomRange(0.5, 1);
    /** 隨機尺寸 */
    particle.scaling = new Vector3(scaling, Scalar.RandomRange(0.5, 1), scaling);
  });
};

/** 初始化 */
stones.initParticles();
stones.setParticles();

return stones;
}
</script>
```

▲ 圖 12-6 加入石頭

助教：「嗯 ... 看起來還是很白捏。(⊙ ⼩ ⊙')」

鱈魚：「OK 的啦，反正重點是狐狸和老鼠嘛。(´▽`)ﾉ _」

12.3 加入老鼠

此章節程式碼可以在以下連結取得。

- https://gitlab.com/deepmind1/animal-party/web/-/tree/feat/game-fox-and-mouse-join-the-mouse

來加入藏在雪下方的老鼠吧！

由於老鼠藏在雪下方，所以實際上我們在雪上用一個方框，提示玩家老鼠的位置。

助教：「所以根本沒有老鼠？ ㄟ('ε'ㄟ)」

鱈魚：「嘿啊，在雪下，看不到。('·ω·`)」

助教：「啥 ... ㄟ('口'ㄟ)」

建立老鼠 Class 檔案。

→ src\games\fox-and-mouse\mouse.ts

```
import {
  Scene, Vector3, AbstractMesh,
  MeshBuilder, PhysicsImpostor, StandardMaterial, Color3,
} from '@babylonjs/core';

import { defaultsDeep } from 'lodash-es';

export interface MouseParam {
```

```
  position?: Vector3;
}

export class Mouse {
  mesh?: AbstractMesh;
  name: string;
  scene: Scene;
  params: Required<MouseParam> = {
    position: new Vector3(0, 0, 0),
  };

  constructor(name: string, scene: Scene, params?: MouseParam) {
    this.name = name;
    this.scene = scene;
    this.params = defaultsDeep(params, this.params);
  }

  async init() {
    const mesh = MeshBuilder.CreateTorus(`mouse-${this.name}`, {
      diameter: 2,
      tessellation: 4,
      thickness: 0.1,
    }, this.scene);

    const material = new StandardMaterial('mouse-material', this.scene);
    material.diffuseColor = new Color3(0, 0, 0);
    material.alpha = 0.3;

    mesh.material = material;
    this.mesh = mesh;
    this.mesh.position = this.params.position;

    return this;
  }

}
```

在場景中建立看看。

src\games\fox-and-mouse\game-scene.vue

```
...
<script setup lang="ts">
...
const { canvas } = useBabylonScene({
  async init({ scene, camera }) {
    ...
    createStones(scene);
    createMice(scene);
    ...
  }
});
...

async function createMice(scene: Scene) {
  const mouse = await new Mouse('0', scene).init();
}
</script>
```

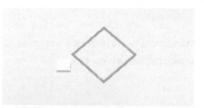

▲ 圖 12-7 老鼠方框

加個動畫吧。

➜ src\games\fox-and-mouse\mouse.ts

```
...

export class Mouse {
  ...
```

```
async init() {
  ...
  // 建立動畫
  const frameRate = 10;
  const alphaAnimation = new Animation(
    'alphaAnimation',
    'alpha',
    frameRate / 2,
    Animation.ANIMATIONTYPE_FLOAT,
    Animation.ANIMATIONLOOPMODE_CYCLE
  );

  const keyFrames = [
    {
      frame: 0,
      value: 0.1
    },
    {
      frame: frameRate / 2,
      value: 0.4
    },
    {
      frame: frameRate,
      value: 0.1
    }
  ];

  alphaAnimation.setKeys(keyFrames);
  material.animations = [alphaAnimation];

  this.scene.beginAnimation(material, 0, frameRate, true);

  ...
  }
}
```

遊戲比的是誰抓的老鼠最大隻,所以我們要新增一個參數,用來表示老鼠尺寸,同時還需要一個變數用來記錄有沒有被抓到。

➔ src\games\fox-and-mouse\mouse.ts

```
...
export interface MouseParam {
  ...
  /** 最小值為 1 */
  size?: number;
}

export class Mouse {
  ...
  params: Required<MouseParam> = {
    ...
    size: 1,
  };

  isCaught = false;

  ...
  caught() {
    this.isCaught = true;

    /** 隱藏動畫 */
    if (!this.mesh?.material) return;
    const {
      animation, frameRate
    } = createAnimation(this.mesh.material, 'alpha', 0);

    this.mesh.material.animations = [animation];
    this.scene.beginAnimation(this.mesh.material, 0, frameRate);
  }
}
```

現在我們只要隨機產生老鼠即可完成老鼠的部分了,但是這裡有個小問題,位置座標不能直接隨機產生,隨機產生可能會讓老鼠重疊,所以先讓我們建立一個可以隨機產生座標又不會重疊的 function。

➜ src\common\utils.ts

```
...
interface GetRandomPositionsParam {
  /** 個軸向偏移量 */
  offset: {
    x: number;
    y: number;
    z: number;
  },
  /** 每個座標之間最小距離 */
  minDistance: number,
  /** 數量 */
  length: number,
  /** 座標原點，預設 new Vector3(0, 0, 0) */
  origin?: Vector3,
}
/** 取得指定數量的隨機座標 */
export function getRandomPositions(param: GetRandomPositionsParam) {
  const {
    offset: { x, y, z },
    minDistance,
    length,
    origin = new Vector3(0, 0, 0),
  } = param;

  const result: Vector3[] = flow([
    () => {
      const positions: Vector3[] = [];

      /** 紀錄嘗試次數 */
      let counter = 0;
      do {
        if (counter > 5000) {
          throw new Error(' 無法計算座標 ');
        }

        const position = new Vector3(
          random(-x, x, true),
          random(-y, y, true),
```

```
        random(-z, z, true),
      );

      /** 檢查是否與任一現存座標重疊 */
      const invalid = positions.some(
        (existPosition) => Vector3.Distance(position, existPosition) < minDistance
      );
      /** 無效，進入下一輪 */
      if (invalid) {
        counter++;
        continue;
      }

      /** 儲存有效座標 */
      positions.push(position);
      /** 歸零嘗試次數 */
      counter = 0;
    } while (positions.length < length);

    return positions;
  },
  /** 將目前中心點平移至指定 origin */
  (positions: Vector3[]) => positions.map((position) => position.add(origin)),
])();

  return result;
}
```

讓我們利用 getRandomPositions 產生老鼠吧,同時調整一下場景邊界與石頭數量。

➜ src\games\fox-and-mouse\game-scene.vue

```
...

<script setup lang="ts">
...
import { getRandomPositions } from '../../common/utils';
...
```

```
/** 遊戲場景邊界 */
const sceneBoundary = {
  x: 16,
  z: 10,
}
...
function createStones(scene: Scene) {
  ...
  stones.addShape(stone, 20);
  ...
}

async function createMice(scene: Scene) {
  /** 老鼠數量，未來會改成依照玩家數量產生 */
  const quantity = 10;
  const { x, z } = sceneBoundary;

  const positions = getRandomPositions({
    offset: {
      x,
      y: 0,
      z,
    },
    minDistance: 5,
    length: quantity,
  });

  const task = positions.map((position, i) =>
    new Mouse(`mouse-${i}`, scene, {
      position,
      size: (i + 1) * 100,
    }).init()
  );

  const results = await Promise.allSettled(task);
}
</script>
```

調整一卜方框的顏色。

➔ src\games\fox-and-mouse\mouse.ts

```
...
export class Mouse {
  ...
  async init() {
    ...
    material.diffuseColor = new Color3(0.8, 0.8, 0.5);
    ...
  }
  ...
}
```

▲ 圖 12-8　隨機產生老鼠方框

再來讓我們加入狐狸吧。(´„•ω•„)

12.4　狐狸登場

此章節程式碼可以在以下連結取得。

- https://gitlab.com/deepmind1/animal-party/web/-/tree/feat/game-fox-and-mouse-the-fox

狐狸控制概念基本上和企鵝相同，首先讓我們新增狐狸檔案。

➡ src\games\fox-and-mouse\fox.ts

```typescript
import {
  Scene, Vector3, AbstractMesh, SceneLoader,
} from '@babylonjs/core';

import { defaultsDeep } from 'lodash-es';

export interface FoxParam {
  position?: Vector3;
  ownerId: string;
}

export class Fox {
  mesh?: AbstractMesh;
  name: string;
  scene: Scene;
  param: Required<FoxParam> = {
    position: new Vector3(0, 0, 0),
    ownerId: '',
  };
```

```
  constructor(name: string, scene: Scene, param?: FoxParam) {
    this.name = name;
    this.scene = scene;
    this.param = defaultsDeep(param, this.param);
  }

  async init() {
    const result = await SceneLoader.ImportMeshAsync('', '/fox-and-mouse/', 'fox.glb',
  this.scene);

    return this;
  }
}
```

在場景中引入一隻狐狸。

➜ src\games\fox-and-mouse\game-scene.vue

```
...

<script setup lang="ts">
...
const { canvas } = useBabylonScene({
  async init({ scene, camera }) {
    ...
    createFoxes(scene);
    ...
  }
});
...
async function createFoxes(scene: Scene) {
  const positions = getSquareMatrixPositions(1, 1);

  const task = positions.map((position, i) =>
    new Fox(`fox-${i}`, scene, {
      position,
      ownerId: '',
    }).init()
```

```
  );

  const results = await Promise.allSettled(task);
}
</script>
```

▲ 圖 12-9 狐狸登場

接著初始化狐狸模型動畫。

➔ src\games\fox-and-mouse\fox.ts

```
...
interface AnimationMap {
  idle?: AnimationGroup,
  walk?: AnimationGroup,
  pounce?: AnimationGroup,
  dive?: AnimationGroup,
}
...
export class Fox {
  ...
  private animation: AnimationMap = {
    idle: undefined,
    walk: undefined,
```

```
    pounce: undefined,
    dive: undefined,
  };

  ...
  private initAnimation(animationGroups: AnimationGroup[]) {
    animationGroups.forEach((group) => group.stop());

    const findAnimation = (name: keyof AnimationMap) => animationGroups.find(
      (group) => group.name === name
    );

    this.animation = {
      idle: findAnimation('idle'),
      walk: findAnimation('walk'),
      pounce: findAnimation('pounce'),
      dive: findAnimation('dive'),
    }
  }

  async init() {
    const result = await SceneLoader.ImportMeshAsync('', '/fox-and-mouse/', 'fox.glb',
this.scene);
    this.initAnimation(result.animationGroups);

    return this;
  }
}
```

助教：「獲得一隻冷靜的豬了！(´▽`)/ 」

鱈魚：「是狐狸啦。... ('◉ ⋒ ◉`) 」

助教：「啊 ... 抱歉捏。(⊙ ⌇⋎⌇ ⊙`) 」

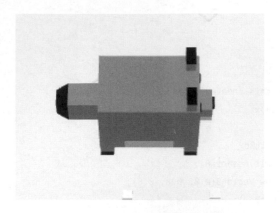

▲ 圖 12-10　初始化狐狸動畫

加上 hit box 並調整一下狐狸尺寸吧。

→ src\games\fox-and-mouse\fox.ts

```
...
export class Fox {
  ...
  private createHitBox() {
    const hitBox = MeshBuilder.CreateBox(`${this.name}-hit-box`, {
      width: 3, depth: 1.5, height: 2
    });
    hitBox.position = new Vector3(0, 1, 0);
    // 設為半透明方便觀察
    hitBox.visibility = 0.5;

    /** 使用物理效果 */
    const hitBoxImpostor = new PhysicsImpostor(
      hitBox,
      PhysicsImpostor.BoxImpostor,
      { mass: 1, friction: 0.7, restitution: 0.7 },
      this.scene
    );

    hitBox.physicsImpostor = hitBoxImpostor;
    return hitBox;
  }
```

```
async init() {
  ...
  // 產生 hitBox
  const hitBox = this.createHitBox();
  this.mesh = hitBox;

  // 將狐狸綁定至 hitBox
  const fox = result.meshes[0];
  fox.scaling = new Vector3(0.8, 0.8, 0.8);
  fox.setParent(hitBox);

  this.mesh.position = this.param.position;

  return this;
  }
}
```

▲ 圖 12-11　狐狸 hit box

hit box 確認 OK 後，就可以隱藏了，現在讓我們加上移動用的 method。

➜ src\games\fox-and-mouse\fox.ts

```
...

export class Fox {
  ...
  /** 指定移動方向與力量 */
  walk(direction: Vector3) {
```

```
    if (!this.mesh) {
      throw new Error(' 未建立 Mesh');
    }

    /** 單位向量化，統一速度 */
    const velocity = direction.normalize().scale(3);
    this.mesh.physicsImpostor?.setLinearVelocity(velocity);
  }
}
```

並在場景加入鍵盤事件後，開啟物理引擎。

➜ src\games\fox-and-mouse\game-scene.vue

```
...
<script setup lang="ts">
import * as CANNON from 'cannon-es';
import { compact, range } from 'lodash-es';
...
const { canvas } = useBabylonScene({
  async init({ scene, camera }) {
    const physicsPlugin = new CannonJSPlugin(true, 8, CANNON);
    scene.enablePhysics(new Vector3(0, -9, 0), physicsPlugin);
    ...
  }
});

...

async function createFoxes(scene: Scene) {
  const positions = getSquareMatrixPositions(5, 1, new Vector3(0, 1, 0));

  const task = positions.map((position, i) =>
    new Fox(`fox-${i}`, scene, {
      position,
      ownerId: '',
    }).init()
  );
```

```
const results = await Promise.allSettled(task);

/** 去除建立失敗的狐狸 */
const foxes = compact(results.map(
  (result) => result.status === 'fulfilled' ? result.value : undefined,
));

window.addEventListener('keydown', (event) => {
  switch (event.key) {
    case '8': {
      foxes[0].walk(new Vector3(0, 0, -1));
      break;
    }
    case '4': {
      foxes[0].walk(new Vector3(1, 0, 0));
      break;
    }

    case '2': {
      foxes[0].walk(new Vector3(0, 0, 1));
      break;
    }
    case '6': {
      foxes[0].walk(new Vector3(-1, 0, 0));
      break;
    }
  }
})
}
</script>
```

因為地板沒有加入物理效果，狐狸往地心奔去了！ ᐅ(˚∀˚)ᐅ

地板追加物理效果。

➡ src\games\fox-and-mouse\game-scene.vue

```
...
<script setup lang="ts">
...
```

```
function createSnowfield(scene: Scene) {
  ...
  snowfield.physicsImpostor = new PhysicsImpostor(
    snowfield,
    PhysicsImpostor.PlaneImpostor,
    { mass: 0, friction: 0, restitution: 0 },
    scene
  );

  return snowfield;
}
...
</script>
```

現在按下對應數字鍵，狐狸會移動了，不過一動起來就停不下來，我們需要在放開按鍵的時候將速度設為 0。

➜ src\games\fox-and-mouse\game-scene.vue

```
...
<script setup lang="ts">
...
async function createFoxes(scene: Scene) {
  ...
  window.addEventListener('keyup', (event) => {
    switch (event.key) {
      case '8': case '4': case '2': case '6': {
        foxes[0].walk(new Vector3(0, 0, 0));
        break;
      }
    }
  })
}
...
</script>
```

現在狐狸可以自由移動了，讓狐狸轉起來吧。

邏輯和企鵝轉向一模一樣，只是方向稍微不太一樣。

➜ src\games\fox-and-mouse\fox.ts

```
...
export class Fox {
  ...
  /** 取得力與人物的夾角 */
  private getDirectionAngle(vector: Vector3) {
    if (!this.mesh) {
      throw new Error('未建立 Mesh');
    }

    const forceVector = vector.normalize();
    /** 人物面相 -X 軸方向 */
    const characterVector = new Vector3(-1, 0, 0);
    const deltaAngle = Math.acos(Vector3.Dot(forceVector, characterVector));

    /** 反餘弦求得角度範圍為 0~180 度，需要自行判斷負角度部分。
     *  向量 Z 軸分量為負時，表示夾角為負。
     */
    if (forceVector.z < 0) {
      return deltaAngle * -1;
    }

    return deltaAngle;
  }
  private rotate(angle: number) {
    if (!this.mesh) return;

    /** 若角度超過 180 度，則先直接切換至兩倍補角處，讓轉向更自然 */
    const currentAngle = this.mesh.rotation.y;
    if (Math.abs(angle - currentAngle) > Math.PI) {
      const supplementaryAngle = Math.PI * 2 - Math.abs(currentAngle);
      if (currentAngle < 0) {
        this.mesh.rotation = new Vector3(0, supplementaryAngle, 0);
      } else {
        this.mesh.rotation = new Vector3(0, -supplementaryAngle, 0);
```

```
      }
   }

   const { animation, frameRate } = createAnimation(this.mesh, 'rotation', new
Vector3(0, angle, 0), {
      speedRatio: 3,
   });

   this.scene.beginDirectAnimation(this.mesh, [animation], 0, frameRate);
 }

 ...

 /** 指定移動方向與力量 */
 walk(direction: Vector3) {
   ...

   /** 長度為 0 不用旋轉 */
   if (velocity.length() > 0) {
     const targetAngle = this.getDirectionAngle(velocity);
     this.rotate(targetAngle);
   }
 }
}
```

現在讓我們使用狀態機來管理狐狸狀態吧。

首先定義狀態列舉。

➜ src\games\fox-and-mouse\fox.ts

```
...
enum State {
  /** 可自由動作 */
  IDLE = 'idle',
  /** 走路中 */
  WALK = 'walk',
  /** 抓老鼠 */
  POUNCE = 'pounce',
```

```
  /** 插在地上 */
  DIVE = 'dive',
}
...
```

目前考慮 idle 和 walk 狀態。

- idle 觸發 walk 事件後轉移至 walk 狀態。

- walk 狀態時，沒有持續觸發 walk 事件將於 500 毫秒後自動返回 idle 狀態。

➜ src\games\fox-and-mouse\fox.ts

```
...
import { createMachine, interpret } from 'xstate';
...

export class Fox {
  ...
  /** 儲存被解釋後的狀態機服務 */
  private stateService;

  constructor(name: string, scene: Scene, param?: FoxParam) {
    ...

    const stateMachine = createMachine(
      {
        id: 'fox',
        initial: State.IDLE,
        context: {},
        states: {
          [State.IDLE]: {
            on: {
              [State.WALK]: State.WALK,
            },
          },
          [State.WALK]: {
            after: [
              {
                delay: 500,
```

```
        target: State.IDLE,
      },
    ],
    on: {
      [State.WALK]: State.WALK,
    },
  },
},
predictableActionArguments: true,
},
{
  actions: { },
}
);


this.stateService = interpret(stateMachine);
this.stateService.onTransition((content) => {
  console.log('content : ', content.value)
});
this.stateService.start();
}
...
}
```

現在按下數字鍵就會在 console 中看到狀態變化的 log。

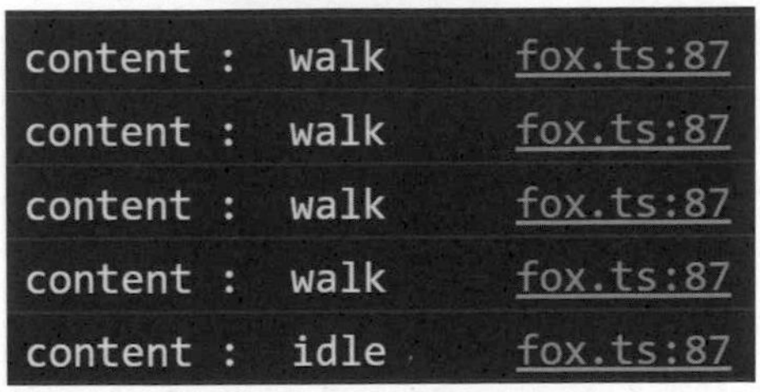

```
content :   walk        fox.ts:87
content :   walk        fox.ts:87
content :   walk        fox.ts:87
content :   walk        fox.ts:87
content :   idle        fox.ts:87
```

▲ 圖 12-12　狀態變化 log

在加入動畫之前，我們先將先前企鵝遊戲中，用來混合動畫的 method 抽離獨立到 utils 中吧。

➜ src\common\utils.ts

```
...
/** 動畫混合
 * 讓目前播放動畫的權重從 1 到 0，而目標動畫從 0 到 1。
 * 利用 Generator Function 配合 babylon 的 [runCoroutineAsync API](https://doc.
babylonjs.com/features/featuresDeepDive/events/coroutines)
 * 迭代，達成融合效果。
 */
export function* animationBlending(
  fromAnimation: AnimationGroup,
  toAnimation: AnimationGroup,
  loop = true,
  step = 0.1
) {
  let currentWeight = 1;
  let targetWeight = 0;

  toAnimation.play(loop);

  while (targetWeight < 1) {
    targetWeight += step;
    currentWeight -= step;

    toAnimation.setWeightForAllAnimatables(targetWeight);
    fromAnimation.setWeightForAllAnimatables(currentWeight);
    yield;
  }

  toAnimation.play(loop);
  fromAnimation.stop();
}
```

　　新增 oldState 儲存上次狀態，方便後續處理動畫。

➜ src\games\fox-and-mouse\fox.ts

```
...
export class Fox {
  ...
```

```
  private oldState: `${State}` | undefined = undefined;

  constructor(name: string, scene: Scene, param?: FoxParam) {
    ...
    this.stateService.onTransition((state) => {
      const stateValue = state.value as `${State}`;
      /** 沒變化，提早結束 */
      if (this.oldState === stateValue) return;

      console.log(`stateValue : `, stateValue);
      console.log(`oldState : `, this.oldState);

      this.oldState = stateValue;
    });
    this.stateService.start();
  }
  ...
}
```

現在讓我們新增處理動畫的 method，並在動畫初始化後播放 idle 動畫。

→ src\games\fox-and-mouse\fox.ts

```
...
export class Fox {
  ...
  constructor(name: string, scene: Scene, param?: FoxParam) {
    ...
    this.stateService.onTransition((state) => {
      const stateValue = state.value as `${State}`;
      /** 沒變化，提早結束 */
      if (this.oldState === stateValue) return;

      this.processStateAnimation(stateValue);

      this.oldState = stateValue;
    });
    ...
  }
```

```
  private initAnimation(animationGroups: AnimationGroup[]) {
    ...
    this.animation.idle?.play(true);
  }
  /** 處理狀態動畫
   *
   * 利用 [runCoroutineAsync API](https://doc.babylonjs.com/features/featuresDeepDive/
events/coroutines) 實現
   */
  private processStateAnimation(newState: `${State}`) {
    const targetAnimation = this.animation[newState];

    /** 沒有上一個狀態，直接播放新動畫 */
    if (!this.oldState) {
      targetAnimation?.play();
      return;
    }

    const playingAnimation = this.animation[this.oldState];

    if (!targetAnimation || !playingAnimation) return;

    /** idle、walk 循環播放 */
    const loop = ['idle', 'walk', 'dive'].includes(newState);

    this.scene.onBeforeRenderObservable.runCoroutineAsync(
      animationBlending(playingAnimation, targetAnimation, loop)
    );
  }
  ...
}
```

我們新增一個轉移事件，就是當速度被設為 0 時，會轉移至 idle 狀態。

➔ src\games\fox-and-mouse\fox.ts

```
...
export class Fox {
  ...
  constructor(name: string, scene: Scene, param?: FoxParam) {
```

```
  ...
  const stateMachine = createMachine(
    {
      ...
      states: {
        ...
        [State.WALK]: {
          ...
          on: {
            [State.IDLE]: State.IDLE,
            [State.WALK]: State.WALK,
          },
        },
      },
      ...
    },
    ...
  );
  ...
}
...
/** 指定移動方向與力量 */
walk(direction: Vector3) {
  ...

  /** 單位向量化，統一速度 */
  const velocity = direction.normalize().scale(3);
  this.mesh.physicsImpostor?.setLinearVelocity(velocity);

  /** 長度為 0 不用旋轉 */
  if (velocity.length() > 0) {
    this.stateService.send({ type: State.WALK });

    const targetAngle = this.getDirectionAngle(velocity);
    this.rotate(targetAngle);
  } else {
    this.stateService.send({ type: State.IDLE });
  }
}
}
```

現在我們完成 idle 和 walk 動畫了！

用狀態機處理是不是好理解很多啊。ヽ(*´∼`*)ﾉ

讓我們繼續完成抓老鼠和成功抓到老鼠的狀態吧。

首先來定義狀態：

- idle、walk 狀態都可以藉由 pounce 事件轉移至 pounce 狀態。

- pounce 狀態時觸發 dive 事件會根據內部變數判斷是否抓到老鼠，有抓到前往 dive 狀態，否則前往 idle 狀態。

➡ src\games\fox-and-mouse\fox.ts

```
...
export class Fox {
  /** 抓到的老鼠尺寸 */
  mouseSize = 0;
  ...
  constructor(name: string, scene: Scene, param?: FoxParam) {
    ...

    const stateMachine = createMachine(
      {
        ...
        context: {
          touchedMouse: false,
        },
        states: {
          [State.IDLE]: {
            on: {
              ...
              [State.POUNCE]: State.POUNCE,
            },
          },
          [State.WALK]: {
            ...
            on: {
              ...
```

```
            [State.POUNCE]: State.POUNCE,
          },
        },
        [State.POUNCE]: {
          on: {
            [State.DIVE]: [
              {
                cond: () => this.mouseSize > 0,
                target: State.DIVE,
              },
              {
                target: State.IDLE,
              },
            ]
          },
        },
        [State.DIVE]: {}

      },
      predictableActionArguments: true,
    }
  );
  ...
  }
  ...
}
```

新增 method：

- pounce，用來執行抓老鼠動作。

- setMouseSize：儲存老鼠尺寸。

- getState，取得目前人物狀態。

➜ src\games\fox-and-mouse\fox.ts

```
...
export class Fox {
  ...
  setMouseSize(size: number) {
    this.mouseSize = size;
  }
  /** 抓老鼠 */
  pounce() {
    this.stateService.send({ type: State.POUNCE });
  }

  getState() {
    const { value } = this.stateService.getSnapshot();
    return value as `${State}`;
  }
}
```

調整一下 processStateAnimation，回傳目標動畫，讓我們可以得知動畫是否播放完成。

➜ src\games\fox-and-mouse\fox.ts

```
...
export class Fox {
  ...
  /** 處理狀態動畫
   *
   * 利用 [runCoroutineAsync API](https://doc.babylonjs.com/features/featuresDeepDive/
events/coroutines) 實現
   */
```

```
private processStateAnimation(newState: `${State}`) {
  const targetAnimation = this.animation[newState];

  /** 沒有上一個狀態，直接播放新動畫 */
  if (!this.oldState) {
    targetAnimation?.play();
    return targetAnimation;
  }

  const playingAnimation = this.animation[this.oldState];

  if (!targetAnimation || !playingAnimation) return;

  /** idle、walk 不循環播放 */
  const loop = ['idle', 'walk'].includes(newState);

  this.scene.onBeforeRenderObservable.runCoroutineAsync(
    animationBlending(playingAnimation, targetAnimation, loop)
  );

  return targetAnimation;
}
...
}
```

接著調整一下狀態機服務的 onTransition 部分，如果狀態為 pounce 則偵測動畫是否結束。

➡ src\games\fox-and-mouse\fox.ts

```
...
export class Fox {
  ...
  constructor(name: string, scene: Scene, param?: FoxParam) {
    ...
    this.stateService.onTransition((state) => {
      const stateValue = state.value as `${State}`;
      /** 沒變化，提早結束 */
      if (this.oldState === stateValue) return;
```

```
      const animation = this.processStateAnimation(stateValue);
      if (animation && stateValue === 'pounce') {
        animation.onAnimationEndObservable.addOnce(async () => {
          /** 延遲一下，不要那麼快轉換 */
          await promiseTimeout(500);
          this.stateService.send({ type: State.DIVE });
        });
      }

      this.oldState = stateValue;
    });
    ...
  }
turn targetAnimation;
  }
  ...
}
```

在遊戲場景中使用數字鍵 0，呼叫狐狸的 pounce method。

➔　src\games\fox-and-mouse\game-scene.vue

```
...
<script setup lang="ts">
...
async function createFoxes(scene: Scene) {
  ...
  window.addEventListener('keydown', (event) => {
    switch (event.key) {
      ...
      case '0': {
        foxes[0].pounce();
        break;
      }
    }
  })
  ...
}
```

```
...
</script>
```

現在按下數字鍵 0，狐狸會跳起來抓老鼠了。ㄑ(•ω•)ㄏ

▲ 圖 12-13 狐狸 pounce 動畫

讓我們回到遊戲場景，透過偵測物體碰撞來控制狐狸與老鼠是否抓到與被抓到。

我們讓畫面先產生 2 隻狐狸並回傳建立完成的狐狸，createMice 也要將老鼠回傳。

➜ src\games\fox-and-mouse\game-scene.vue

```ts
...
<script setup lang="ts">
...
async function createMice(scene: Scene) {
  ...
  const results = await Promise.allSettled(task);

  return compact(results.map(
    (result) => result.status === 'fulfilled' ? result.value : undefined,
  ));
}

async function createFoxes(scene: Scene) {
  const positions = getSquareMatrixPositions(5, 2, new Vector3(0, 1, 0));
```

```
  ...

  return foxes;
}
</script>
```

　　現在會發現狐狸會意外飄動，讓我們回到狐狸 Class，加入移動限制。

➜ src\games\fox-and-mouse\fox.ts

```
...
export class Fox {
...
  async init() {
    ...

    // 持續在每個 frame render 之前呼叫
    this.scene.registerBeforeRender(() => {
      hitBox.physicsImpostor?.setAngularVelocity(
        new Vector3(0, 0, 0)
      );

      if (['idle', 'pounce', 'dive'].includes(this.getState())) {
        hitBox.physicsImpostor?.setLinearVelocity(
          new Vector3(0, 0, 0)
        );
      }
    });

    ...
  }

  /** 指定移動方向與力量 */
  walk(direction: Vector3) {
    ...
    if (['pounce', 'dive'].includes(this.getState())) return;

    /** 單位向量化，統一速度 */
    ...
```

```
  }
...
}
```

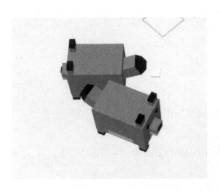

▲　圖 12-14　多隻狐狸

現在讓我們偵測狐狸與老鼠的碰撞偵測。

➜ src\games\fox-and-mouse\game-scene.vue

```
...
<script setup lang="ts">
...
const { canvas } = useBabylonScene({
  async init({ scene, camera }) {
    ...
    const mice = await createMice(scene);
    const foxes = await createFoxes(scene);

    window.addEventListener('keydown', (ev) => {
      ...
    });

    scene.registerBeforeRender(() => {
      detectCollideEvent(foxes, mice);
    });

    emit('init');
  }
```

```
});
...

/** 偵測碰撞事件 */
function detectCollideEvent(foxes: Fox[], mice: Mouse[]) {

}
</script>
```

最後讓我們完成 detectCollideEvent 的內容吧！

➜ src\games\fox-and-mouse\game-scene.vue

```
...
<script setup lang="ts">
...
/** 偵測碰撞事件 */
function detectCollideEvent(foxes: Fox[], mice: Mouse[]) {
  mice.forEach((mouse) => {
    const mouseMesh = mouse.mesh;

    /** 已經被抓到的老鼠不用檢查 */
    if (mouse.isCaught || !mouseMesh) return;

    foxes.forEach((fox) => {
      const foxState = fox.getState();

      /** dive 狀態不需要檢查 */
      if (['dive'].includes(foxState)) return;

      const foxMesh = fox.mesh;
      if (!foxMesh) return;

      /** 抓到了 */
      if (foxState === 'pounce' && foxMesh.intersectsMesh(mouseMesh)) {
        mouse.caught();
        fox.setMouseSize(mouse.params.size);
      }
    });
```

```
  });
}
</script>
```

現在狐狸在老鼠方框上方抓到老鼠後，持續插在地上的動畫了！ XD

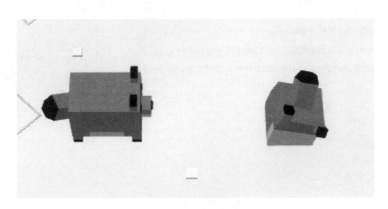

▲ 圖 12-15 抓到老鼠

最後我們只要讓狐狸可以自定義顏色，狐狸部分就大功告成了。

➜ src\games\fox-and-mouse\fox.ts

```
...
export interface FoxParam {
  position?: Vector3;
  ownerId: string;
  color?: Color3;
}

export class Fox {
  ...
  param: Required<FoxParam> = {
    position: new Vector3(0, 0, 0),
    ownerId: '',
    color: new Color3(1, 0.277, 0),
  };
  ...
  async init() {
```

```
  ...
  this.mesh = hitBox;

  /** 找到 body mesh 將材質改顏色 */
  const bodyMesh = result.meshes.find(
    ({ name }) => name === 'body_primitive2'
  );
  /** 確認模型材質為 PBRMaterial */
  if (bodyMesh?.material instanceof PBRMaterial) {
    bodyMesh.material.albedoColor = this.param.color;
  }

  ...
  }
  ...
}
```

改個顏色試試看。

➡ src\games\fox-and-mouse\game-scene.vue

```
...
<script setup lang="ts">
...
async function createFoxes(scene: Scene) {
  ...
  const task = positions.map((position, i) =>
    new Fox(`fox-${i}`, scene, {
      ...
      color: new Color3(0, 0.5, 1),
    }).init()
  );
  ...
}
...
</script>
```

變種狐狸出現了！ヽ(≧∀≦)ノ

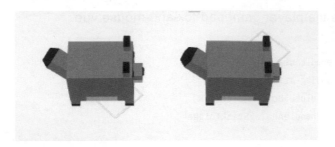

▲ 圖 12-16 狐狸變色

12.5 建立控制搖桿

此章節程式碼可以在以下連結取得。

▲ 此章節程式碼可以在以下連結取得。

- https://gitlab.com/deepmind1/animal-party/web/-/tree/feat/game-fox-and-mouse-gamepad

現在讓我們建立控制狐狸的搖桿吧，基本上和企鵝搖桿一模一樣，只有一個特殊差別，至於是甚麼差別讓我們賣個關子。(´„•ω•„)

新增搖桿頁面。

➡ src\views\the-player-gamepad-fox-and-mouse.vue

```
<template>
  <player-gamepad-container>
    <gamepad-analog-stick
      class="absolute bottom-5 left-8"
      @trigger="handleAnalogStickTrigger"
    />

    <gamepad-btn
      class="absolute bottom-10 right-20"
      size="6rem"
      @trigger="(status) => handleBtnTrigger('a', status)"
    >
      <div class="text-9xl mb-8">
        A
      </div>
    </gamepad-btn>

    <gamepad-btn
      class="absolute top-10 right-20 opacity-90"
      size="3rem"
      icon="done"
      @trigger="(status) => handleBtnTrigger('confirm', status)"
    />
  </player-gamepad-container>
</template>

<script setup lang="ts">
import GamepadBtn from '../components/gamepad-btn.vue';
import GamepadAnalogStick from '../components/gamepad-analog-stick.vue';
import PlayerGamepadContainer from '../components/player-gamepad-container.vue';

import { useLoading } from '../composables/use-loading';
import { useClientPlayer } from '../composables/use-client-player';
import { KeyName } from '../types';

const loading = useLoading();
```

```
const { emitGamepadData } = useClientPlayer();

function init() {
  loading.hide();
}
init();


function handleBtnTrigger(keyName: `${KeyName}`, status: boolean) {
  console.log(`[ handleBtnTrigger ] : `, { keyName, status });

  emitGamepadData([{
    name: keyName,
    value: status,
  }]);
}

function handleAnalogStickTrigger(data: { x: number, y: number }) {
  console.log(`[ handleAnalogStickTrigger ] : `, data);

  emitGamepadData([
    {
      name: 'x-axis',
      value: data.x,
    },
    {
      name: 'y-axis',
      value: data.y,
    }
  ]);
}
</script>
```

在 route 新增搖桿。

➔　src\router\router.ts

```
...
export enum RouteName {
  ...
```

```
    PLAYER_GAMEPAD_FOX_AND_MOUSE = 'player-gamepad-fox-and-mouse',
}

const routes: RouteRecordRaw[] = [
  ...
  {
    path: `/player-gamepad`,
    ...
    children: [
      ...
      {
        path: `fox-and-mouse`,
        name: RouteName.PLAYER_GAMEPAD_FOX_AND_MOUSE,
        component: () => import('../views/the-player-gamepad-fox-and-mouse.vue')
      },
    ]
  },
  ...
]
...
```

　　追加搖桿跳轉邏輯。

➜ src\views\the-player-gamepad.vue

```
...
<script setup lang="ts">
...
/** 遊戲與搖桿頁面對應資料 */
const gamepadMap: Record<GameName, {
  routeName: RouteName,
}> = {
  ...
  'fox-and-mouse': {
    routeName: RouteName.PLAYER_GAMEPAD_FOX_AND_MOUSE,
  },
}
...</script>
```

即使內容邏輯與企鵝搖桿相同，使用 player-gamepad-container 元件後，程式碼是不是精簡很多啊！ヽ(●`∀´●)ﾉ

助教：「所以到底和企鵝搖桿差在哪？(⊙_⊙)？」

鱈魚：「差別在於會震動！(´▽`)ﾉ_」

助教：「震動是有個雞毛用途。(⊙⌣⊙')」

鱈魚：「當玩家移動到老鼠上方時，老鼠越大隻，震動時間就會越久。」

助教：「原來老鼠大小是這樣判斷的喔，還以為你漏掉了。(´･ω･`)」

鱈魚：「你對我是不是有甚麼很深的偏見。('◉⌣◉`)」

老樣子我們需要先取得 Web API 授權才行，來新增授權定義吧。

➔ src\types\game.type.ts

```
...
export interface PlayerPermission {
  gyroscope: PlayerPermissionState;
  vibrate: PlayerPermissionState;
}

export const permissionInfoMap: Record<keyof PlayerPermission, {
  label: string;
}> = {
  'gyroscope': {
    label: ' 陀螺儀 ',
  },
  'vibrate': {
    label: ' 震動回饋 ',
  },
}
...
```

　　並在遊戲選單中將狐狸遊戲加入必要權限。

➡ src\components\game-menu.vue

```
...
<script setup lang="ts">
...
const games: GameInfo[] = [
  ...
  {
    name: 'fox-and-mouse',
    ...
    condition: {
      ...
      requiredPermissions: [
        'vibrate',
      ],
    }
  },
];
...
</script>
...
```

　　前往 permission-card，新增選項。

➡ src\components\permission-card.vue

```
...
<script setup lang="ts">
import { ..., useVibrate } from '@vueuse/core';
...

const { isSupported } = useVibrate();
const vibrateState = computed<`${PlayerPermissionState}`>(
  () => isSupported.value ? 'granted' : 'not-support',
);
...n getStateInfo(value: PlayerPermissionState) {
  return stateInfoMap?.[value] ?? stateInfoMap['not-support'];
}
```

```
const permissions = computed<{
  ...
  onClick?: () => void;
}[]>(() => ([
  ...
  {
    key: 'vibrate',
    icon: 'vibration',
    label: ' 震動回饋 ',
    caption: ' 控制震動馬達，提供震動回饋 ',
    state: vibrateState.value,
    stateInfo: getStateInfo(vibrateState.value),
    /** 點一下震動 */
    onClick: () => vibrate([100, 10, 50])
  },
]));
...
</script>
```

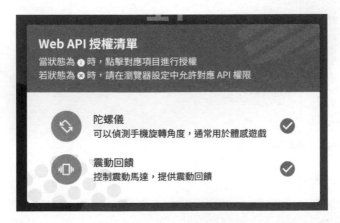

▲ 圖 12-17　震動回饋

現在讓我們實作傳遞震動訊號部分吧。

12.6 搖桿接收資料

和以往遊戲機接收搖桿訊號不同，改成搖桿接收遊戲機來的訊號了！

這裡會發現 player.type 的命名不好，讓我們重新命名一下，改為 gamepad. type，同時調整 index.ts。

➜ src\types\index.ts

```
...
export * from './gamepad.type';
```

新增 game-console.type 定義相關資料吧。

➜ src\types\game-console.type.ts

```
/** 震動資料 */
export interface ConsoleVibrateData {
  name: `vibrate`;
  value: number;
}


type Data = ConsoleVibrateData;

/** 遊戲機資料 */
export interface GameConsoleData {
  /** 目標玩家 ID，undefined 表示給所有玩家 */
  targets?: string[];
  data: Data;
}
```

統一導出。

➜ src\types\index.ts

```
...
export * from './game-console.type';
```

發送資料前，先來定義 socket 事件。

➔ src\types\socket.type.ts

```
...
import { GameConsoleData, GameConsoleState, GamepadData, Player } from '.';
...

interface OnEvents {
  ...
  'game-console:console-data': (data: GameConsoleData) => void;
}

interface EmitEvents {
  ...
  'game-console:console-data': (data: GameConsoleData) => void;
}
...
```

接著分別新增發送資料與接收資料部分吧。

- use-client-game-console，新增 emitConsoleData，對搖桿發送資料。

- use-client-player，新增 onConsoleData，接收遊戲機送來的資料。

➔ src\composables\use-client-game-console.ts

```
import { GameConsoleData, GameConsoleStatus, GameName, GamepadData, Player, Room }
from "../types";
...
export function useClientGameConsole() {
  ...
  async function emitConsoleData(data: GameConsoleData) {
    if (!mainStore.client?.connected) {
      return Promise.reject('client 尚未連線 ');
    }

    mainStore.client.emit('game-console:console-data', data);
  }
```

```
  return {
    ...
    emitConsoleData,
  }
}
```

➜ src\composables\use-client-player.ts

```
...
import { GameConsoleData, GameConsoleState, Player, Room, SignalData } from "../
types";
...

export function useClientPlayer() {
  ...
  const consoleDataHook = createEventHook<GameConsoleData>();

  /** 元件解除安裝前，移除 Listener 以免記憶體洩漏 */
  onBeforeUnmount(() => {
    ...
    mainStore.client?.removeListener('game-console:console-data', consoleDataHook.
trigger);

  });

  ...

  return {
    ...
    /** 接收遊戲機發送資料 */
    onConsoleData: (fn: Parameters<typeof consoleDataHook['on']>[0]) => {
      mainStore.client?.on('game-console:console-data', consoleDataHook.trigger);
      return consoleDataHook.on(fn);
    },

    ...
  }
}
```

網頁端準備完成,現在讓我們回到久違的伺服器端,同步 console 資料吧。

12.6.1 伺服器同步 console 資料

此章節程式碼可以在以下連結取得。

- https://gitlab.com/deepmind1/animal-party/server/-/tree/feat/sync-console-data

第一步老樣子,新增資料定義。

➔ types\game-console.type.ts

```ts
/** 震動資料 */
export interface ConsoleVibrateData {
  name: `vibrate`;
  value: number;
}


type Data = ConsoleVibrateData;

/** 遊戲機資料 */
export interface GameConsoleData {
  /** 目標玩家 ID,undefined 表示給所有玩家 */
  targets?: string[];
  data: Data;
}
```

統一導出。

➜ types\index.ts

```
export * from './socket.type';
export * from './player.type';
export * from './game-console.type';
```

接著新增事件定義。

➜ types\socket.type.ts

```
...
import { GameConsoleData, GamepadData } from 'types';

export interface OnEvents {
  ...
  'game-console:console-data': (data: GameConsoleData) => void;
}

export interface EmitEvents {
  ...
  'game-console:console-data': (data: GameConsoleData) => void;
}
...
```

最後在 game-console.gateway 處理事件。

➜ src\game-console\game-console.gateway.ts

```
...
@WebSocketGateway()
export class GameConsoleGateway {
  ...
  @SubscribeMessage<keyof OnEvents>('game-console:console-data')
  async handleGameConsoleData(socket: ClientSocket, data: GameConsoleData) {
    const client = this.wsClientService.getClient({ socketId: socket.id });
    if (!client) {
      const result: SocketResponse = {
        status: 'err',
```

```
      message: ' 此 socket 不存在 client',
    };
    return result;
  }

  const room = this.roomService.getRoom({ founderId: client.id });
  if (!room) {
    const result: SocketResponse = {
      status: 'err',
      message: 'client 未建立任何房間 ',
    };
    return result;
  }

  /** 發送給所有玩家 */
  const sockets = await this.server.in(room.id).fetchSockets();
  sockets.forEach((socketItem) => {
    socketItem.emit('game-console:console-data', data);
  });

  const result: SocketResponse = {
    status: 'suc',
    message: ' 傳輸遊戲機資料成功 ',
    data: undefined,
  };
  return result;
}}
```

現在伺服器會自動同步遊戲機發出的資料了！(。^▽^)

12.7 一起抓老鼠！

此章節程式碼可以在以下連結取得。

- https://gitlab.com/deepmind1/animal-party/web/-/tree/feat/game-fox-and-mouse-finish

讓我們回到遊戲場景中，把老鼠的尺寸資料發送出來吧。

邏輯很簡單，如果狐狸有碰到老鼠就發出尺寸資料，否則發出 0。

不過由於碰撞偵測是每秒 60 次，傳輸老鼠尺寸資料不需要那麼頻繁，以免塞爆傳輸，所以我們必須先替每個玩家建立 throttle 後的傳輸 function。

→ src\games\fox-and-mouse\game-scene.vue

```ts
<script setup lang="ts">
...
import { ..., throttle } from 'lodash-es';
...
/** 為每個玩家建立 throttled emit */
type ThrottledEmitMouseSizeMap = Record<string, (size: number) => void>
const throttledEmitMouseSizeMap: ThrottledEmitMouseSizeMap = gameConsole.players.
value.reduce(
  (result, player) => {
    result[player.clientId] = throttle((size: number) => {
      gameConsole.emitConsoleData({
        targets: [player.clientId],
        data: {
          name: 'vibrate',
```

```
      value: size,
    }
  });
}, 500);

return result;
},
{} as ThrottledEmitMouseSizeMap
);
...
</script>
```

調整一下 detectCollideEvent 內容。

➡ src\games\fox-and-mouse\game-scene.vue

```
<script setup lang="ts">
...
/** 偵測碰撞事件 */
function detectCollideEvent(foxes: Fox[], mice: Mouse[]) {
  foxes.forEach((fox) => {
    const foxState = fox.getState();

    /** dive 狀態不需要檢查 */
    if (['dive'].includes(foxState)) return;

    const foxMesh = fox.mesh;
    if (!foxMesh) return;

    /** 抓到的老鼠尺寸 */
    let caughtSize = 0;
    mice.forEach((mouse) => {
      const mouseMesh = mouse.mesh;
      /** 已經被抓到的老鼠不用檢查 */
      if (mouse.isCaught || !mouseMesh) return;

      const isIntersects = foxMesh.intersectsMesh(mouseMesh),

      /** 抓到了 */
```

```
      if (foxState === 'pounce' && isIntersects) {
        mouse.caught();
        fox.setMouseSize(mouse.params.size);
      }

      if (isIntersects) {
        caughtSize = mouse.params.size;
      }
    })

    /** 發送尺寸 */
    throttledEmitMouseSizeMap[fox.param.ownerId]?.(caughtSize);
  });
}
...
</script>
```

現在我們把狐狸與老鼠的數量都改為依照實際玩家數量產生吧。

➜ src\games\fox-and-mouse\game-scene.vue

```
<script setup lang="ts">
...
async function createMice(scene: Scene) {
  /** 老鼠數量，比玩家數量多 3 個 */
  const quantity = gameConsole.players.value.length + 3;
  ...
}
async function createFox(id: string, position: Vector3, scene: Scene) {
  /** 依照玩家 ID 取得對應顏色名稱並轉換成 rgb */
  const codeName = gameConsole.getPlayerCodeName(id);
  const color = getPlayerColor({ codeName });
  const hex = getPaletteColor(color);
  const rgb = textToRgb(hex);

  const fox = await new Fox(`fox-${id}`, scene, {
    ownerId: id,
    position,
    color: new Color3(rgb.r / 255, rgb.g / 255, rgb.b / 255),
```

```
  }).init();

  return fox;
}
async function createFoxes(scene: Scene) {
  const players = gameConsole.players.value;
  const positions = getSquareMatrixPositions(5, players.length, new Vector3(0, 1, 0));

  const tasks = players.map((player, i) =>
    createFox(player.clientId, positions[i], scene),
  );
  ...
}
...
</script>
```

初始化偵測玩家搖桿資料並新增控制狐狸功能。

➡ src\games\fox-and-mouse\game-scene.vue

```
...
<script setup lang="ts">
...

const emit = defineEmits<{
  ...
  (e: 'back-to-lobby'): void;
}>();
...


const { canvas } = useBabylonScene({
  async init({ engine, scene, camera }) {
    ...
    initGamepadEvent(foxes);


    ...
  }
});
```

```
...

function initGamepadEvent(foxes: Fox[]) {
  gameConsole.onGamepadData((data) => {
    const { playerId } = data;

    /** 找到對應的狐狸 */
    const target = foxes.find(
      ({ param }) => param.ownerId === playerId
    );
    if (!target) return;

    ctrlFox(target, data);
  });
}

/** 根據 key 取得資料 */
const findSingleData = curry((keys: SignalData[], name: `${KeyName}`): SignalData |
 undefined =>
  keys.find((key) => key.name === name)
);

/** 控制指定狐狸 */
function ctrlFox(fox: Fox, data: GamepadData) {
  const { keys } = data;
  const findData = findSingleData(keys);

  // 確認按鍵
  const confirmData = findData('confirm');
  if (confirmData && isGameOver.value) {
    emit('back-to-lobby');
    isGameOver.value = false;
    return;
  }

  // 抓老鼠
  const aData = findData('a');
  if (aData) {
```

```
    fox.pounce();
    return;
  }

  // 移動按鍵
  const xData = findData('x-axis');
  const yData = findData('y-axis');

  const x = xData?.value ?? 0;
  const y = yData?.value ?? 0;

  if (typeof x === 'number' && typeof y === 'number') {
    /** 搖桿向左時 x 為負值，而往左是螢幕的 +x 方向，所以要反轉 */
    fox.walk(new Vector3(-x, 0, y));
  }
}
</script>
```

人物都就定位後，讓我們完成遊戲完整邏輯吧。

首先是偵測遊戲是否結束。

➡ src\games\fox-and-mouse\game-scene.vue

```
...
<script setup lang="ts">
...
const isGameOver = ref(false);
/** 遊戲場景邊界 */
...
const { canvas } = useBabylonScene({
  async init({ engine, scene, camera }) {
    ...
    scene.registerBeforeRender(() => {
      ...
      detectGameOver(foxes, engine);
    });

    ...
  }
```

```
  });

  ...
  /** 偵測遊戲是否結束 */
  async function detectGameOver(foxes: Fox[], engine: Engine) {
    const anyEmptyHanded = foxes.some(({ mouseSize }) => mouseSize === 0);

    /** 有人還沒抓到老鼠，遊戲尚未結束 */
    if (anyEmptyHanded) return;

    /** 等待一下再結束 */
    await promiseTimeout(2000);

    engine.stopRenderLoop();
    isGameOver.value = true;
  }
</script>
```

新增計算排名的 function。

→ src\games\fox-and-mouse\game-scene.vue

```
...
<script setup lang="ts">
...
const players: Fox[] = [];
function getRankedIdList(foxes: Fox[]) {
  const result = cloneDeep(foxes)
    /** 從大到小 */
    .sort((a, b) => b.mouseSize - a.mouseSize)
    .map(({ param }) => param.ownerId);

  return result;
}

const { canvas } = useBabylonScene({
  async init({ engine, scene, camera }) {
    ...
    const foxes = await createFoxes(scene);
```

```
    players.push(...foxes);
    ...
  }
});
...
</script>
```

最後讓我們引入排名元件。

➜ src\games\fox-and-mouse\game-scene.vue

```
<template>
  <div class="overflow-hidden">
    ...
    <q-dialog
      v-model="isGameOver"
      persistent
    >
      <player-leaderboard :id-list="getRankedIdList(players)">
        <div class="text-xl text-gray-400 p-5 text-center">
          按下 <q-icon name="done" /> 回到大廳
        </div>
      </player-leaderboard>
    </q-dialog>
  </div>
</template>

<script setup lang="ts">
...
import PlayerLeaderboard from '../../components/player-leaderboard.vue';
...
</script>
```

完成狐狸遊戲機頁面。

➜ src\views\the-game-console-fox-and-mouse.vue

```
<template>
  <game-scene
    class=" w-full h-full"
    @init="handleInit"
  />
</template>

<script setup lang="ts">
import { ref } from 'vue';

import GameScene from '../games/fox-and-mouse/game-scene.vue';

import { useLoading } from '../composables/use-loading';

const loading = useLoading();

function handleInit() {
  loading.hide();
}
</script>

<style scoped lang="sass">
</style>
```

只差搖桿震動功能就大功告成了！、(≧∀≦)、

讓我們在搖桿中偵測 console 資料。

➜ src\views\the-player-gamepad-fox-and-mouse.vue

```
<template>
...
</template>

<script setup lang="ts">
...
```

```
import { useVibrate } from '@vueuse/core';
...
const { emitGamepadData, onConsoleData, } = useClientPlayer();
const mainStore = useMainStore();
const { vibrate, stop } = useVibrate();
...
const mouseSize = ref(0);
function startVibrate() {
  const size = mouseSize.value;
  if (mouseSize.value === 0) {
    stop();
    return;
  }

  vibrate([size, 500]);

  const nextTime = size + 500;
  setTimeout(() => {
    startVibrate();
  }, nextTime);
}
watch(mouseSize, () => {
  startVibrate();
});

onConsoleData(({ targets, data }) => {
  if (
    !targets?.includes(mainStore.clientId) ||
    data.name !== 'vibrate'
  ) return;

  mouseSize.value = data.value;
});
</script>
```

到目前為止，狐狸遊戲邏輯都完成了！✧*｡٩(ˊ◡ˋ*) و✧*｡

▲ 圖 12-18　狐狸遊戲完成

最後讓我們完成教學模式的部分。

首先實現遊戲模式邏輯，新增元件的 props。

→ src\games\fox-and-mouse\game-scene.vue

```
...
<script setup lang="ts">
...
interface Props {
  mode?: `${GameSceneMode}`;
}
const props = withDefaults(defineProps<Props>(), {
  mode: 'normal',
});
...
</script>
```

showcase 模式我們讓狐狸可以移動，但是不能抓老鼠。

➔ src\games\fox-and-mouse\game-scene.vue

```
...
<script setup lang="ts">
...
/** 控制指定狐狸 */
function ctrlFox(fox: Fox, data: GamepadData) {
  ...
  // 抓老鼠
  const aData = findData('a');
  if (aData) {
    if (props.mode === 'showcase') {
      return;
    }
    fox.pounce();
    return;
  }

  ...
}
</script>
```

training 模式的部分，直接發出 game-over 事件，由父元件負責重啟遊戲。

➔ src\games\fox-and-mouse\game-scene.vue

```
<template>
  <div class="overflow-hidden">
    ...
    <q-dialog
      :model-value="isGameOver && props.mode === 'normal'"
      persistent
    >
      ...
    </q-dialog>
  </div>
</template>
```

```ts
<script setup lang="ts">
...
const emit = defineEmits<{
  ...
  (e: 'game-over'): void;
}>();
...
/** 偵測遊戲是否結束 */
async function detectGameOver(foxes: Fox[], engine: Engine) {
  ...
  /** 等待一下再結束 */
  await promiseTimeout(2000);

  if (!isGameOver.value) {
    emit('game-over');
  }

  isGameOver.value = true;
}
...
</script>
```

現在讓我們調整遊戲頁面，概念和小雞遊戲相同。

新增教學卡片。

➜ src\games\fox-and-mouse\tutorial-card.vue

```vue
<template>
  <q-card class="flex flex-col rounded-[2rem] shadow-sm overflow-hidden text-sky-900">
    <q-card-section class=" relative text-white text-4xl text-center bg-sky-600 p-7
!rounded-none">
      遊戲說明

      <base-polygon
        class=" absolute top-0 left-0 -translate-x-1/4 -translate-y-1/2"
        opacity="0.2"
        rotate="30deg"
      />
```

```
    <base-polygon
      class=" absolute bottom-0 right-0 translate-x-1/4 translate-y-1/2"
      opacity="0.2"
      fill="spot"
      shape="pentagon"
      rotate="30deg"
    />
  </q-card-section>

<q-card-section class=" text-xl p-10 text-center">
    仔細感受雪下的動靜，抓到最大隻的老鼠吧！
</q-card-section>

<q-card-section class="section">
  <div class=" w-1/3 flex flex-center">
    <div class="pad ">
      <div class="thumb thumb-action" />
    </div>
  </div>

  <p class=" text-center flex-1">
      移動類比搖桿控制狐狸移動
  </p>
</q-card-section>

<q-card-section class="section">
  <div class=" w-1/3 flex flex-center">
    <div class="mouse" />
  </div>


  <p class=" text-center flex-1">
    踩在方框上，搖桿震動時間越長，表示老鼠越大隻
  </p>
</q-card-section>

<q-card-section class="section">
  <div class=" w-1/3 flex flex-center">
    <div class="button click">
```

```
        <div class=" text-white text-5xl">
          A
        </div>
      </div>
    </div>


    <p class=" text-center flex-1">
      按下 A 抓老鼠
    </p>
  </q-card-section>

  <q-card-section class="section justify-center flex-1">
    <p>
      練習完成後，按下
      <q-icon
        name="done"
        color="white"
        class="p-2 bg-[#000]/60 rounded-full"
      />
      準備完成
    </p>
  </q-card-section>
  </q-card>
</template>

<script setup lang="ts">
import BasePolygon from '../../components/base-polygon.vue';
</script>

<style scoped lang="sass">
.section
  @apply text-xl p-6 px-10 flex items-center flex-nowrap gap-6

.pad, .button
  @apply rounded-full flex justify-center items-center
  background: rgba(#111, 0.6)

.pad
```

```
    width: 7rem
    height: 7rem
.button
    width: 6rem
    height: 6rem

.thumb
    width: 40%
    height: 40%
    background: white
    border-radius: 9999px
    opacity: 0.8

.mouse
    width: 4rem
    height: 4rem
    border: 3px solid rgba(#222, 0.4)
    transform: rotate(45deg)
    animation: mouse 2s infinite

@keyframes mouse
    0%, 100%
        opacity: 1
    50%
        opacity: 0.4


.thumb-action
    animation: thumb-action 3s ease-in-out infinite

@keyframes thumb-action
    0%, 100%
        transform: translate(0%, 0%)
    25%
        transform: translate(0%, 50%)
    50%
        transform: translate(-50%, 0%)
    75%
        transform: translate(0%, -50%)
```

```
.click
  animation: click 1s ease-in-out infinite

@keyframes click
  0%, 20%, 100%
    transform: scale(1)
  40%
    transform: scale(0.98)
    opacity: 0.8


</style>
```

調整遊戲頁面。

➜ src\games\fox-and-mouse\tutorial-card.vue

```
<template>
  <div class="w-full h-full bg-white">
    <transition
      name="opacity"
      mode="out-in"
    >
      <!-- 練習 -->
      <div
        v-if="sceneMode === 'training'"
        class="w-full h-full flex bg-sky-100"
      >
        <div class=" relative flex flex-col flex-nowrap w-[70%]">
          <game-scene
            mode="training"
            class=" w-full h-[90%] rounded-br-3xl"
            @init="handleInit"
          />

          <!-- 玩家頭像 -->
          <transition-group
            tag="div"
            class=" absolute bottom-0 w-full overflow-hidden px-10 pointer-events-none"
            name="avatar"
```

```
        >
          <player-list-avatar
            v-for="player in players"
            :key="player.clientId"
            :player="player"
            :code-name="player.codeName"
            class="player"
            :class="{ 'ready': player.ok }"
          />
        </transition-group>
      </div>

      <!-- 遊戲說明 -->
      <div class="p-4 flex-1 ">
        <tutorial-card class="w-full h-full" />
      </div>
    </div>

    <!-- 正式 -->
    <div
      v-else
      class=" w-full h-full"
    >
      <game-scene
        :mode="sceneMode"
        class=" absolute w-full h-full"
        @back-to-lobby="handleBackToLobby"
      />

      <countdown-overlay @done="handleTimeout" />
    </div>
  </transition>
  </div>
</template>

<script setup lang="ts">
import { computed, ref, watch } from 'vue';
import { RouteName } from '../router/router';
import { GameSceneMode } from '../types';
```

```
import GameScene from '../games/fox-and-mouse/game-scene.vue';
import TutorialCard from '../games/fox-and-mouse/tutorial-card.vue';
import PlayerListAvatar from '../components/player-list-avatar.vue';
import CountdownOverlay from '../components/countdown-overlay.vue';

import { useLoading } from '../composables/use-loading';
import { useRouter } from 'vue-router';
import { useClientGameConsole } from '../composables/use-client-game-console';
import { whenever } from '@vueuse/core';

const gameConsole = useClientGameConsole();
const loading = useLoading();
const router = useRouter();

const sceneMode = ref<`${GameSceneMode}`>('training');

function handleInit() {
  loading.hide();
}

async function handleBackToLobby() {
  await loading.show();
  router.push({
    name: RouteName.GAME_CONSOLE_LOBBY
  });
}

function handleTimeout() {
  sceneMode.value = 'normal';
}

/** 紀錄準備完成玩家 */
const readiedPlayerIdList = ref<string[]>([]);

const players = computed(() => {
  return gameConsole.players.value.map((player) => {
    const codeName = gameConsole.getPlayerCodeName(player.clientId);
    const ok = readiedPlayerIdList.value.includes(player.clientId);
```

```
    return {
      ...player,
      codeName,
      ok,
    }
  });
});

whenever(
  () => players.value.every(({ ok }) => ok),
  () => {
    sceneMode.value = 'showcase';
  }
);

gameConsole.onGamepadData((data) => {
  const lastDatum = data.keys.at(-1);
  if (lastDatum?.value || lastDatum?.name !== 'confirm') return;

  readiedPlayerIdList.value.push(data.playerId);
});
</script>

<style scoped lang="sass">
.player
  transform: translateY(0%)
  transition-duration: 0.4s
  &.ready
    opacity: 0.5
    transform: translateY(10%)
    &::after
      position: absolute
      content: 'OK'
      color: white
      text-shadow: 0 0 6px rgba(#000, 0.4)
      left: 50%
      bottom: 15%
      font-size: 2rem
```

```
      transform: translateX(-50%)
</style>
```

與小雞的教學模式不同，這裡我們要實現 game over 時，重啟遊戲邏輯。

使用 nanoid 產生隨機字串，並使用 key 強制更新遊戲場景元件。

➜ src\games\fox-and-mouse\tutorial-card.vue

```
<template>
  <div class="w-full h-full bg-white">
    <transition ... >
      <!-- 練習 -->
      <div
        v-if="sceneMode === 'training'"
        class="w-full h-full flex bg-sky-100"
      >
        <div class=" ...">
          <transition
            name="opacity"
            mode="out-in"
          >
            <game-scene
              :key="gameId"
              mode="training"
              class=" w-full h-[90%] rounded-br-3xl"
              @init="handleInit"
              @game-over="handleGameOver"
            />
          </transition>

          <!-- 玩家頭像 -->
          ...
        </div>

        <!-- 遊戲說明 -->
        ...
      </div>

      <!-- 正式 -->
```

```
    ...
    </transition>
  </div>
</template>

<script setup lang="ts">
...
import { nanoid } from 'nanoid';
...
/** 用來更新教學遊戲場景 */
const gameId = ref('');
function handleGameOver() {
  gameId.value = nanoid();
}
...
</script>
...
```

現在我們完成狐狸遊戲了，大家可以一起抓老鼠了！✧*。٩(´ ꒳ ` *)و✧*。

▲ 圖 12-19 狐狸遊戲完成

鱈魚：「細心的讀者們一定會發現教學頁面的部份又重複啦，不過這裡就不抽離元件了。ᕦ(ﾟ∀ﾟ)ᕤ」

助教：「又偷懶？('◉ ᗢ ◉`)」

鱈魚：「才不是哩，是因為篇幅有限，派對該告了一個段落了。(´·ω·`)」

助教和鱈魚：「感謝各位讀者的陪伴，派對準備收攤，未來有機會再見囉！下台一鞠躬！`(°▽、°)`(°▽、°)」

第13章

結束是另一個開始

　　一路走來，我們堆疊了很多功能，接下來讓我們根據反饋，重構部分功能，讓體驗更好吧！ヽ(*´∀`*)ﾉ

13.1 重構企鵝遊戲

此章節程式碼可以在以下連結取得。

- https://gitlab.com/deepmind1/animal-party/web/-/tree/feat/refactor-the-first-penguin

我們把企鵝遊戲加入加入教學頁面。

首先將遊戲教學頁面中重複的內容抽離為單一元件。

➜ src\components\game-console-scene-container.vue

```
<template>
  <div class="w-full h-full bg-white">
    <transition ... >
        <!-- 練習 -->
        <div
          v-if="props.sceneMode === 'training'"
          class="w-full h-full flex bg-sky-100"
        >
          <div class=" ... ">
            <div class=" ... overflow-hidden">
              <slot name="training-scene" />
            </div>
            <!-- 玩家頭像 -->
            ...
          </div>

          <!-- 遊戲說明 -->
          <div class="p-4 flex-1 ">
            <slot name="tutorial-card" />
          </div>
        </div>

        <!-- 正式 -->
        <div
          v-else
          class=" w-full h-full">
        >
          <slot name="normal-scene" />
        </div>
    </transition>
  </div>
</template>

<script setup lang="ts">
import { computed, ref, watch } from 'vue';
import { GameSceneMode } from '../types';
```

```
import PlayerListAvatar from '../components/player-list-avatar.vue';

import { useClientGameConsole } from '../composables/use-client-game-console';
import { whenever } from '@vueuse/core';

interface Props {
  sceneMode?: `${GameSceneMode}`;
}
const props = withDefaults(defineProps<Props>(), {
  sceneMode: 'training',
});

const emit = defineEmits<{
  (e: 'all-ready'): void;
}>();

const gameConsole = useClientGameConsole();

/** 紀錄準備完成玩家 */
const readiedPlayerIdList = ref<string[]>([]);

const players = computed(() => {
  ...
});

whenever(
  () => players.value.every(({ ok }) => ok),
  () => emit('all-ready')
);

gameConsole.onGamepadData((data) => {
  ...
});
</script>

<style scoped lang="sass">
...
</style>
```

接著使用 use-babylon-scene，取代原本建立場景程式。

➜ src\games\the-first-penguin\game-scene.vue

```
<template>
...
</template>

<script setup lang="ts">
...
import { useBabylonScene } from '../../composables/use-babylon-scene';
...

const gameConsole = useClientGameConsole();

const { canvas } = useBabylonScene({
  createCamera(scene: Scene) {
    const camera = new ArcRotateCamera(
      'camera',
      Math.PI / 2,
      Math.PI / 4,
      34,
      new Vector3(0, 0, 2),
      scene
    );

    return camera;
  },

  async init({ scene, engine }) {
    const physicsPlugin = new CannonJSPlugin(true, 8, CANNON);
    scene.enablePhysics(new Vector3(0, -9.81, 0), physicsPlugin);

    createSea(scene);
    createIce(scene);

    const players = gameConsole.players.value;
    const positions = getSquareMatrixPositions(
      5, players.length, new Vector3(0, 10, 0)
    );
```

```
    const tasks = players.map(({ clientId }, index) =>
      createPenguin(clientId, index, scene, {
        position: positions[index],
      })
    );
    const result = await Promise.allSettled(tasks)
    result.forEach((data) => {
      if (data.status !== 'fulfilled') return;
      penguins.push(data.value);
    });

    initGamepadEvent();

    /** 持續運行指定事件 */
    scene.registerAfterRender(() => {
      detectCollideEvents(penguins);
      detectOutOfBounds(penguins);
      detectWinner(penguins, engine);
    });

    emit('init');
  },
});
...
async function createPenguin(id: string, index: number, scene: Scene, params:
CreatePenguinParams) {
  ...
  const penguin = await new Penguin(`penguin-${index}`, scene, ...).init();

  return penguin;
}
...

/** 偵測是否有贏家 */
function detectWinner(penguins: Penguin[], engine: Engine) {
  ...
}
...
</script>
```

```
<style scoped lang="sass">
...

</style>
```

在企鵝遊戲中實現遊戲模式，新增 props 與 emit 事件。

➔ src\games\the-first-penguin\game-scene.vue

```
...
<script setup lang="ts">
...
interface Props {
  mode?: `${GameSceneMode}`;
}
const props = withDefaults(defineProps<Props>(), {
  mode: 'normal',
});

const emit = defineEmits<{
  (e: 'init'): void;
  (e: 'back-to-lobby'): void;
  (e: 'game-over'): void;
}>();
...
</script>
...
```

新增模式邏輯，showcase 無法控制企鵝。

➔ src\games\the-first-penguin\game-scene.vue

```
...
<script setup lang="ts">
...
function initGamepadEvent() {
  gameConsole.onGamepadData((data) => {
    if (props.mode === 'showcase') return;
    ...
  });
```

```
}
...
</script>
...
```

　　training 模式與狐狸遊戲相同，發出 game-over 事件並將 backToLobby function 內容改為發出 back-to-lobby 事件。

➡ src\games\the-first-penguin\game-scene.vue

```
<template>
...
</template>

<script setup lang="ts">
...

/** 偵測是否有贏家 */
function detectWinner(penguins: Penguin[], engine: Engine) {
  ...

  if (!isGameOver.value) {
    emit('game-over');
  }

  winnerCodeName.value = gameConsole.getPlayerCodeName(winnerId);
  isGameOver.value = true;
}
...
async function backToLobby() {
  isGameOver.value = false;

  emit('back-to-lobby');
}
</script>

<style scoped lang="sass">
...
</style>
```

現在調整一下遊戲機頁面，加入教學。

➜ src\views\the-game-console-the-first-penguin.vue

```html
<template>
  <game-console-scene-container
    :scene-mode="sceneMode"
    @all-ready="handleAllReady"
  >
    <template #training-scene>
      <transition
        name="opacity"
        mode="out-in"
      >
        <game-scene
          :key="gameId"
          mode="training"
          class=" w-full h-full"
          @init="handleInit"
          @game-over="handleGameOver"
        />
      </transition>
    </template>

    <template #normal-scene>
      <game-scene
        :mode="sceneMode"
        class="absolute w-full h-full"
        @back-to-lobby="handleBackToLobby"
      />

      <countdown-overlay @done="handleTimeout" />
    </template>
  </game-console-scene-container>
</template>

<script setup lang="ts">
import { ref } from 'vue';
import { GameSceneMode } from '../types';
import { nanoid } from 'nanoid';
import { RouteName } from '../router/router';
```

```
import GameConsoleSceneContainer from '../components/game-console-scene-container.
vue';
import GameScene from '../games/the-first-penguin/game-scene.vue';
import CountdownOverlay from '../components/countdown-overlay.vue';

import { useLoading } from '../composables/use-loading';
import { useRouter } from 'vue-router';

const loading = useLoading();
const router = useRouter();

const sceneMode = ref<`${GameSceneMode}`>('training');

function handleInit() {
  loading.hide();
}

async function handleBackToLobby() {
  await loading.show();
  router.push({
    name: RouteName.GAME_CONSOLE_LOBBY
  });
}

function handleTimeout() {
  sceneMode.value = 'normal';
}
function handleAllReady() {
  sceneMode.value = 'showcase';
}

/** 用來更新教學遊戲場景 */
const gameId = ref('');
async function handleGameOver() {
  gameId.value = nanoid();
}
</script>
...
```

程式碼是不是變得簡潔很多呢？(ㄒ ㄥ ㄒ)✧

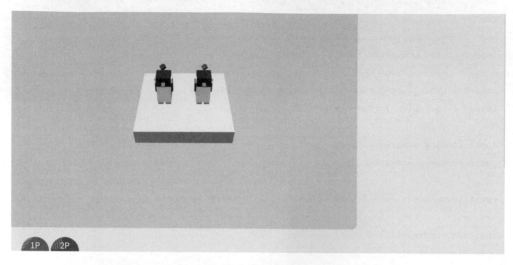

▲ 圖 13-1 企鵝遊戲加入教學

最後讓我們完成教學卡片。

➜ src\games\the-first-penguin\tutorial-card.vue

```
<template>
  <q-card class="flex flex-col rounded-[2rem] shadow-sm overflow-hidden text-sky-900">
    <q-card-section class=" relative text-white text-4xl text-center bg-sky-600 p-7
!rounded-none">
      遊戲說明

      <base-polygon
        class=" absolute top-0 left-0 -translate-x-1/4 -translate-y-1/2"
        opacity="0.2"
        rotate="30deg"
      />
      <base-polygon
        class=" absolute bottom-0 right-0 translate-x-1/4 translate-y-1/2"
        opacity="0.2"
        fill="spot"
        shape="pentagon"
        rotate="30deg"
```

```
  />
</q-card-section>

<q-card-section class=" text-xl p-10 text-center">
  誰都不想第一個下水，努力把別人撞下海中吧！
</q-card-section>

<q-card-section class="section">
  <div class=" w-1/3 flex flex-center">
    <div class="pad ">
      <div class="thumb thumb-action" />
    </div>
  </div>

  <p class=" text-center flex-1">
    移動類比搖桿控制企鵝移動
  </p>
</q-card-section>

<q-card-section class="section">
  <div class=" w-1/3 flex flex-center">
    <div class="button click">
      <div class=" text-white text-5xl">
        A
      </div>
    </div>
  </div>

  <p class=" text-center flex-1">
    按下 A 發動旋轉攻擊，可以全方位擊退其他企鵝
  </p>
</q-card-section>

<q-card-section class="section justify-center flex-1">
  <p>
    練習完成後，按下
    <q-icon
      name="done"
```

```
          color="white"
          class="p-2 bg-[#000]/60 rounded-full"
        />
        準備完成
      </p>
    </q-card-section>
  </q-card>
</template>

<script setup lang="ts">
import BasePolygon from '../../components/base-polygon.vue';
</script>

<style scoped lang="sass">
.section
  @apply text-xl p-6 px-10 flex items-center flex-nowrap gap-6

.pad, .button
  @apply rounded-full flex justify-center items-center
  background: rgba(#111, 0.6)

.pad
  width: 7rem
  height: 7rem
.button
  width: 6rem
  height: 6rem

.thumb
  width: 40%
  height: 40%
  background: white
  border-radius: 9999px
  opacity: 0.8

.mouse
  width: 4rem
  height: 4rem
  border: 3px solid rgba(#222, 0.4)
  transform: rotate(45deg)
```

```
    animation: mouse 2s infinite

@keyframes mouse
  0%, 100%
    opacity: 1
  50%
    opacity: 0.4



.thumb-action
  animation: thumb-action 3s ease-in-out infinite

@keyframes thumb-action
  0%, 100%
    transform: translate(0%, 0%)
  25%
    transform: translate(0%, 50%)
  50%
    transform: translate(-50%, 0%)
  75%
    transform: translate(0%, -50%)

.click
  animation: click 1s ease-in-out infinite

@keyframes click
  0%, 20%, 100%
    transform: scale(1)
  40%
    transform: scale(0.98)
    opacity: 0.8

</style>
```

最後加入教學卡片。

➜ src\views\the-game-console-the-first-ponguin.vue

```
<template>
  <game-console-scene-container ... >
```

```
    ...
    <template #tutorial-card>
      <tutorial-card class="w-full h-full" />
    </template>
  </game-console-scene-container>
</template>

<script setup lang="ts">
...
import TutorialCard from '../games/the-first-penguin/tutorial-card.vue';
...
</script>
```

現在企鵝遊戲也有完整的教學引導了！✧*。٩(´ロ`*)و✧*。

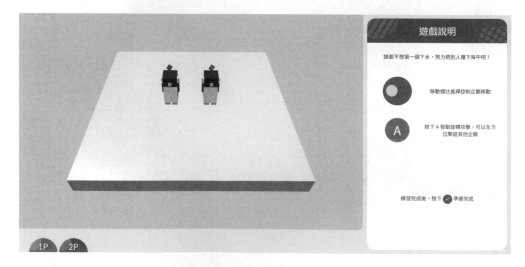

▲ 圖 13-2 企鵝遊戲加入教學卡片

讀者們也可以嘗試看看將所有遊戲的教學頁面都改用 game-console-scene-container 重構看看喔，可以深刻地感受到元件化的優點。(´▽`)/

13.2 取消搖桿方向限制

此章節程式碼可以在以下連結取得。

- https://gitlab.com/deepmind1/animal-party/web/-/tree/feat/gamepad-rwd

本來是想重現 joy-con 那種在特定遊戲下要直握或橫握的效果，但是實際上除了徒增困擾之外，一點用都沒有。XD

 Tips

如果螢幕方向鎖定的 Web API 沒有被棄用的話，就可以實現這個願望了。
。·˚·(つ д ˚)·˚·

- https://developer.mozilla.org/en-US/docs/Web/API/Screen/lockOrientation

現在讓我們將搖桿改成直握或橫握皆可，首先是大廳搖桿。

取消 orientation 限制後，微調按鈕位置。

➡ src\views\the-player-gamepad-lobby.vue

```
<template>
  <player-gamepad-container>
    ...
    <gamepad-btn
      class="absolute bottom-10 right-10 opacity-90"
      ...
    />
```

```
    <gamepad-btn
      class="absolute top-10 right-10 opacity-90"
      ...
    />

    ...
  </player-gamepad-container>
</template>

<script setup lang="ts">
...
</script>
```

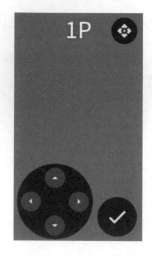

▲ 圖 13-3　直向大廳搖桿

　　接著是企鵝遊戲搖桿與狐狸遊戲搖桿，由於兩個搖桿外觀一模一樣，所以一起調整。

　　同時將企鵝遊戲搖桿使用 player-gamepad-container 元件重構。

➜ src\views\the-player-gamepad-the-first-penguin.vue

➜ src\views\the-player-gamepad-fox-and-mouse.vue

```
<template>
  <player-gamepad-container>
```

```
    ...
    <gamepad-btn
      class="absolute bottom-10 right-10"
      ...
    >
      ...
    </gamepad-btn>

    <gamepad-btn
      class="absolute top-10 right-10 opacity-90"
      ...
    />
  </player-gamepad-container>
</template>

<script setup lang="ts">
...
</script>
```

▲ 圖 13-4　直向企鵝遊戲與狐狸遊戲搖桿

小雞遊戲的搖桿本來就是直握，所以不用調整。(′,,•ω•,,)

第 14 章

後記

感謝所有讀者看到這裡，其實派對還沒結束，只要保持赤子之心，繼續更新遊戲，我們就能一直嗨下去。✧*。٩(´ロ`*)و✧*。

在這次專案中，盡可能保持簡單，所以省略了許多功能，若大家有興趣可以加入更多細節，例如：

- 玩家網頁意外斷線後，重啟網頁，會詢問玩家是否重新加入先前遊戲室。

- 玩家可自訂名稱與頭像。

- 新增遊戲排名，每個玩家可以依據排名得分。

- 新增玩家個人資料系統，玩家可以查看玩過甚麼遊戲、累積得分、累積遊玩時間與成就獎章等等資訊。

- 更多遊戲。ᕕ(ﾟ∀ﾟ｡)ᕗ

- 更多種讀取畫面。(´,,•ω•,,)

等等更多可能性！ヽ(●´∀`●)ﾉ

最後讓我再次感謝大家，希望以後還有機會再見！(´･ﾛ･`)/ゝ

MEMO